T0306092

Advanced Risk Analysis in Engineering Enterprise Systems

STATISTICS: Textbooks and Monographs

STATISTICS: Textbooks and Monographs

Recent Titles

Computational Methods in Statistics and Econometrics, *Hisashi Tanizaki*

Applied Sequential Methodologies: Real-World Examples with Data Analysis, *edited by Nitis Mukhopadhyay, Sujay Datta, and Saibal Chattopadhyay*

Handbook of Beta Distribution and Its Applications, *edited by Arjun K. Gupta and Saralees Nadarajah*

Item Response Theory: Parameter Estimation Techniques, Second Edition, *edited by Frank B. Baker and Seock-Ho Kim*

Statistical Methods in Computer Security, *edited by William W. S. Chen*

Elementary Statistical Quality Control, Second Edition, *John T. Burr*

Data Analysis of Asymmetric Structures, *Takayuki Saito and Hiroshi Yadohisa*

Mathematical Statistics with Applications, *Asha Seth Kapadia, Wenyaw Chan, and Lemuel Moyé*

Advances on Models, Characterizations and Applications, *N. Balakrishnan, I. G. Bairamov, and O. L. Gebizlioglu*

Survey Sampling: Theory and Methods, Second Edition, *Arijit Chaudhuri and Horst Stenger*

Statistical Design of Experiments with Engineering Applications, *Kamel Rekab and Muzaffar Shaikh*

Quality by Experimental Design, Third Edition, *Thomas B. Barker*

Handbook of Parallel Computing and Statistics, *Erricos John Kontoghiorghes*

Statistical Inference Based on Divergence Measures, *Leandro Pardo*

A Kalman Filter Primer, *Randy Eubank*

Introductory Statistical Inference, *Nitis Mukhopadhyay*

Handbook of Statistical Distributions with Applications, *K. Krishnamoorthy*

A Course on Queueing Models, *Joti Lal Jain, Sri Gopal Mohanty, and Walter Böhm*

Univariate and Multivariate General Linear Models: Theory and Applications with SAS, Second Edition, *Kevin Kim and Neil Timm*

Randomization Tests, Fourth Edition, *Eugene S. Edgington and Patrick Onghena*

Design and Analysis of Experiments: Classical and Regression Approaches with SAS, *Leonard C. Onyiah*

Analytical Methods for Risk Management: A Systems Engineering Perspective, *Paul R. Garvey*

Confidence Intervals in Generalized Regression Models, *Esa Uusipaikka*

Introduction to Spatial Econometrics, *James LeSage and R. Kelley Pace*

Acceptance Sampling in Quality Control, *Edward G. Schilling and Dean V. Neubauer*

Applied Statistical Inference with MINITAB®, *Sally A. Lesik*

Nonparametric Statistical Inference, Fifth Edition, *Jean Dickinson Gibbons and Subhabrata Chakraborti*

Bayesian Model Selection and Statistical Modeling, *Tomohiro Ando*

Handbook of Empirical Economics and Finance, *Aman Ullah and David E. A. Giles*

Randomized Response and Indirect Questioning Techniques in Surveys, *Arijit Chaudhuri*

Applied Time Series Analysis, *Wayne A. Woodward, Henry L. Gray, and Alan C. Elliott*

Advanced Risk Analysis in Engineering Enterprise Systems, *C. Ariel Pinto and Paul R. Garvey*

Advanced Risk Analysis in Engineering Enterprise Systems

C. Ariel Pinto
Old Dominion University
Norfolk, Virginia, USA

Paul R. Garvey
The MITRE Corporation
Bedford, Massachusetts, USA

CRC Press is an imprint of the
Taylor & Francis Group, an **informa** business
A CHAPMAN & HALL BOOK

Approved for Public Release (09-0449, 10-2944). Distribution unlimited.

CRC Press
Taylor & Francis Group
6000 Broken Sound Parkway NW, Suite 300
Boca Raton, FL 33487-2742

International Standard Book Number: 978-1-4398-2614-0 (Hardback)

Library of Congress Cataloging-in-Publication Data

Pinto, Cesar Ariel.
Advanced risk analysis in engineering enterprise systems / Cesar Ariel Pinto, Paul R. Garvey.
p. cm.
Includes bibliographical references and index.
ISBN 978-1-4398-2614-0 (hardback)
1. Reliability (Engineering) 2. Technology--Risk assessment. 3. Systems engineering. I. Garvey, Paul R., 1956- II. Title.

TA169.P56 2012
620.0068'1--dc23 2011043038

Visit the Taylor & Francis Web site at
http://www.taylorandfrancis.com

and the CRC Press Web site at
http://www.crcpress.com

To my parents Alberto and Aida for everything a son can wish for and to my wife Mylene and children Marlo and Ela for being my inspiration.

C. Ariel Pinto

To the memory of my daughter Katrina, to my wife Maura and daughters Alana and Kirsten. To my parents Eva and Ralph.

Ad Majorem Dei Gloriam

Paul R. Garvey

Contents

Preface xv
Acknowledgments... xix
Authors xxi

1. Engineering Risk Management...1
1.1 Introduction 1
1.1.1 Boston's Central Artery/Tunnel Project 2
1.2 Objectives and Practices...6
1.3 New Challenges 12
Questions and Exercises...13

2. Perspectives on Theories of Systems and Risk 15
2.1 Introduction 15
2.2 General Systems Theory 15
2.2.1 Complex Systems, Systems-of-Systems, and Enterprise Systems...20
2.3 Risk and Decision Theory..24
2.4 Engineering Risk Management 36
Questions and Exercises...39

3. Foundations of Risk and Decision Theory 41
3.1 Introduction 41
3.2 Elements of Probability Theory 41
3.3 The Value Function...63
3.4 Risk and Utility Functions..81
3.4.1 vNM Utility Theory...81
3.4.2 Utility Functions 85
3.5 Multiattribute Utility—The Power Additive Utility Function 97
3.5.1 The Power-Additive Utility Function 97
3.5.2 Applying the Power-Additive Utility Function...98
3.6 Applications to Engineering Risk Management 101
3.6.1 Value Theory to Measure Risk...102
3.6.2 Utility Theory to Compare Designs.................................. 114
Questions and Exercises...119

4. A Risk Analysis Framework in Engineering Enterprise Systems.....125
4.1 Introduction 125
4.2 Perspectives on Engineering Enterprise Systems 125

4.3 A Framework for Measuring Enterprise Capability Risk 129
4.4 A Risk Analysis Algebra 133
4.5 Information Needs for Portfolio Risk Analysis 149
4.6 The "Cutting Edge" 150
Questions and Exercises 151

5. An Index to Measure Risk Corelationships 157
5.1 Introduction 157
5.2 RCR Postulates, Definitions, and Theory 158
5.3 Computing the RCR Index 164
5.4 Applying the RCR Index: A Resource Allocation Example 171
5.5 Summary 174
Questions and Exercises 174

6. Functional Dependency Network Analysis 177
6.1 Introduction 177
6.2 FDNA Fundamentals 178
6.3 Weakest Link Formulations 186
6.4 FDNA (α, β) Weakest Link Rule 191
6.5 Network Operability and Tolerance Analyses 215
6.5.1 Critical Node Analysis and Degradation Index 222
6.5.2 Degradation Tolerance Level 227
6.6 Special Topics 237
6.6.1 Operability Function Regulation 237
6.6.2 Constituent Nodes 239
6.6.3 Addressing Cycle Dependencies 245
6.7 Summary 247
Questions and Exercises 249

7. A Decision-Theoretic Algorithm for Ranking Risk Criticality 257
7.1 Introduction 257
7.2 A Prioritization Algorithm 257
7.2.1 Linear Additive Model 258
7.2.2 Compromise Models 259
7.2.3 Criteria Weights 262
7.2.4 Illustration 265
Questions and Exercises 269

8. A Model for Measuring Risk in Engineering Enterprise Systems 271
8.1 A Unifying Risk Analytic Framework and Process 271
8.1.1 A Traditional Process with Nontraditional Methods 271
8.1.2 A Model Formulation for Measuring Risk in Engineering Enterprise Systems 272

8.2 Summary 279
Questions and Exercises 279

9. Random Processes and Queuing Theory 281
9.1 Introduction 281
9.2 Deterministic Process 282
9.2.1 Mathematical Determinism 283
9.2.2 Philosophical Determinism 284
9.3 Random Process 284
9.3.1 Concept of Uncertainty 286
9.3.2 Uncertainty, Randomness, and Probability 287
9.3.3 Causality and Uncertainty 289
9.3.4 Necessary and Sufficient Causes 291
9.3.5 Causalities and Risk Scenario Identification 291
9.3.6 Probabilistic Causation 293
9.4 Markov Process 298
9.4.1 Birth and Death Process 300
9.5 Queuing Theory 300
9.5.1 Characteristic of Queuing Systems 302
9.5.2 Poisson Process and Distribution 303
9.5.3 Exponential Distribution 304
9.6 Basic Queuing Models 304
9.6.1 Single-Server Model 304
9.6.2 Probability of an Empty Queuing System 306
9.6.3 Probability That There Are Exactly *N* Entities Inside the Queuing System 307
9.6.4 Mean Number of Entities in the Queuing System 308
9.6.5 Mean Number of Waiting Entities 308
9.6.6 Average Latency Time of Entities 308
9.6.7 Average Time of an Entity Waiting to Be Served 309
9.7 Applications to Engineering Systems 310
9.8 Summary 315
Questions and Exercises 316

10. Extreme Event Theory 323
10.1 Introduction to Extreme and Rare Events 323
10.2 Extreme and Rare Events and Engineering Systems 324
10.3 Traditional Data Analysis 325
10.4 Extreme Value Analysis 327
10.5 Extreme Event Probability Distributions 329
10.5.1 Independent Single-Order Statistic 331
10.6 Limit Distributions 334
10.7 Determining Domain of Attraction Using Inverse Function 336
10.8 Determining Domain of Attraction Using Graphical Method 341

10.8.1 Steps in Visual Analysis of Empirical Data 341
10.8.2 Estimating Parameters of GEVD 345
10.9 Complex Systems and Extreme and Rare Events 347
10.9.1 Extreme and Rare Events in a Complex System 348
10.9.2 Complexity and Causality 349
10.9.3 Complexity and Correlation 349
10.9.4 Final Words on Causation 350
10.10 Summary 351
Questions and Exercises 351

11. Prioritization Systems in Highly Networked Environments 357
11.1 Introduction 357
11.2 Priority Systems 357
11.2.1 PS Notation 358
11.3 Types of Priority Systems 363
11.3.1 Static Priority Systems 363
11.3.2 Dynamic Priority Systems 365
11.3.3 State-Dependent DPS 365
11.3.4 Time-Dependent DPS 371
11.4 Summary 375
Questions and Exercises 375
Questions 376

12. Risks of Extreme Events in Complex Queuing Systems 379
12.1 Introduction 379
12.2 Risk of Extreme Latency 379
12.2.1 Methodology for Measurement of Risk 381
12.3 Conditions for Unbounded Latency 386
12.3.1 Saturated PS 388
12.4 Conditions for Bounded Latency 389
12.4.1 Bounded Latency Times in Saturated Static PS 389
12.4.2 Bounded Latency Times in a Saturated SDPS 392
12.4.3 Combinations of Gumbel Types 394
12.5 Derived Performance Measures 395
12.5.1 Tolerance Level for Risk 395
12.5.2 Degree of Deficit 397
12.5.3 Relative Risks 398
12.5.4 Differentiation Tolerance Level 400
12.5.5 Cost Functions 401
12.6 Optimization of PS 403
12.6.1 Cost Function Minimization 404
12.6.2 Bounds on Waiting Line 404
12.6.3 Pessimistic and Optimistic Decisions in Extremes 406
12.7 Summary 410
Questions and Exercises 411

Appendix Bernoulli Utility and the St. Petersburg Paradox 415
A.1.1 The St. Petersburg Paradox 415
A.1.2 Use Expected Utility, Not Expected Value 417
Questions and Exercises 419

References 421

Index 429

Preface

Engineering today's systems is a challenging and complex task. Increasingly, systems are engineered by bringing together many separate systems, which together provide an overall capability that is otherwise not possible. Many systems no longer physically exist within clearly defined boundaries, are characterized by their ubiquity and lack of specification, and are unbounded, for example, the Internet.

More and more communication systems, transportation systems, and financial systems connect across domains and seamlessly interface with an uncountable number of users, information repositories, applications, and services. These systems are an enterprise of people, processes, technologies, and organizations. Enterprise systems operate in network-centric ways to deliver capabilities through richly interconnected networks of information and communication technologies.

Engineering enterprise systems is an emerging discipline. It encompasses and extends traditional systems engineering to create and evolve webs of systems and systems-of-systems. In addition, the engineering management and management sciences communities need new approaches for analyzing and managing risk in engineering enterprise systems. The aim of this book is to present advances in methods designed to address this need.

This book is organized around a set of advanced topics in risk analysis that apply to engineering enterprise systems. They include:

A framework for modeling and measuring engineering risks

Capability portfolio risk analysis and management

Functional dependency network analysis (FDNA)

Extreme-event theory

Prioritization systems in highly networked enterprise environments

Measuring risks of extreme latencies in complex queuing networks

The first three topics address the engineering management problem of how to represent, model, and measure risk in large-scale, complex systems engineered to function in enterprise-wide environments. An analytical framework and computational model is presented. In addition, new protocols that capture dependency risks and risk corelationships (RCRs) that may exist in an enterprise are developed and presented. These protocols are called the RCR index and the FDNA approach.

Extreme and rare event risks are of increased concern in systems that function at an enterprise scale. Uncertainties in system behavior are intensified in those that operate in highly networked, globally connected environments. The realization of extreme latencies in delivering time-critical data, applications, or services in these environments can have catastrophic consequences. As such, the last three topics listed above address extreme and rare events and how these considerations can be captured in engineering enterprise systems.

Chapter 1 presents an introduction to engineering risk management. This chapter discusses the nature of risk and uncertainty and their considerations in engineering systems. The objectives of engineering risk management are described along with an overview of modern practices. New perspectives on managing risk in engineering systems-of-systems and enterprise systems are discussed.

Chapter 2 offers perspectives on the theories of systems and risk. This chapter introduces literature foundational to general systems theory, risk and decision theory, and their application to engineering risk management.

Chapter 3 presents foundations of risk and decision theory. Topics include an introduction to probability theory (the language of risk) and decision-making under uncertainty. Value and utility functions are introduced, along with how they apply to the analysis of risk in engineering systems.

Chapter 4 introduces enterprise systems. The discussion includes planning for their engineering and the environments within which they operate. Chapter 4 applies the concepts from the preceding chapters and presents an analytical framework for modeling and measuring risk in engineering enterprise systems. This chapter shows how modeling risk in the enterprise problem space can be represented by a supplier–provider metaphor using mathematical graphs.

Chapters 5 and 6 address the topic of capturing and analyzing dependencies in engineering enterprise systems, from a capability portfolio perspective. Chapter 5 presents a new management metric called the risk corelationship (RCR) index. The RCR index measures risk inheritance between supplier programs and its ripple effects across a capability portfolio. The index identifies and captures horizontal and vertical impacts of risk inheritance, as it increases the threat that risks on one supplier program may adversely affect others and ultimately their contributions to their associated capabilities.

Chapter 6 presents a new analytic technique called Functional Dependency Network Analysis (FDNA). FDNA is an approach that enables management to study and anticipate the ripple effects of losses in supplier-program contributions on dependent capabilities before risks that threaten the suppliers are realized.

The RCR index identifies the supplier programs that face high levels of risk inheritance in delivering their contributions to capability. FDNA identifies whether the level of operability loss, if such risks occur, remains within acceptable margins. Together, the RCR index and FDNA enables management to target risk resolution resources to the supplier programs that face high risk and are most critical to operational capabilities.

Chapter 7 presents an advanced decision-theoretic ranking algorithm that captures each risk's multiconsequential impacts and dependencies that may exist in an enterprise. The algorithm offers a logical and rational basis for addressing the "choice problem" of selecting which capability risks to lessen, or eliminate, as a function of their criticality to the enterprise.

Chapter 8 brings together the concepts developed in the preceding chapters into a coherent recipe for representing, modeling, and measuring risk in engineering large-scale, complex systems designed to function in enterprise-wide environments. This chapter offers decision-makers formal ways to model and measure enterprise-wide risks, their potential multiconsequential impacts, dependencies, and their rippling effects within and beyond enterprise boundaries.

Chapter 9 presents a discussion on random processes and queuing theory as a way to understand chance behaviors in systems. Techniques from these approaches are applied to show how systems under tight resource constraints can be viewed from the queuing perspective.

Chapter 10 presents an introduction to extreme and rare events, the significance of extremes in managing risks in complex systems, and the use of extreme-event probability distributions in assessing risks of rare events.

Chapter 11 discusses the elements of a networked environment, including advanced exploration of the role of priorities in managing the performance of a network. Finally, the two major types of priority discipline models—static and dynamic—and their roles in designing and managing networks are described.

Chapter 12 presents risks of extreme events in complex queuing systems and discusses the role of priority disciplines in managing queuing systems. In addition, this chapter presents tools and techniques for managing extremes, particularly in supersaturated queuing systems based on knowledge of their priority disciplines.

This book is appropriate for advanced risk analysis studies in the engineering systems and engineering management communities. Readers need a background in systems science, systems engineering, and closely related fields. Mathematical competence in differential and integral calculus, risk and decision theory, random processes, and queuing theory is recommended. However, key concepts from these subjects are presented in this book. This facilitates understanding the application of these concepts in the topic areas described. Exercises are provided in each chapter to further the understanding of theory and practice.

Acknowledgments

The authors gratefully acknowledge the following individuals for their encouragement to pursue the research and scholarship in this book.

James H. Lambert, PE, D.WRE, Ph.D., associate director, in the Center for Risk Management of Engineering Systems and research associate professor, in the Department of Systems and Information Engineering at the University of Virginia.

Resit Unal, Ph.D., professor and chair of the Department of Engineering Management and Systems Engineering, at Old Dominion University.

Charles B. Keating, Ph.D., professor in the Department of Engineering Management and Systems Engineering and director in the National Centers for System of Systems Engineering (NCSOSE) at Old Dominion University.

Joost Reyes Santos, Ph.D., assistant professor in the Department of Engineering Management and Systems Engineering at The George Washington University.

C. Ariel Pinto
Paul R. Garvey

Authors

C. Ariel Pinto, Ph.D., is with the faculty of Engineering Management & Systems Engineering at Old Dominion University. He obtained undergraduate and graduate degrees in industrial engineering from the University of the Philippines, and a Ph.D. in systems engineering with concentration in engineering risk analysis from the University of Virginia. He worked in the areas of risk assessment, analysis, modeling, and management while at the Center for Risk Management of Engineering Systems at the University of Virginia, and at the Software Industry Center at Carnegie Mellon University. Dr. Pinto is a co-founder of the Emergent Risk Initiative at Old Dominion University. He is a member of the American Society for Engineering Management, the Society for Risk Analysis, and the International Council in Systems Engineering.

Paul R. Garvey, Ph.D., is Chief Scientist for the *Center for Acquisition and Systems Analysis* at The MITRE Corporation. Dr. Garvey is widely published in the application of risk and decision analytic methods to engineering systems problems. He is the author of two other textbooks published by Chapman-Hall/CRC Press. They are *Probability Methods for Cost Uncertainty Analysis* (2000) and *Analytical Methods for Risk Management* (2008). Dr. Garvey completed his undergraduate and graduate degrees in mathematics from Boston College and Northeastern University. He earned a Ph.D. in engineering management, with a concentration in engineering systems risk and decision analysis from Old Dominion University.

1

Engineering Risk Management

1.1 Introduction

Risk is a driving consideration in decisions that determine how engineering systems are developed, produced, and sustained. Critical to these decisions is an understanding of risk and how it affects the engineering and management of systems. What do we mean by risk?

In general, risk means the possibility of loss or injury. Risk is an event that, if it occurs, has unwanted consequences. In the context of engineering management, risk can be described as answering the question, "What can go wrong with my system or any of its parts?" (Kaplan and Garrick, 1981). In the past 300 years, a theory of risk has grown from connections between the theories of probability and economics.

In probability theory, risk is defined as the chance an unwanted event occurs (Hansson, 2008). In economics, risk is characterized by the way a person evaluates the monetary worth of participation in a lottery or a gamble—any game in which the monetary outcome is determined by chance. We say a person is risk-averse if he/she is willing to accept with certainty an amount of money less than the expected amount he/she might receive from a lottery.

There is a common, but subtle, inclusion of loss or gain in these definitions of risk. Probability theory studies risk by measuring the chances unwanted events occur. What makes an event unwanted? In economics, this question is answered in terms of a person's monetary perspective or value structure. In general, "unwanted" is an adjective that needs human interpretation and value judgments specific to a situation.

Thus, the inclusion of probability and loss (or gain) in the definition of risk is important. Defining risk by these two fundamental dimensions enables trade-offs between them with respect to decision making and course-of-action planning. This is essential in the systems engineering community, which traditionally considers risk in terms of its probability and consequence (e.g., cost, schedule, and performance impacts). Understanding these dimensions and their interactions often sets priorities for whether, how, and when risks are managed in the engineering of systems.

What does it mean to manage risk? From a systems engineering perspective, risk management is a formal process used to continuously identify, analyze, and adjudicate events that, if they occur, have unwanted impacts on a system's ability to achieve its outcome objectives (Garvey, 2008). Applied early, risk management can expose potentially crippling areas of risk in the engineering of systems. This provides management the time to define and implement corrective strategies. Moreover, risk management can bring realism to technical and managerial decisions that define a system's overall engineering strategy.

Successfully engineering today's systems requires deliberate and continuous attention to managing risk. Managing risk is an activity designed to improve the chance that these systems will be completed within cost, on time, and will meet safety and performance requirements.

Engineering today's systems is more sophisticated and complex than ever before. Increasingly, systems are engineered by bringing together many separate systems that, as a whole, provide an overall capability that is not possible otherwise. Many systems no longer physically exist within clearly defined boundaries and specifications, which is a characteristic of traditional systems. Today, systems are increasingly characterized by their ubiquity and lack of specifications. They operate as an enterprise of dynamic interactions between technologies and users, which often behaves in unpredictable ways.

Enterprise systems involve and evolve webs of users, technologies, systems, and systems-of-systems through environments that offer cross-boundary access to a wide variety of resources, systems, and information repositories. Examples of enterprise systems include the transportation networks, a university's information infrastructure, and the Internet.

Enterprise systems create value by delivering capabilities that meet user needs for increased flexibility, robustness, and scalability over time rather than by specifying, *a priori*, firm and fixed requirements. Thus, enterprise system architectures must always be open to innovation, at strategic junctures, which advances the efficacy of the enterprise and its delivery of capabilities and services to users.

Engineering enterprise systems involve much more than discovering and employing innovative technologies. Engineering designs must be adaptable to the evolving demands of user enclaves. In addition, designs must be balanced with respect to expected performance while they are continuously risk-managed throughout an enterprise system's evolution.

Engineers and managers must develop a holistic understanding of the social, political, and economic environments within which an enterprise system operates. Failure to fully consider these dimensions, as they influence engineering and management decisions, can be disastrous. Consider the case of Boston's Central Artery/Tunnel (CA/T) project, informally known as the "Big Dig."

1.1.1 Boston's Central Artery/Tunnel Project

Boston's Central Artery/Tunnel (CA/T) project began in 1991 and was completed in 2007. Its mission was to rebuild the city's main transportation infrastructure such that more than 10 hours of daily traffic congestion would be markedly reduced.

At its peak, the Big Dig involved 5000 construction personnel and more than 100 separate engineering contracts, and its expenditure rate reached $3 million a day. The CA/T project built 161 lane miles of highway in a 7.5 mile corridor (half in tunnels) and included 200 bridges and 4 major highway interchanges (Massachusetts Turnpike Authority, *Big Dig*).

The Big Dig was an engineering and management undertaking on an enterprise scale—a public works project that rivaled in complexity with the Hoover Dam (Stern 2003). From the lens of history, design and engineering risks, though significant, were dwarfed by the project's social, political, environmental, and management challenges. Failure to successfully address various aspects of these challenges led to a $12 billion increase in completion year costs and to serious operational safety failures—one which caused loss of life.

Case studies of the CA/T project will be written for many years. The successes and failures of Boston's Big Dig offer a rich source for understanding the risks associated with engineering large-scale, complex enterprise systems. The following discussion summarizes key lessons from the Big Dig and relates them to similar challenges faced in other enterprise engineering projects.

Research into the management of risk for large-scale infrastructure projects is limited, but some findings are emerging from the engineering community. A study by Reilly and Brown (2004) identified three significant areas of risk that persistently threaten enterprise-scale infrastructure projects such as the Big Dig. These areas are as follows.

System Safety: Experience from the Big Dig

This area refers to the risk of injury or catastrophic failure with the potential for loss of life, personal injury, extensive materiel and economic damage, and loss of credibility of those involved (Reilly and Brown, 2004).

On July 10, 2006, 12 tons of cement ceiling panels fell onto a motor vehicle traveling through one of the new tunnels. The collapse resulted in a loss of life. The accident occurred in the D-Street portal of the Interstate 90 connector tunnel in Boston to Logan Airport. One year later, the National Transportation Safety Board (NTSB) determined that "the probable cause of the collapse was the use of an epoxy anchor adhesive with poor creep resistance, that is, an epoxy formulation that was not capable of sustaining long-term loads" (NTSB, 2007). The safety board summarized its findings as follows:

> Over time, the epoxy deformed and fractured until several ceiling support anchors pulled free and allowed a portion of the ceiling to collapse. Use of an inappropriate epoxy formulation resulted from the failure of Gannett Fleming, Inc., and Bechtel/Parsons Brinckerhoff to identify potential creep in the anchor adhesive as a critical long-term failure mode and to account for possible anchor creep in the design, specifications, and approval process for the epoxy anchors used in the tunnel.

> The use of an inappropriate epoxy formulation also resulted from a general lack of understanding and knowledge in the construction community about creep in adhesive anchoring systems. Powers Fasteners, Inc. failed to provide the Central Artery/Tunnel project with sufficiently complete, accurate, and detailed information about the suitability of the company's Fast Set epoxy for sustaining long-term tensile loads. Contributing to the accident was the failure of Powers Fasteners, Inc., to determine that the anchor displacement that was found in the high occupancy vehicle tunnel in 1999 was a result of anchor creep due to the use of the company's Power-Fast Fast Set epoxy, which was known by the company to have poor long-term load characteristics. Also contributing to the accident was the failure of Modern Continental Construction Company and Bechtel/Parsons Brinckerhoff, subsequent to the 1999 anchor displacement, to continue to monitor anchor performance in light of the uncertainty as to the cause of the failures. The Massachusetts Turnpike Authority also contributed to the accident by failing to implement a timely tunnel inspection program that would likely have revealed the ongoing anchor creep in time to correct the deficiencies before an accident occurred.
>
> **(NTSB/HAR-07/02, 2007)**

Design, Maintainability, and Quality: Experience from the Big Dig

This area refers to the risk of not meeting design, operational, maintainability, and quality standards (Reilly and Brown, 2004).

In many ways a system's safety is a reflection of the integrity of its design, maintainability, and quality. In light of the catastrophic failure just described, of note is the article "Lessons of Boston's Big Dig" by Gelinas (2007) in the *City Journal*. The author writes:

> As early as 1991, the state's Inspector General (IG) warned of the "increasingly apparent vulnerabilities ... of (Massachusetts's) long-term dependence on a consultant" whose contract had an "open-ended structure" and "inadequate monitoring." The main deficiency, as later IG reports detailed, was that Bechtel and Parsons—as "preliminary designer," "design coordinator," "construction coordinator," and "contract administrator"—were often in charge of checking their own work. If the team noticed in managing construction that a contract was over budget because of problems rooted in preliminary design, it didn't have much incentive to speak up.
>
> **(Gelinas, 2007)**

Cost–Schedule Realism: Experience from the Big Dig

This area refers to the risks of significant increases in project and support costs and of significant delays in project completion and start of revenue operations (Reilly and Brown, 2004).

The completion cost of the Big Dig was $14.8 billion. Its original estimate was $2.6 billion. The project's completion cost was 470% larger than its original estimate. If the impacts of unwanted events are measured by cost, then the risks realized by the Big Dig were severe.

Numerous investigations have been made into the reasons why the Big Dig's costs increased to this magnitude. A key finding was lack of cost–schedule realism in the project's initial stages. This was driven by many factors such as incompleteness in cost scope, ignoring the impacts of inflation, overreliance on long-term federal political support (at the expense of building local political and community advocacy), and failure to incorporate risk into cost–schedule estimates.

Sadly, news reported concerns on cost overruns as a factor that contributed to the collapse of cement ceiling tiles in the new tunnel. Consider the following excerpt from the *City Journal's* article "Lessons of Boston's Big Dig" (Gelinas, 2007):

> This problem of murky responsibility came up repeatedly during the Big Dig, but most tragically with the ceiling collapse. Designers engineered a lightweight ceiling for the tunnel in which Milena del Valle died. But Massachusetts, annoyed by cost overruns and cleanliness problems on a similar ceiling, and at the suggestion of federal highway officials, decided to fit the new tunnel with a cheaper ceiling, which turned out to be heavier.
>
> Realizing that hanging concrete where no built-in anchors existed to hold it would be a difficult job, the ceiling's designer, a company called Gannett Fleming, called for contractors to install the ceiling with an unusually large built-in margin for extra weight. Shortly after contractors installed the ceiling using anchors held by a high strength epoxy (as Gannett specified) workers noticed it was coming loose.
>
> Consultants and contractors decided to take it apart and reinstall it. Two years later, after a contractor told Bechtel that "several anchors appear to be pulling away from the concrete," Bechtel directed it to "set new anchors and retest." After the resetting and retesting, the tunnel opened to traffic, with fatal consequences.
>
> **(Gelinas 2007)**

The paper "Management and Control of Cost and Risk for Tunneling and Infrastructure Projects" (Reilly and Brown, 2004) offers reasons from Fred Salvucci (former Massachusetts Secretary of Transportation) for the project's schedule slip and cost growth:

> The reasons had much to do with Governmental policies, local and national politics, new requirements not planned for in the beginning and, political and management transitions that disrupted continuity. Technical complexity was a factor—but it was not the major cause of the schedule slip and cost growth.
>
> **(Reilly and Brown, 2004)**

In summary, the engineering community should study and learn from the successes and failures of Boston's CA/T project. The technical and engineering successes of the Big Dig are truly noteworthy, but sadly so are its failures. Project failures often trace back to judgments unduly influenced by cost, schedule, and sociopolitical pressures. Adherence to best practices in the management of engineering projects is often minimized by these pressures.

Clearly, the emergence of enterprise systems makes today's engineering practices even more challenging than before. Projects at this scale, as experienced in the Big Dig, necessitate the tightest coupling of engineering, management, and sociopolitical involvement in unprecedented ways so that success becomes the norm and failure the exception. Risks can never be eliminated. However, their realization and consequences can be minimized by the continuous participation of independent boards, stakeholder communities, and well-defined lines of management authority.

1.2 Objectives and Practices

Engineering risk management is a core program management process. The objectives of engineering risk management are the early and continuous identification, management, and resolution of risks such that engineering a system is accomplished within cost, delivered on time, and meets performance requirements (Garvey, 2008). Why is engineering risk management important? There are many reasons. The following are key considerations.

- An engineering risk management program fosters the early and continuous identification of risks so that options can be considered and actions implemented before risks seriously threaten a system's performance objectives.
- Engineering risk management enables risk-informed decision-making and course-of-action planning throughout a program's development life cycle, particularly when options, alternatives, or opportunities need to be evaluated.
- An engineering risk management program enables identified risk events to be mapped to a project's work breakdown structure. From this, the cost of their ripple effects can be estimated. Thus, an analytical justification can be established between a project's risk events and the amount of risk reserve (or contingency) funds that may be needed.
- The analyses derived from an engineering risk management program will help identify where management should consider allocating limited (or competing) resources to the most critical risks in an engineering system project.

- Engineering risk management can be designed to provide management with situational awareness in terms of a project's risk status. This includes tracking the effectiveness of courses of action and trends in the rate at which risks are closed with those newly identified and those that remain unresolved.

What are risks? Risks are events that, if they occur, cause unwanted change in the cost, schedule, or technical performance of an engineering system. The occurrence of risk is an event that has negative consequences to an engineering system project. Risk is a probabilistic event; that is, risk is an event that may occur with probability p or may not occur with probability $(1-p)$.

Why are risks present? Pressures to meet cost, schedule, and technical performance are the practical realities in engineering today's systems. Risk is present when expectations in these dimensions push what is technically or economically feasible. Managing risk is managing the inherent contention that exists within and across these dimensions, as shown in Figure 1.1.

What is the goal of engineering risk management? As mentioned earlier, the goal is to identify cost, schedule, and technical performance risks early and continuously, such that control of any of these dimensions is not lost or the consequences of risks, if they occur, are well understood.

Risk management strives to enable risk-informed decision making throughout an engineering system's life cycle. Engineering risk management process and practice vary greatly from very formal to very informal. The degree of formality is governed by management style, commitment, and a project team's attitude toward risk identification, analysis, and management. Next, we present two basic definitions:

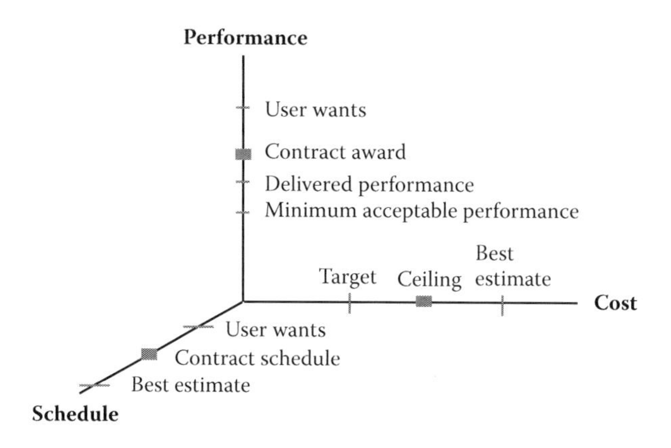

FIGURE 1.1
Pressures on a program manager's decision space. (Adapted from Garvey, P. R., *Analytical Methods for Risk Management: A Systems Engineering Perspective*, Chapman Hall/CRC Press, Taylor & Francis Group (UK), London, 2008.)

Definition 1.1: Risk

Risk is an event that, if it occurs, adversely affects the ability of an engineering system project to achieve its outcome objectives (Garvey, 2008).

From Definition 1.1, a risk event has two aspects. The first is its occurrence probability.* The second is its impact (or consequence) to an engineering system project, which must be nonzero.

A general expression for measuring risk is given by (Garvey, 2008).†

$$Risk = f(Probability, Consequence) \tag{1.1}$$

Definition 1.2: Uncertainty

An event is uncertain if there is indefiniteness about its outcome (Garvey, 2008).

There is a distinction between the definition of risk and the definition of uncertainty. Risk is the chance of loss or injury. In a situation that includes favorable and unfavorable events, risk is the probability an unfavorable event occurs. Uncertainty is the indefiniteness about the outcome of a situation. Uncertainty is sometimes classified as aleatory or epistemic.

Aleatory derives from the Latin word *aleatorius* (gambler). Aleatoric uncertainty refers to inherent randomness associated with some events in the physical world (Ayyub, 2001). For example, the height of waves is aleatoric. Epistemic is an adjective that means *of or pertaining to knowledge*. Epistemic uncertainty refers to uncertainty about an event due to incomplete knowledge (Ayyub, 2001). For example, the cost of engineering a future system is an epistemic uncertainty.

We analyze uncertainty for the purpose of measuring risk. In an engineering system, the analysis might involve measuring the risk of failing to achieve performance objectives, overrunning the budgeted cost, or delivering the system too late to meet users' needs.

Why is the probability formalism used in risk management? Because risk is a potential event, probability is used to express the chance that the event will occur. However, the nature of these events is such that objectively derived measures of occurrence probabilities are typically not possible. Risk management necessarily relies (in part) on probabilities that stem from expert judgment, which are known as measures of belief or subjective probabilities. Are such measures valid?

In 1933, Russian mathematician Kolmogorov established a definition of probability in terms of three axioms (Kolmogorov, 1956). He defined

---

* If *A* is a *risk event*, then the probability that *A* occurs must be strictly greater than zero and strictly less than one.

† Some authors present a general expression for risk that includes *Scenario* as a third aspect, with *Scenario* addressing the question *"what can happen?"*. Including *Scenario* in the general expression for risk is known as Kaplan's triplet, which is discussed in Chapter 2.

probability as a measure that is independent of objective or subjective interpretations of probability. Known as the axiomatic definition, it is the view of probability accepted today.

By this definition, it is assumed that for each random event A, in a sample space Ω, there is a real number $P(A)$ that denotes the probability of A. In accordance with Kolmogorov's axioms, probability is a numerical measure that satisfies the following:

Axiom 1 $0 \le P(A) \le 1$ for any event A in Ω

Axiom 2 $P(\Omega) = 1$

Axiom 3 For any sequence of mutually exclusive events $A_1, A_2, \ldots$ defined on Ω, $P(\bigcup_{i=1}^{\infty} A_i) = \sum_{i=1}^{\infty} P(A_i)$. For any *finite sequence* of mutually exclusive events $A_1, A_2, \ldots, A_n$ defined on Ω, Axiom 3 becomes: $P(\bigcup_{i=1}^{n} A_i) = \sum_{i=1}^{n} P(A_i)$

The first axiom states that the probability of any event is a nonnegative number in the interval 0–1. The second axiom states that a *sure event* is certain to occur. In probability theory, the sample space Ω is referred to as the *sure event*; therefore, we have $P(\Omega)$ equal to 1. The third axiom states that for any infinite or finite sequence of mutually exclusive events, the probability of at least one of these events occurring is the sum of their respective probabilities.

Thus, it is possible for probability to reflect a measure of belief in an event's occurrence. For instance, an engineer might assign a probability of 0.70 to the event "the radar software for the Advanced Air Traffic Control System (AATCS) will not exceed 100K source instructions." Clearly, this event is nonrepeatable. The AATCS cannot be built n times (and under identical conditions) to objectively determine if this probability is indeed 0.70. When an event such as this appears, its probability may be subjectively assigned.

Subjective probabilities should be based on available evidence and previous experience with similar events. They must be plausible and consistent with Kolmogorov's axioms and the theorems of probability.

What about consequence? What does consequence mean and how can it be measured? As mentioned earlier, a risk event's consequence is typically expressed in terms of its impact on an engineering system's cost, schedule, and technical performance. However, there are often other important dimensions to consider. These include programmatic, political, and economic impacts.

Consequence can be measured in many ways. Common measurement methods include techniques from value or utility function theory, which are presented later in this book. These formalisms enable risk events that impact a project in different types of units (e.g., dollars, months, and processing speed) to be compared along normalized, dimensionless scales. This is

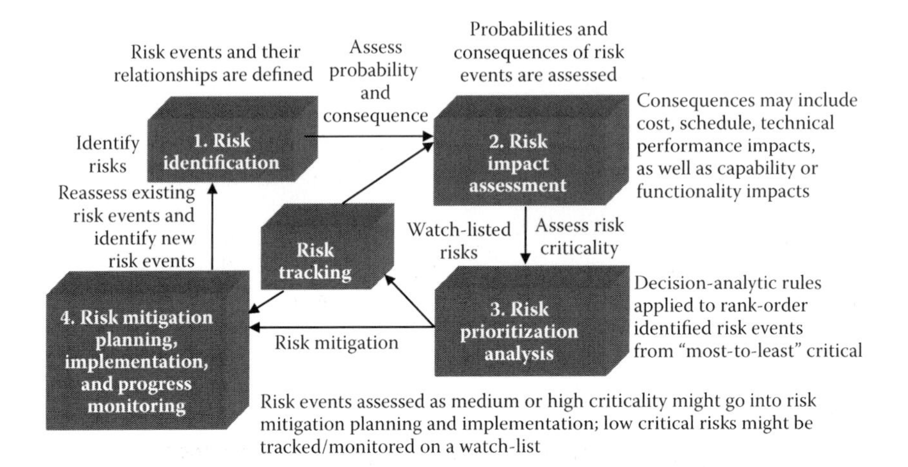

FIGURE 1.2
Steps common to a risk management process.

especially necessary when risk events are rank-ordered or prioritized on the basis of their occurrence probabilities and consequences.

Assessing a risk event's occurrence probability and its consequence is only a part of the overall process of managing risk in an engineering system project. In general, risk management can be characterized by the process illustrated in Figure 1.2. The following describes each step in this process.

Risk Identification

Risk identification is the critical first step of the risk management process. Its objective is the early and continuous identification of risks to include those within and external to the engineering system project. As mentioned earlier, these risks are events that, if they occur, have negative impacts on the project's ability to achieve its performance objectives.

Risk Impact (Consequence) Assessment

In this step, an assessment is made of the impact each risk event could have on the engineering system project. Typically, this includes how the event could impact cost, schedule, and technical performance objectives. Impacts are not limited to these criteria. Additional criteria such as political or economic consequences may also require consideration—discussed later in this book.

An assessment is also made of the probability that each risk event will occur. As mentioned previously, this often involves subjective probability assessments, particularly if circumstances preclude a direct evaluation of probability by objective methods.

Risk Prioritization Analysis

In this step, the overall set of identified risk events, their impact assessments, and their occurrence probabilities are processed to derive a ranking of the most-to-least critical risks. Decision analytic techniques such as utility theory, value function theory, and ordinal methods are formalisms often used to derive this ranking.

A major purpose for prioritizing (or ranking) risks is to form a basis for allocating critical resources. These resources include the assignment of additional personnel or funding (if necessary) to focus on resolving risks deemed most critical to the engineering system project.

Risk Mitigation Planning and Progress Monitoring

This step involves the development of mitigation plans designed to manage, eliminate, or reduce risk to an acceptable level. Once a plan is implemented, it is continually monitored to assess its efficacy with the intent to revise its courses of action if needed.

Systems engineering practices often necessitate the use of historical experience and expert judgments. In recognition of this, the analytical methods developed herein derive from formalisms designed for situations in which the availability of quantitative data is the exception rather than the rule. Specifically, value and utility function theory will be used to represent and measure risk and its effects on engineering systems. These formalisms originate from the von Neumann and Morgenstern axioms of expected utility theory (von Neumann and Morgenstern, 1944) and from modern works on preference theory (Keeney and Raiffa, 1976). Thoughts on these formalisms are given by R. L. Keeney in his book *Value-Focused Thinking: A Path to Creative Decision Making* (1992). Keeney writes:

> The final issue concerns the charge that value (utility) models are not scientific or objective. With that, I certainly agree in the narrow sense. Indeed values are subjective, but they are undeniably a part of decision situations. Not modeling them does not make them go away. It is simply a question of whether these values are included implicitly and perhaps unknowingly in a decision process or whether there is an attempt to make them explicit and consistent and logical. In a broader sense, the systematic development of a model of values is definitely scientific and objective. It lays out the assumptions on which the model is based, the logic supporting these assumptions, and the basis for data (that is, specific value judgments). This makes it possible to appraise the implications of different value judgments. All of this is very much in the spirit of scientific analysis. It certainly seems more reasonable—even more scientific—to approach important decisions with the relevant values explicit and clarified rather than implicit and vague.
>
> **(Keeney, 1992)**

This view reflects the philosophy and the analytic school of thought in this book. It is in this spirit that the formalisms herein were developed to address the very real and complex management problems in engineering today's advanced enterprise systems.

1.3 New Challenges

As mentioned earlier, today's systems are increasingly characterized by their ubiquity and lack of specification. They are an enterprise of systems and systems-of-systems. Through the use of advanced network and communications technologies, these systems continuously operate to meet the demands of globally distributed and uncountable users and communities.

Engineering enterprise systems is an emerging discipline that encompasses and extends traditional systems engineering to create and evolve webs of systems and systems-of-systems. They operate in a network-centric way to deliver capabilities via services, data, and applications through richly interconnected networks of information and communications technologies.

More and more defense systems, transportation systems, and financial systems connect across boundaries and seamlessly interface with users, information repositories, applications, and services. These systems are an enterprise of people, processes, technologies, and organizations.

Thoughts on how to design, engineer, and manage enterprise systems are at the cutting edge of modern systems thinking and engineering. Lack of clearly defined boundaries and diminished hierarchical control are significant technical and managerial challenges. Along with this, the engineering management community needs to establish methods for identifying, analyzing, and managing risks in systems engineered to operate in enterprise contexts.

What makes managing risks in engineering enterprise systems more challenging than that in engineering traditional systems? How does the delivery of capability to users affect how risks are identified and managed in engineering enterprise systems?

With regard to the first question, the difference is principally a matter of scope. From a high-level perspective, the basic risk management process (shown in Figure 1.2) is the same. The challenge comes from implementing and managing this process across a large-scale, complex enterprise – in which contributing systems may be in different stages of maturity and managers, users, and stakeholders may have different capability needs and priorities.

With regard to the second question, an enterprise system is often planned and engineered to deliver capabilities through a series of time-phased increments or evolutionary builds. Thus, risks can originate from many different sources and threaten enterprise capabilities at different points of time. Furthermore, these risks must align to the capabilities they potentially affect, and the scope of their consequences must be understood. In addition, the extent to which enterprise risks may have unwanted collateral effects on other dependent capabilities must be carefully examined.

A final distinguishing challenge in engineering enterprise systems is not only their technologies but also the way users interface with them and each other. Today, the engineering and social science communities are joining in ways not previously seen when planning and evolving the design, development, and operation of enterprise systems (Allen et al., 2004).

The materials in this book present formal methods that promote a holistic understanding of risks in engineering enterprise systems, their potential consequences, dependencies, and rippling effects across the enterprise space. Ultimately, risk management in this context aims to establish and maintain a complete view of risks across the enterprise so that capabilities and performance objectives are achieved via risk-informed resource and investment decisions.

Questions and Exercises

1. In this chapter, engineering risk management was described as a program management process and one that, at its best, is *indistinguishable* from program management.
   (A) Discuss how one might institute protocols to ensure that risk management and program management are inseparable disciplines in the design and engineering of systems.
   (B) What leadership qualities are needed in the management environment to accomplish (A)?
2. The aim of engineering risk management was described as the *early and continuous* identification, management, and resolution of risks such that the engineering of a system is accomplished within cost, delivered on time, and meets performance. Discuss protocols needed on an engineering system project to ensure *early and continuous* identification of risks throughout a project's life cycle.
3. Given Definition 1.1, if A is a *risk event*, then why must the probability of A be strictly greater than 0 and strictly less than 1?
4. This chapter discussed how pressures to meet cost, schedule, and technical performance are the practical realities in engineering today's systems. Risk becomes present, in large part, because expectations in these dimensions push what is technically or economically

feasible. Discuss ways an engineering manager might lessen or guard against these pressures within and across these dimensions. See Figure 1.1.

5. Figure 1.2 presented a general risk management process for engineering system projects. Discuss how this process might be tailored for use in a risk management program designed for engineering an *enterprise system* that consists of a web of systems and systems-of-systems.
6. Review the discussion on Boston's CA/T project in Section 1.1. Enumerate what you consider to be key lessons from this project, as they relate to risks and their consequences. For each lesson in your list, write what you would do to ensure that such risks are minimized (or eliminated) if you were the risk manager on a project with similar challenges.

2

Perspectives on Theories of Systems and Risk

2.1 Introduction

This chapter introduces literature foundational to general systems theory, risk and decision theory, and their application to engineering risk management. Modern literature on engineering systems and risk has only begun to address the complexities and multidisciplinary nature of this topic. However, foundational perspectives on ways to view this space exist in the literature and in the scientific community. Many of these perspectives originate from general systems theory (von Bertalanffy, 1968), a topic discussed in Section 2.2.

The major disciplines described in this book are presented in Figure 2.1 in the form of a literature map. The three axes shown represent engineering systems, risk and decision theory, and engineering risk management. General systems theory provides a foundation for how aspects of these disciplines are applied to the approaches presented in this book.

2.2 General Systems Theory

We begin this discussion with a definition of the word *system*. A system is an *interacting mix of elements forming an intended whole greater than the sum of its parts*. These elements may include people, cultures, organizations, policies, services, techniques, technologies, information/data, facilities, products, procedures, processes, and other human-made (or natural) entities. The whole is sufficiently cohesive to have an identity distinct from its environment (White, 2006). General systems theory is a view that systems are "everywhere" (von Bertalanffy, 1968). Natural laws and social behaviors are parts of highly complex, interdependent systems and elements. General systems theory is a philosophy that considers how to view and pursue scientific inquiry. Its concepts provide a basis for the analytic approaches in this book.

General systems theory is a phrase coined 40 years ago; however, systems and systems thinking have long been part of man's history. Anthropological

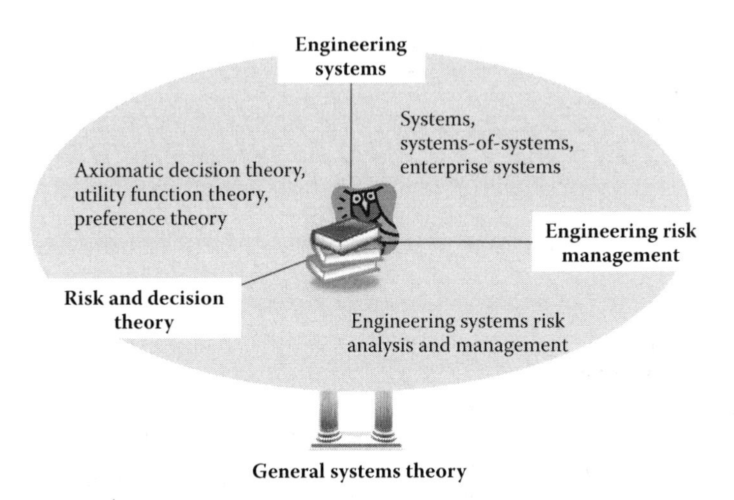

FIGURE 2.1
Literature map dimensions.

evidence reveals the creation of hunting systems by Paleolithic human cultures that existed more than 50,000 years ago.

Cro-Magnon artifacts demonstrate the increasing sophistication of their hunting devices and hunting systems to capture large and dangerous game from safer distances. One such device was a thrower. The thrower operated like a sling shot. When inserted with a spear and thrown, the device increased the spear's speed, range, and lethality. This enabled hunters to attack from distances and so lessened their risk of injury. These were learned from empirical observations and not, at that time, from formal understanding of the laws of motion.

Let us review this from the perspective of general systems theory. Three elements were integrated to make the Cro-Magnon's weapon: the arrowhead; a long, thick stick; and the thrower device. Independently, these elements were ineffective as a weapon for capturing prey. However, when integrated as a whole system, the spear's potential was deadly. The word "potential" is because the spear itself is inert; a human thrower is needed for the spear's effectiveness to be realized.

In this sense, the human thrower is the weapon system, and the spear is the weapon. When multiple human throwers are engaged in a coordinated attack, they operate as a system of weapon systems engaged on a target. Here, the spears are the individual weapon systems that operate together to launch an even more powerful assault on a target than realized by a single human thrower acting as a single weapon system.

It took systems thinking to integrate the weapon's three elements. It took even more of it to improve the weapon's effectiveness by launching weapons simultaneously at targets through group attack strategies. This is one of the many examples of systems, systems thinking, and even system-of-systems thinking in early human culture. It highlights a view that systems are not only everywhere but have always been everywhere. They are ubiquitous throughout nature and society. In 1968, Karl Ludwig von Bertalanffy authored the book "General Systems Theory: Foundations, Development, Applications." He conjectured a theory of systems as "a general science of wholeness" where "the whole is more than the sum of its parts" (von Bertalanffy, 1968). These ideas were understood well by our ancestors. Next, we fast forward from Paleolithic times to the Industrial Revolution.

Historians generally associate the Industrial Revolution with mid- to late-18th century England. Here, mechanical innovations moved agriculture-based economies to economies driven by mass production of manufactured goods. As a result, society experienced dramatic population shifts from rural farms to cities where factories and factory jobs were plenty.

Historians refer to the Industrial Revolution in two phases. The first phase involved mechanical innovations that replaced manual labor with machine-driven mass production of goods. The second phase put many of these innovations into complex applications, including some that required the use of electricity.

Consider steam power. In the first phase of the Industrial Revolution, steam powered many types of manufacturing machines that were operated in factories—especially in factories where waterpower was absent. However, steam power was soon recognized as the driving power for shipping and railway systems. Ultimately, these innovations led to electromechanical technologies that enabled wide-scale transportation systems to be built and operated across the expanse of a nation's land and sea territories. Thus, one can trace the beginning of modern day engineering systems to those innovations, inventions, and processes that appeared more than a century ago.

Today, engineering systems continue to advance, but they do so within another revolution—the digital or information age. Unlike engineering systems built during the height of the Industrial Revolution, today's systems are focused less on enabling the mass production of physical goods and more on enabling global connectivity. With this, engineering systems now make possible the instantaneous transport of digital information around the world.

With an understanding of the past and a perspective on today, Figure 2.2 presents the literature foundational to this book. A chronology of modern scholarship on systems theory and the engineering of systems, systems-of-systems, and enterprise systems is shown. We begin with Bertalanffy's seminal work on general systems theory.

Over 40 years ago, Karl Ludwig von Bertalanffy proposed a general theory of systems to explain fundamental commonalities that seem to underlie natural and sociobehavioral phenomena. He theorized that natural phenomena and social behavior at their elemental levels are systems comprising entities that operate and interact in open and continually dynamic ways.

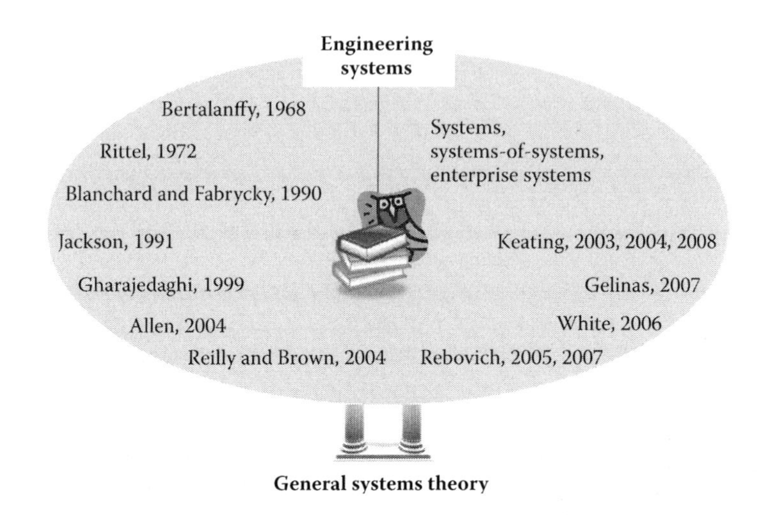

FIGURE 2.2
Literature map: engineering systems.

Bertalanffy (1968) argues that a closed system,* such as an urn containing red and blue marbles, eventually tends toward a state of most probable distribution. This reflects a tendency toward maximum disorder. A system might be closed at certain macrolevels of organization with behaviors predictable with certainty; however, at lower and lower levels of organization, the system eventually becomes more and more open, disordered, and with behaviors not predictable with certainty. "Thus a basic problem posed to modern science is a general theory of organization" (von Bertalanffy, 1968) with general systems theory as a framework within which the behavior and interaction of entities operating within an organization can be discovered.

Bertalanffy (1968) regarded general systems theory as a "general science of wholeness." He saw the incompleteness of trying to understand "observable phenomena" as a collection of entities that could be studied "independently of each other." Systems are fundamentally organizations made of entities "not understandable by investigation of their respective parts in isolation" (von Bertalanffy, 1968) but in how they assemble, react, and interact as a whole. Thus, the behavior of a system (an organization) is not simply the sum of the behavior of its parts.

Bertalanffy's insights were profound. He foresaw not only the challenges in engineering today's complex systems but also ways to view and study their dynamics, interactions, and behaviors. Others also expressed views consistent with those of Bertalanffy when planning and designing engineering systems. For example, Rittel (1972) wrote on the importance of "grasping the whole of

* A closed system is isolated from its environment (von Bertalanffy, 1968). An open system is one that is not closed.

a system" rather than viewing it in "piecemeal," and since a system has many "facets," planning its design is necessarily multidisciplinary.

Bertalanffy and Rittel recognized the need for many different specialties to work together in the engineering of systems as they evolved from single-purpose machines, such as steam engines, to highly complex machines such as space vehicles (von Bertalanffy, 1968; Rittel, 1972). Throughout most of the Industrial Revolution, single-purpose machines were built by engineers trained in their underlying technologies. These machines were comprised of components that were similar to each other and were within the scope of the engineer's training. As the industrial age moved to its second phase, and beyond, engineering systems required the assemblage of more and more different technologies. Today, many different specialties are needed to successfully design, build, and field information age systems.

What are these specialties? These are indeed not only the traditional engineering sciences but also include management, economics, cost analysis, and other analytical areas such as reliability and logistics analyses, modeling and simulation, and human factors. In the past, authors like Blanchard and Fabrycky (1990) brought these specialties and related areas into the modern study of systems engineering. As a result, systems engineering has become the principal discipline from which we address the depth and breadth of the sociotechnical challenges in engineering today's advanced systems.

What is meant by sociotechnical challenges? They are challenges that range from how a system affects society to how society affects a system. With regard to the former, we discussed how the Industrial Revolution changed society from an agricultural economy to one driven by automation and the mass production of goods. With regard to the latter, in Chapter 1 we discussed how social and political attitudes affected technical decisions on Boston's Central Artery/Tunnel (CA/T) project. Thus, sociotechnical challenges stem from how systems interact with people and how people interact with systems.

Social interactions with systems, as enabled by their technologies, are innumerable. They produce desirable and undesirable effects. For example, the ability to easily purchase goods from around the world via networked systems and services led to the emergence of cyber-based commerce and the economic opportunities that it provides. Unfortunately, opportunities often come with risks. Consider the risks posed by cybercrime to electronic commerce. Cybercrime is a socially initiated undesirable behavior that intentionally exploits the vulnerabilities in a system's technologies.

In both cases, electronic commerce and cybercrime illustrate a property called emergent behavior. Emergent properties derive from "the whole of a system and not the properties of its parts; nor can it be deduced from the properties of its parts" (Gharajedaghi, 1999). Emergent behavior has always been possible in systems. However, emergent behaviors in industrial age systems could be anticipated and addressed better than in systems engineered in the current age. Emergent behaviors in today's systems are often so subtle, or originate so deeply in layers of architecture, that their effects or origins can go unnoticed. Thus, there is a persistence of uncertainty and unpredictability in the performance and behavior of information age systems. Why is this? A simple answer is because of networks and networked computing. But that is where simplicity ends in this problem space.

2.2.1 Complex Systems, Systems-of-Systems, and Enterprise Systems

The computer was a closed system before the advent of networks in ways similar to single-purpose machines of the Industrial Revolution. Network technologies brought isolated computers into an open system of globally connected machines in which information dissemination and collaboration is nearly instantaneous.

Networking became the enabling technology of information age systems. Using this technology, separate and autonomous computers could now form into systems of networked computers, and computing grew in scale, complexity, and purpose.

Thus, in today's literature, the terms *complex systems, systems-of-systems,* and *enterprise systems* are commonly found. What do these terms mean and how are they related? It is important to note that the systems engineering community is not settled on answers to these questions. These are cutting-edge topics in engineering systems and systems research. However, a convergence of thought is beginning to emerge. We begin with the term *complex system.* Keating et al. (2003) describe complex systems as those having attributes characterized by Jackson (1991), which are given below:

- Large number of variables or elements and rich interactions among elements
- Difficulty in identifying attributes and emergent properties
- Loosely organized (structured) interaction among elements
- Probabilistic, as opposed to deterministic, behavior in the system
- System evolution and emergence over time
- Purposeful pursuit of multiple goals by system entities or subsystems (pluralistic)
- Possibility of behavioral influence or intervention in the system
- Largely open to the transport of energy, information, or resources from/across the system boundary to the environment.

Examples of complex systems include space shuttles, nuclear power plants, and magnetic resonance imaging scanners. More recently, and consistent with the above, White (2006) defines a complex system as "an open system with continually cooperating and competing elements—a system that continually evolves and changes its behavior according to its own condition and its external environment. Changes between states of order and chaotic flux are possible. Relationships between elements are imperfectly known and difficult to understand, predict, or control."

Engineering systems today are challenged when complex systems become more and more networked in ways that create metasystems—*systems-of-systems* "comprised of multiple embedded and interrelated autonomous complex subsystems" (Keating, 2004). Similarly, White (2006) defines a system-of-systems as "a collection of systems that function to achieve a purpose not achievable by the individual systems acting independently. Each system can operate independently and accomplish its own separate purpose." In a system-of-systems, their whole is indeed more than the sum of their parts; however, it cannot exist without them.

A system-of-systems is formed by the integration of multiple subsystems, where each subsystem can be a complex system. Examples of systems-of-systems include the national airspace system, the international earth observation program known as GEOSS (global earth observation system-of-systems) and navigation systems such as the global positioning system. Building systems-of-systems like these is an enormous engineering and management challenge. Engineering systems of networked systems-of-systems is an even greater challenge. It is the newest demand being faced by today's systems engineering community.

Systems of networked systems-of-systems are sometimes called *enterprise systems*. Enterprise systems, such as the Internet, are the cutting edge of information age computing and global communications.

The literature on engineering enterprise systems is very young. However, scholarship has begun to emerge from academia and industry. Writings by Allen et al. (2004) and Rebovich (2005) reflect thought trends from academic and industry perspectives, respectively.

In the monograph "Engineering Systems: An Enterprise Perspective," Allen et al. (2004) reflect on the nature of an enterprise and its effects on design and engineering solutions. They state, "Such designs are no longer purely technical. In many cases, the enterprise issues are far more difficult than the technical ones to solve; moreover, there must be adaptation on both sides of the relationship between system and enterprise." Moreover, Allen identifies the critical and sometimes orthogonal relationships and goals of the multiple stakeholders in the design of an enterprise system. Allen et al. (2004) write:

> An enterprise perspective on system design makes us aware of the fact that most such designs engage multiple stakeholders. These can range from shareholders to suppliers to members of the workforce to customers to society. What impact can this far-reaching effect have on system design? First of all, stakeholders' interests are not always in alignment.
>
> System design may have to take this into account, balancing the interests of the various stakeholders. As a result, the design process is far more complex than one would be led to believe from the

> engineering science model that we teach to undergraduate engineering students. The best technical solution to a design may very well not be the best overall solution. In fact, it seldom is, and there may not even be a best technical design. Take for example the current F-35 aircraft design. With several customers, each having different missions for this system, the designers cannot optimize the design for any one of the customers' desires. In addition, since recruiting customers in different countries often means engaging suppliers from those countries, adaptations may need to be made in the design to match the capabilities of those suppliers.

Allen's insights echoed Bertalanffy's insights that recognized that systems, such as the F-35 or Boston's Big Dig (discussed in Chapter 1) are fundamentally organizations made of entities (e.g., people) understood by "studying them not in isolation" but by studying how they assemble, react, and interact as a whole. Rebovich (2005, 2007) and other systems thinkers at MITRE offer a view on what is meant by an enterprise and what is fundamentally different. MITRE (2007) writes the following:

> By enterprise we mean a network of interdependent people, processes and supporting technology not fully under control of any single entity. In business literature an enterprise frequently refers to an organization, such as a firm or government agency; in the computer industry it refers to any large organization that uses computers.
>
> Our definition emphasizes the interdependency of individual systems and even systems-of-systems. We include firms, government agencies, large information-enabled organizations and any network of entities coming together to collectively accomplish explicit or implicit goals. This includes the integration of previously separate units. The enterprise displays new behaviors that emerge from the interaction of the parts.

What is fundamentally different?

> A mix of interdependency and unpredictability, intensified by rapid technology change, is driving the need for new systems engineering techniques. When large numbers of systems are networked together to achieve some collaborative advantage, interdependencies spring up among the systems. Moreover, when the networked systems are each individually adapting to both technology and mission changes, then the environment for any given system becomes essentially unpredictable. The combination of massive interdependencies and unpredictability is fundamentally different. Systems engineering success is defined not for an individual known system, but for the network of constantly changing systems.
>
> **(MITRE, 2007)**

From this, it is inferred that a key differentiator of an enterprise system is diminished control over its engineering by a centralized authority.

Centralized or hierarchical control over design decisions is a feature in engineering systems-of-systems and traditional, well-bounded systems (e.g., an airplane and an automobile). Systems-of-systems are, in most cases, engineered in accordance with stated specifications. These may be shaped by multiple stakeholders, but they are managed by a centralized authority with overall responsibility for engineering and fielding the system-of-systems.

This is not the case in engineering enterprise systems. An enterprise system is not characterized by firm and fixed specifications under the control of a centralized authority and agreed to by all participants at different organizational levels. The envelope that captures stakeholders affected by, or involved with, an enterprise system is so broad that centralized or hierarchical control over its engineering is generally not possible and perhaps not even desirable.

Given these challenges and considerations, how is engineering an enterprise planned? The short answer, given what we have seen so far, is through a continual and evolutionary development of capability. What is meant by capability? Based on their experiences to date, planners of enterprise systems define capability as the ability to achieve an effect to a standard under specified conditions using multiple combinations of means and ways to perform a set of tasks (Office of the Secretary of Defense, 2005).

An enterprise is essentially a society of connected users with competing needs, interests, and behaviors. Thus, an enterprise system is characterized more by the capabilities it must field than by the specifications within which it must operate. Moreover, capabilities are constrained by the readiness of technology, availability of suppliers of technology, and the operational limits of the systems and systems-of-systems that enable them.

An enterprise system must be adaptable to evolving missions, changing capability needs, and the dynamics of human behaviors that interact within and across the enterprise. Rebovich (2007) writes:

> Enterprise capabilities evolve through largely unpredictable technical and cultural dimensions. Enterprise capabilities are implemented by the collective effort of organizations whose primary interests, motivations, and rewards come from successfully fielding system capabilities.

Rebovich (2007) further writes:

> Enterprise engineering is an emerging discipline for developing enterprise capabilities. It is a multidisciplinary approach that takes a broad perspective in synthesizing technical and nontechnical (political, economic, organizational, operational, social and cultural) aspects of an enterprise capability.
>
> Enterprise engineering is directed towards enabling and achieving enterprise-level and cross-enterprise operations outcomes. Enterprise engineering is based on the premise that an enterprise is a collection of entities that want to succeed and will adapt to do so. The implication

> of this statement is that enterprise engineering processes are more about shaping the space in which organizations develop systems so that an organization innovating and operating to succeed in its local mission will—automatically and at the same time—innovate and operate in the interest of the enterprise.
>
> Enterprise engineering processes are focused more on shaping the environment, incentives and rules of success in which classical engineering takes place. Enterprise engineering coordinates, harmonizes and integrates the efforts of organizations and individuals through processes informed or inspired by natural evolution and economic markets. Enterprise engineering manages largely through interventions (innovations) instead of (rigorous/strict) controls.

The literature on systems and systems theory will continue to evolve. Systems science is endless. The more we explore, the more our present day understanding is shaped and further challenged. The more we advance in technology and global connectedness, the more open, complex, and virtual the enabling systems will become.

Engineering and managing the development of enterprise systems necessitates, as never before, an openness and adaptability of process, practice, and procedure. Engineering methodologies that are appropriate today might not be appropriate tomorrow. Engineering management practices that scale today might not scale tomorrow. Because we cannot see beyond our line of sight, we should reserve judgment on the finality of any one process or practice at this stage of understanding.

It is with this view that the advanced analytical methods presented in this book were designed. The analytic philosophy was to approach risk analysis in the enterprise space from a *whole systems* perspective. A perspective with roots in the writings of Bertalanffy (1968) and influenced by recognizing that the whole of an enterprise is not only more than the sum of its parts—but is continually shaped, expanded, or diminished by them.

2.3 Risk and Decision Theory

The literature on risk and decision theory is vast. Its intellectual foundations are deeply rooted in mathematics and economics. Risk and decision theory has been a field of study for at least 300 years and one with a rich history of cross-domain applications. Engineering, management, and behavioral sciences all apply and advance aspects of risk and decision theory.

The study of risk is the study of chance and the study of choice. Risk is the chance that an unwanted event occurs. Taking a risk is a choice to gamble on an event whose outcome is uncertain. Risk is the probability that an unfavorable outcome is realized. However, a favorable or an unfavorable outcome is a personal determination—one that is influenced by a person's view of value or worth.

Probability theory is the formalism to study chance. Decision theory is the formalism to study choice. Together, they provide the formalism to study risk. The importance of combining the study of chance and the study of choice was recognized by Swiss mathematician Daniel Bernoulli (1738) in his essay "Exposition of a New Theory on the Measurement of Risk."

The following presents a brief review of the literature foundational to risk and decision theory as it applies to this book. Figure 2.3 identifies these authors. We will begin with Bernoulli and his seminal 1738 essay, which proposed a mathematical relationship between chance and choice, and end with Bertalanffy's insights on the importance of this topic to general systems theory.

Daniel Bernoulli published one of the most influential essays on the theory of risk and its measurement.* He formed the idea that valuing monetary loss or gain from a gamble or lottery should be measured in the context of a player's personal circumstance and *existing wealth*. It was the first time a person's affluence was directly considered in how they value an amount of money won or lost, instead of just its absolute numerical sum (e.g., Figure 2.4).

> To do this the determination of the value of an item must not be based on its price, but rather on the utility it yields. The price of the item is dependent only on the thing itself and is equal for everyone; the utility, however, is dependent on the particular circumstances of the person making the estimate. Thus, there is no doubt that *a gain of one thousand ducats is more significant to a pauper than to a rich man though both gain same amount*.
>
> **(Bernoulli, 1738)**

> Meanwhile, let us use this as a fundamental rule: If the utility of each possible profit expectation is multiplied by the number of ways in which it can occur, and we then divide the sum of these products by the total number of possible cases, a mean utility [*moral expectation*] will be

* Daniel Bernoulli's essay was part of correspondences on a problem that became known as the St. Petersburg paradox, described in Appendix A. The paradox was one of five problems posed by Daniel's cousin Nicolas Bernoulli (1687–1759) to Pierre Raymond de Montmort (1678–1719)—a French mathematician who wrote a treatise on probability theory and games of chance. In a 1728 letter to Nicolas Bernoulli, Swiss mathematician Gabriel Cramer (1704–1752) independently developed concepts similar to those in Daniel Bernoulli's essay. Like Daniel Bernoulli, Cramer wrote "*mathematicians estimate money in proportion to its quantity, and men of good sense in proportion to the usage that they may make of it*." Cramer went on to propose a square root function to represent "proportion of usage," where Daniel Bernoulli derived a logarithmic function. Recognition of Cramer's thoughts as remarkably similar to his own is acknowledged by Daniel Bernoulli at the close of his 1738 essay (Bernoulli, 1738).

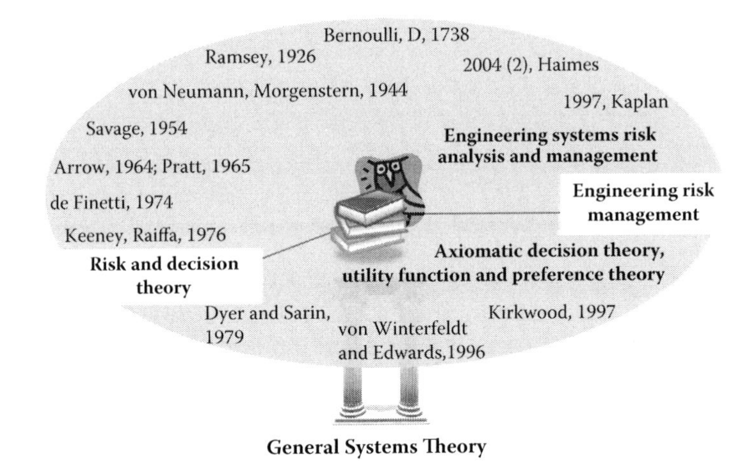

FIGURE 2.3
Literature map: risk and decision theory, engineering management.

FIGURE 2.4
A Swiss Silver Thaler, Zurich, 1727.* (From CNG coins. http://www.cngcoins.com, permission is granted to copy, distribute, and/or modify this document under the terms of the GNU Free Documentation license, Version 1.2.)

> obtained, and the profit which corresponds to this utility will equal the value of the risk in question.
>
> **(Bernoulli, 1738, para. 4)**

> Thus, it becomes evident that no valid measurement of the *value of a risk can be obtained without consideration being given to its utility*, that is to say, the utility of whatever gain accrues to the individual or, conversely, how much profit is required to yield a given utility. However it hardly seems plausible to make any precise generalizations since the utility of an item may change with circumstances. *Thus, though a poor man generally obtains more utility than does a rich man from an equal gain.*
>
> **(Bernoulli, 1738)**

* MONETA REIPUBLICÆ TIGURINÆ, oval city coat-of-arms within ornate frame supported by lions rampant, one holding palm, the other sword. DOMINI CONSERVA NOS IN PACE, aerial view of city along the Limmat River with boats; date in ornate cartouche.

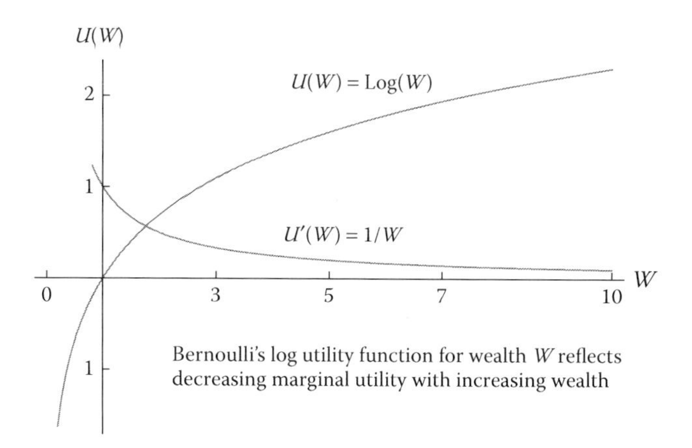

FIGURE 2.5
Bernoulli's log utility function.

With this, Bernoulli introduced the idea of *expected utility theory* and the logarithmic utility function (Figure 2.5) to represent decision-making under uncertainty. It was a formalism that directly captured personal or subjective measures of value (or worth) into a risk calculus.

As seen in Figure 2.5, Bernoulli's log utility function is concave. Concave functions always appear "hill-like." The log utility function exhibits a property known as diminishing marginal utility.* This means that for every unit increase in wealth, there is a corresponding decrease in the rate of additional utility with respect to that change in wealth.

Concave utility functions are always associated with a risk-averse person. A risk-averse person is willing to accept, with certainty, an amount of money less than the expected winnings that might be received from a lottery or gamble. Bernoulli's risk measurement theory assumed that all persons are risk-averse. This assumption was reasonable given the socioeconomic realities of eighteenth century Europe.

Despite the newness of Bernoulli's theory, it would be 200 years before John von Neumann and Oskar Morgenstern (1944) extended its ideas to a set of axioms known as the *axioms of expected utility theory*.† The axioms of expected utility theory state conditions that must exist for rational decision-making in the presence of uncertainty.

Subject to these conditions, a rational individual will choose (or prefer) the option from a set of options (with uncertain outcomes) with maximum expected utility. With this, von Neumann and Morgenstern define a utility

---

* Also known as "Bernoulli's *increasing-at-a-decreasing-rate* thesis, which economists would later term diminishing marginal utility of wealth" (Fishburn, 1989).

† The axioms of expected utility are sometimes called the *axioms of choice* or the *preference axioms*.

function over options with uncertain outcomes (e.g., lotteries and gambles) instead of over wealth as defined by Bernoulli.

Before presenting these axioms, it is important to mention that decision theorists treat individual preferences as *primitives*. In decision theory, a primitive is one that is not derived from other conditions (Garvey, 2008). Decision theory is a calculus that operates on primitives to interpret which option among competing options is the rational choice instead of interpreting why an individual prefers one option more than others.

How are preferences expressed? This can be illustrated by the following two examples:

(1) A person strictly prefers red more than black (red $\succ$ black)*

(2) A person *weakly* prefers five *A-widgets* more than nine *B-widgets* (5 *A-widgets* $\succeq$ 9 *B-widgets*)

The principal axioms of von Neumann-Morgenstern (vNM) expected utility theory are as follows:

Completeness Axiom Given lottery A and lottery B, a person can state that A is strictly preferred to B (i.e., $A \succ B$) or B is strictly preferred to A (i.e., $B \succ A$), or the person is indifferent to them (i.e., $A \sim B$).

Transitivity Axiom If a person prefers lottery A more than lottery B and lottery B more than lottery C, then lottery A is preferred to lottery C.

Continuity Axiom If a person prefers lottery A more than lottery B and lottery B more than lottery C, then there is a probability p that this person is *indifferent* between receiving lottery B with *certainty* and receiving a compound lottery† with probability p of receiving lottery A and probability $(1-p)$ of receiving lottery C.

The continuity axiom means that a person is willing to act on an event that has a favorable or unfavorable outcome if the probability that the unfavorable outcome occurs is reasonably small. Another way to view this axiom is as follows: a slight change in an outcome's occurrence probability p does not change a person's preference ordering of these outcomes. The continuity axiom implies that a continuous utility function exists that represents a person's preference relation.

Independence (Substitution) Axiom If a person prefers lottery A more than lottery B, then a compound lottery that produces lottery A with probability p

---

* The notation $\succ$ is a preference relation notation. Here, $A \succ B$ means a person *strictly* prefers outcome A more than outcome B; $A \succeq B$ means a person *weakly* prefers outcome A more than outcome B (the outcome from A is at least as good as the outcome from B); $A \sim B$ means a person is *indifferent* between outcome A or outcome B.

† A compound lottery is one whose possible outcomes are themselves simple lotteries; a simple lottery is a gamble or risky prospect whose outcomes are determined by chance. Lotteries are discussed further in Chapter 3.

and lottery C with probability $(1 - p)$ is preferred to a compound lottery that produces lottery B with probability p and lottery C with probability $(1 - p)$.

The independence axiom means that a person's preference order of preference for any pair of lotteries, A and B, is preserved when A and B are mixed with a third lottery C, provided lottery A and lottery B are produced with probability p and lottery C is produced with probability $(1 - p)$.

Thus, in vNM utility theory, if an individual's preferences obey these axioms, then they act rationally in choosing the option that has maximum expected utility from a set of options whose outcomes are uncertain. The vNM axiomatization of utility furthered a formal theory of risk with respect to choices under uncertainty. From this, the existence of utility functions could be claimed, and their shapes could be associated with a person's attitude for risk-averse, risk-seeking, or risk-neutral behavior.

In a situation where gains are preferred to losses, a risk-averse person is one who is willing to accept a gain with certainty that is less than the expected amount received from a lottery. The opposite characterizes a risk-seeking person. A risk-seeking person is one who is willing to accept a loss that is greater than the expected amount received from a lottery. A risk-neutral person is one who is neither risk-averse nor risk-seeking. Such a person would be willing to accept a gain or a loss equal only to the expected amount received from a lottery.

There is a class of utility functions that model attitudes with respect to risk-averse, risk-seeking, and risk-neutral behaviors. A family of such functions is shown in Figure 2.6.

Concave utility functions model risk-averse attitudes. When gains are preferred to losses, concave utility functions show a slowing rate of increase in utility for every unit increase in x. Convex utility functions model risk-seeking attitudes. When gains are preferred to losses, convex utility functions show a quickening rate of increase in utility for every unit increase

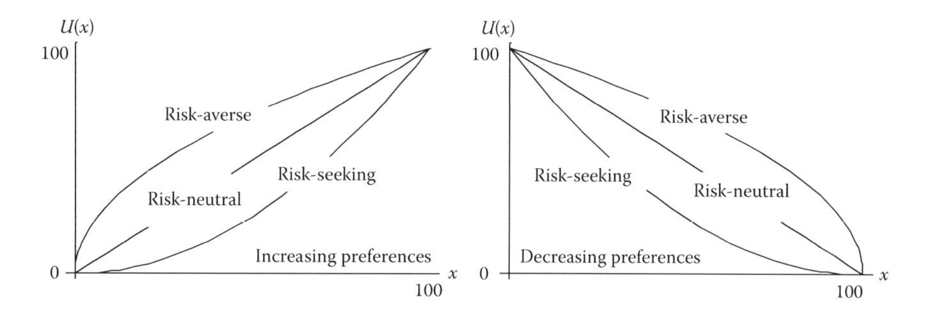

FIGURE 2.6
Families of risk attitude or utility functions. (Adapted from Garvey, P. R., *Analytical Methods for Risk Management: A Systems Engineering Perspective*, Chapman Hall/CRC Press, Taylor & Francis Group (UK), London, 2008.)

in x. Linear utility functions model risk-neutral attitudes. Here, for every unit increase in x there is a constant rate of increase in utility, regardless of whether gains are preferred to losses.

From this, utility theorists began to measure a person's degree of risk averseness by the steepness of their utility function. From calculus, the second derivative of a function provides information about its curvature. Because a vNM utility function $U(x)$ is monotonic and continuous on a close interval $a \le x \le b$, from differential calculus, $U(x)$ is concave if $U''(x) < 0$ and convex if $U''(x) > 0$ for all x such that $a \le x \le b$. So, can $U''(x)$ provide a measure of a person's *degree* of risk averseness that can be compared with another? Not by itself.

A key property of vNM utility functions is that they are unique up to an affine transformation. They are cardinal functions. Preference differences between points along their curves have meaning in accordance with cardinal interval scales.* Unfortunately, $U''(x)$ is not invariant under an affine transformation. However, two theoretical economists K. J. Arrow (Arrow, 1965) and J. W. Pratt (Pratt, 1964) created a measure that used $U''(x)$ and preserved the preference structure of $U(x)$. This became known as the Arrow–Pratt measure of absolute risk aversion. The measure is defined as follows:

$$R_A = -\frac{U''(x)}{U'(x)}$$

Using the Arrow–Pratt measure of absolute risk aversion, the Bernoulli log utility function has decreasing absolute risk aversion; that is,

$$R_A = -\frac{U''(W)}{U'(W)} = -\frac{(-1/W^2)}{1/W} = \frac{1}{W} = U'(W)$$

if $U(W) = \log(W)$ and W denotes wealth. Thus, the Bernoulli log utility function has decreasing absolute risk aversion and decreasing marginal utility with increasing wealth, as shown in Figure 2.5. This means that the amount of wealth a person is willing to risk increases as wealth increases.

* An interval scale is a measurement scale in which attributes are assigned numbers such that differences between them have meaning. The zero point on an interval scale is chosen for convenience and does not necessarily represent the absence of the attribute being measured. Examples of interval scales are the Fahrenheit (F) or Celsius (C) temperature scales. The mathematical relationship between these scales is an *affine transformation*; that is, $F = (9/5) * C + 32$. The zero point in a temperature scale does not mean the absence of temperature. In particular, 0°C is assigned as the freezing point of water. Because distances between numbers in an interval scale have meaning, addition and subtraction of interval scale numbers is permitted; however, because the zero point is arbitrary, multiplication and division of interval scale numbers is not permitted. For example, we can say that 75°F is 25°F hotter than 50°F; but, we cannot say 75°F is 50% hotter than 50°F. However, ratios of differences can be expressed meaningfully; for example, one difference can be one-half or twice or three times another.

The works of Daniel Bernoulli, John von Neumann, Oskar Morgenstern, and others brought about ways to study rational decisions relative to risk and risk taking. A theory of risk emerged in which a person's choice to engage in events with uncertain outcomes could be represented by bounded and monotonic functions called utility functions—mathematical expressions that capture preferences, measures of worth, or degrees of risk aversion *unique* to an individual.

In many ways, utility theory as a basis for a theory of risk was a revolution in the growth of mathematical thought. Before Daniel Bernoulli's 1738 essay on the St. Petersburg paradox, mathematics was principally applied to problems in natural sciences. By 1738, however, mathematics was intersecting with problems in social sciences and most prominently with the study of economics. Economics provided the ideal problem environment to evolve theories of rational choice, as reflected in a person's decision to invest in options with uncertain outcomes.

Around the same time von Neumann and Morgenstern were forming an axiomatic basis for a theory of rational choice, mathematicians were revisiting views on the nature of probability and its meaning as a measure. As mentioned earlier, the study of risk is the study of chance and the study of choice; thus, the dual concepts of probability and choice are integral to the theory of risk.

In 1926, F. P. Ramsey of the University of Cambridge wrote the chapter "Truth and Probability," which appears as Chapter VII in the book *The Foundations of Mathematics, and other Logical Essays* (Ramsey, 1931). In his unpublished work, Ramsey produced some of the earliest arguments and proofs on the logical consistency of subjective utility and subjective probability as measures of value and chance; the latter, which he proved, follows the laws of probability.

Ramsey wrote that measuring a person's degree of belief in the truth of a proposition can be determined from the odds a person would accept when gambling on an uncertain outcome. Thus, Ramsey connected an individual's decision to engage in a bet with their previous knowledge or experience on whether the outcome would likely be in their favor or in their disfavor. This is essentially a lottery, which later became the fundamental concept of von Neumann and Morgenstern's theory of rational choice.

Ramsey's view of probability became increasingly a modern interpretation. Independent of Ramsey's essay, Italian mathematician B. de Finetti (1974) went so far to say that "probability does not exist" (Nau, 2002)—meaning that probability has only a subjective meaning (de Finetti, 1974).

> This definition neatly inverts the objectivistic theory of gambling, in which probabilities are taken to be intrinsic properties of events (e.g., propensities to happen and long-run frequencies) and personal betting rates are later derived from them. Of course, subjective probabilities may be informed by classical, logical, or frequentist reasoning in the special cases where they apply.
>
> **(Nau, 2002)**

Like Ramsey, de Finetti associated probability with the "rate at which an individual is willing to bet on the occurrence of an event. Betting rates are the primitive measurements that reveal your probabilities or someone else's probabilities, which are the only probabilities that really exist." (Nau, 2002). Thus, de Finetti, like Ramsey, viewed probability as one dependent on the state of a person's knowledge.

In 1954, L. J. Savage further extended the ideas of Ramsey, von Neumann and Morgenstern, and de Finetti to ultimately form a Bayesian approach to statistical theory (Savage, 1954). In particular, Savage described a relationship between probability and preference as follows:

> Moreover, a utility function u is unique up to a positive affine (linear) transformation, and the subjective probability π is unique. The relation between probability and preference revealed by the representation is
>
> $$\pi(A) > \pi(B) \underset{\text{iff}}{\Leftrightarrow} \{x \text{ if } A, y \text{ if not } A\} \succ \{x \text{ if } B, y \text{ if not } B\}$$
>
> whenever outcome x is preferred to outcome y. For, Savage, you regard A as more probable than B if you would rather bet on A than B for the preferred outcome.
>
> **(Fishburn, 1989)**

Despite continued debates on the interpretation of probability throughout the early 20th century, the issue was essentially settled in 1933. About 10 years before the postulation of the vNM axioms of utility, Russian mathematician, A. N. Kolmogorov (1956) presented a definition of probability in terms of three axioms. Introduced formally in Chapter 1, the first axiom states that the probability of any event is a nonnegative number in the interval 0–1. The second axiom states that a *sure* or *certain event* has probability equal to 1. The third axiom states that for any sequence of mutually exclusive events, the probability of at least one of these events occuring is the sum of their respective probabilities. Thus, probability is only a numerical measure that behaves according to these axioms. This encompassed all competing interpretations on its nature and allowed objective and subjective probabilities to be part of the "Laplacian" calculus.

Modern Decision Theory

Decision theory has much of its modern theoretical basis in the classic text "Decisions with Multiple Objectives: Preferences and Value Tradeoffs" by R. L. Keeney and H. Raiffa (1976). In this work, Keeney and Raiffa extended the ideas of value and vNM expected utility theory into a modern theory of preference. Preference theory has become the theoretical foundation for most of the risk analysis methods of today's engineering systems, as well as for the work in this book.

Howard Raiffa has written extensively on subjective probability theory, the need for consistency with Kolmogorov's axioms, and its role in Bayesian statistical inference—particularly in decision-making under conditions of uncertainty. Raiffa introduced concepts of preferential and utility independence—which are key to examining trade-offs between alternatives and their performance across multiple criteria. The study of trade-offs led to a theory of multiattribute utility—whose extensive development Raiffa credits to Ralph Keeney, his doctoral student at that time.

Where trade-offs under conditions of uncertainty are captured by multiattribute utility, trade-offs under conditions of certainty are captured by multiattribute value theory. Keeney and Raiffa (1976) write:

> The multiattribute value problem is one of value tradeoffs. If there is no uncertainty in the problem, if we know the multiattribute consequence of each alternative, the essence of the issue is, *How much achievement on objective 1 is the decision-maker willing to give up to improve achievement on objective 2 by some fixed amount*? If there is uncertainty in the problem, the tradeoff issue remains, but difficulties are compounded because it is not clear what the consequences of each alternative will be. The tradeoff issue often becomes a personal value question and, in those cases, it requires the subjective judgment of the decision-maker. There may be no right or wrong answers to these value questions and, naturally enough, different individuals may have different value structures.

Although, Keeney and Raiffa extended the theoretical foundations of vNM utility theory, it was Krantz et al. (1971) and Dyer and Sarin (1979) who developed value functions as formalisms to capture a person's *strength of preference*. A value function is a real-valued function defined over an evaluation criterion (or attribute) that represents an alternative's (or option's) measure of goodness over the levels of the criterion. A measure of "goodness" reflects a decision-maker's judged value in the performance of an alternative (or option) across the levels of a criterion (or attribute).

Similar to a utility function, a value function* is usually designed to vary from 0 to 1 over the range of levels (or scores) for a criterion. In practice, the value function for a criterion's least preferred level (or score), or the least preferred option or alternative, takes the value 0. The value function for a criterion's most preferred level (or score), or the most preferred option or alternative, takes the value 1.

Dyer and Sarin (1979) introduced the concept of a *measurable value function*. A measurable value function is one in which the value difference between any two levels (or scores) within a criterion (or attribute) represents

* A utility function is a value function, but a value function is not necessarily a utility function (Keeney and Raiffa, 1976).

a decision-maker's strength of preference between them—which is also referred to as preference difference.

A measurable value function* is monotonic in preferences. In this, value differences represent relative strength of preference. Large value differences between options (alternatives) indicate that the difference in preference between them is greater than that between other options (alternatives). Furthermore, the numerical amount of this difference represents the relative amount of preference difference. The concept of value differences is also a "primitive concept" in decision theory; that is, it is a concept not derived from other conditions.

One way to address trade-off problems is found in "Strategic Decision Making: Multiobjective Decision Analysis with Spreadsheets" (Kirkwood, 1997). In this work, Kirkwood writes

> If a decision-maker is multiattribute risk averse, then it is necessary to determine a utility function to convert values calculated using a multiattribute value function into utilities. This utility function can then be used to rank alternatives that have uncertainty about their outcomes.
>
> **(Kirkwood, 1997)**

Kirkwood (1997) presents a utility function known as the *power-additive utility function*. It is an exponential utility function that is a function of a multiattribute value function. With the power-additive utility function, Kirkwood connects utility theory to preference theory and to the concept of multiattribute risk averseness (or risk tolerance). Although the power-additive utility function has many useful theoretical properties, its strengths are in its practical aspects. Its shape is fully determined by a single parameter that reflects the risk averseness of a decision-maker. This parameter is known as *multiattribute risk tolerance* ρ_m.

One way to determine ρ_m is for the decision-maker to select the value that reflects his or her risk attitude. Where increasing preferences apply, an extremely risk-averse decision-maker might select ρ_m in the interval $0.05 \leq \rho_m \leq 0.15$. A less risk-averse decision-maker might select ρ_m in the interval $0.15 < \rho_m < 1$. As ρ_m becomes increasingly large the decision-maker becomes increasingly risk-neutral, and the power-additive utility function approaches a straight line. Here, the expected value of the value function can be used to rank alternatives.† Figure 2.7 presents families of power-additive utility functions for various ρ_m and for increasing preferences.

* The vertical axis of a measurable value function is a cardinal interval scale measure of the strength of a decision-maker's preferences. For this reason, a measurable value function is also referred to as a cardinal value function [refer to Dyer and Sarin (1976) and Kirkwood (1997)].

† This is where the expected value of an outcome would equal its expected utility; hence, either decision rule would be a rational basis for the choice under consideration.

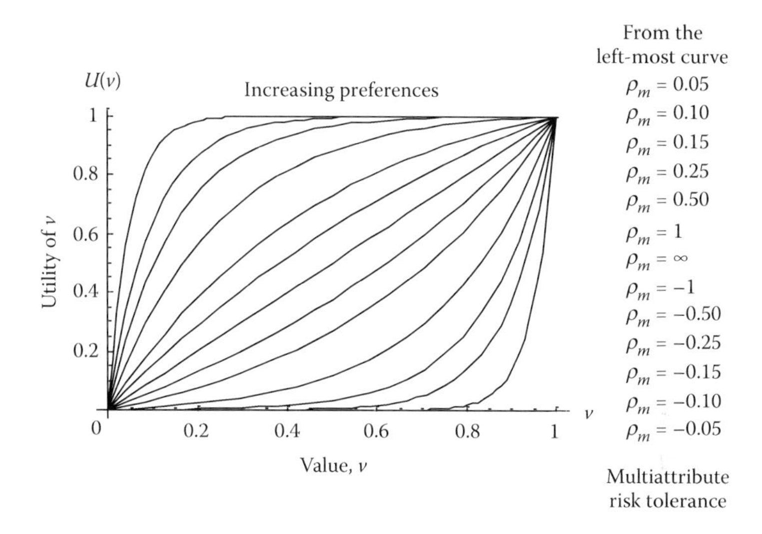

FIGURE 2.7
Families of power-additive utility functions. (Adapted from Garvey, P. R., *Analytical Methods for Risk Management: A Systems Engineering Perspective*, Chapman Hall/CRC Press, Taylor & Francis Group (UK), London, 2008.)

Preference theory and subjective expected utility theory grew in prominence as a mathematical foundation for a theory of rational choice. With this, it became increasingly necessary to understand its conjunction with human behavior relative to decision-making under uncertainty.

By the 1950s, decision theory was joined with behavioral science through the works of Ward Edwards and his graduate student Detlof von Winterfeldt. Two influential articles written by Edwards in 1954 and 1961 introduced vNM utility theory to the field of psychology.

From these papers, an entirely new branch of study called behavioral decision theory emerged. Behavioral scientists began to study whether human behavior followed the views of vNM utility theory. Today, the findings remain somewhat mixed. Nonetheless, Edwards brought a behavioral science view to the topic of rational choice and human decision-making. His many contributions included identifying that persons have preference structures for probabilities and not just for utilities (von Winterfeldt and Edwards, 1986). Consider the following from Fishburn (1989):

> In early work on the psychology of probability, Edwards observed that people's betting behavior reveals preferences among probabilities. For example, given monetary gambles with equal expected values, subjects consistently liked bets with win probability 1/2 and avoided bets with win probability 3/4. Moreover, these probability preferences were reversed in the loss domain, were insensitive to the amounts involved, and could not be explained by curved utility functions.

2.4 Engineering Risk Management

The intellectual groundwork on general systems theory and modern decision theory has natural extensions and applications to managing risks in engineering systems. In general systems theory, vNM utility theory is recognized as concerned with the "behavior of supposedly rational players to obtain maximal gains and minimal losses by appropriate strategies against other players" (Bertalanffy, 1968). Engineering risk management is practiced in similar ways.

To be effective, engineering risk management relies on the behavior of rational decision makers to field systems that maximally achieve performance outcomes, while minimizing failure by taking appropriate risk mitigation strategies against events that threaten success. To achieve this, engineering risk management is best practiced from a "whole systems" perspective—whether it is for traditional systems, systems-of-systems, or enterprise systems.

Successfully engineering today's systems requires deliberate and continuous attention to the management of risk. Managing risk is an activity designed to improve the chance that these systems will be completed within cost, on time, and will meet safety and performance objectives.

As mentioned earlier, the study of risk is the study of chance and choice. In engineering a system, risk is the chance an event occurs with unwanted consequences to the system's cost, schedule, or performance. Furthermore, choices must be made on where to allocate resources to manage risks such that, if they occur, their consequences to the system are eliminated or reduced to acceptable levels.

Until the mid-1970s, risk analyses in engineering systems were often informal and characterized by *ad hoc* collections of qualitative approaches. However, by 1975, qualitative approaches began to be replaced with increasingly insightful quantitative methods. One such method became known as quantitative risk assessment (QRA).*

A founder of QRA was Stan Kaplan, an engineer and applied mathematician who first used the technique to analyze risks associated with engineering and operating nuclear power plants. He gave a definition of risk, which was consistent with that in past scholarship but in a context specific to engineering systems. Kaplan (1997) stated that risk analyses in engineering systems involve answering three questions (Kaplan, 1997). They are as follows:

- *"What can happen?"*
- *"How likely is that to happen?"*
- *"If it happens what are the consequences"?*

* Quantitative risk assessment (QRA) is also known as probabilistic risk assessment (PRA).

These questions became known as Kaplan's triplet and it is represented by the expression

$$Risk = \langle Scenario, Probability, Consequence \rangle$$

where *Scenario, Probability,* and *Consequence* reflect Kaplan's first, second, and third questions, respectively.

A hallmark of QRA is the idea of evidence-based decision making. Here, Kaplan writes that when dealing with an expert, we should never ask for his opinion. Instead, we want his experience, his information, and his evidence (Kaplan, 1997). This includes expert-driven evidence-based probabilities, the second component of Kaplan's triplet. The impetus for this aspect of QRA is rooted in the views of probability expressed by Ramsey, de Finetti, and Savage, as well as from E. T. Jaynes, who wrote extensively on probability as a theory of logic.

> Probabilities need not correspond to physical causal influences or propensities affecting mass phenomena. Probability theory is far more useful if we recognize that probabilities express fundamentally logical inferences pertaining to individual cases.
>
> **(Jaynes, 1988)**

> In our simplest everyday inferences, in or out of science, it has always been clear that two events may be physically independent without being logically independent; or put differently, they may be logically dependent without being physically dependent. From the sound of raindrops striking my window pane, I infer the likely existence of clouds over-head
>
> $$P(Clouds \mid Sound) \approx 1$$
>
> although the sound of raindrops is not a physical causative agent producing clouds. From the unearthing of bones in Wyoming we infer the existence of dinosaurs long ago:
>
> $$P(Dinosaus \mid Bones) \approx 1$$
>
> although the digging of the bones is not the physical cause of the dinosaurs. Yet conventional probability theory cannot account for such simple inferences, which we all make constantly and which are obviously justified. As noted, it rationalizes this failure by claiming that probability theory expresses partial physical causation and does not apply to the individual case.
>
> **(Jaynes, 1988)**

> But if we are to be denied the use of probability theory not only for problems of reasoning about the individual case; but also for problems where the cogent information does not happen to be about a physical cause or a frequency, we shall be obliged to invent arbitrary *ad hockeries* for dealing with virtually all real problems of inference; as indeed the orthodox school of thought has done. Therefore, if it should turn out that probability theory used as logic is, after all, the unique, consistent tool for dealing with such problems, a viewpoint which denies this applicability on ideological grounds would represent a disastrous error of judgment, which deprives probability theory of virtually all its real value and even worse, deprives science of the proper means to deal with its problems.
>
> **(Jaynes, 1988)**

The QRA approach emphasized the importance of scenario-driven risk analyses and the integration of probability and consequence measures with cost-benefit-risk trade-offs to derive optimal risk reduction choices among competing courses of action. Early QRA applications focused on quantifying risks to public safety by certain types of engineering systems, such as nuclear power systems. As QRA methods improved, so did the breadth of their applications to other engineering systems.

The text "Risk Modeling, Assessment, and Management" (Haimes, 2004) was a major contribution in the extension of risk analysis methods to the engineering systems community. Innovations by Haimes included extensions of Keeney-Raiffa decision theory to enable the study of trade-offs between risks and multiconsequential impacts to an engineering system. Haimes pioneered engineering risk management methods into project management processes and practices.

The literature relevant to this book has deep roots in philosophies of logic, probability, utility theory, and general systems theory. Despite, at times, discordant views on issues such as *Does probability exist? Can utility functions represent human preference?* the bodies of work from these disciplines continue to be advanced in theory and application. This book presents these advancements as they pertain to addressing the challenges of managing risk in engineering enterprise systems.

In summary, N. W. Dougherty, president of the American Society for Engineering Education (1954–1955), once said "the ideal engineer is a composite ... he is not a scientist, he is not a mathematician, he is not a sociologist or a writer; but he may use the knowledge and techniques of any or all of these disciplines in solving engineering problems (Dougherty, 1972)." That was true then and is even truer in engineering today's sophisticated, complex, and highly networked engineering systems.

Questions and Exercises

1. Think of real-world examples of a complex system, a system-of-systems, and an enterprise system. Describe their similarities and key distinctions.
2. Study the ideas on general systems theory by Ludwig von Bertalanffy. Write an essay that identifies the characteristics of today's enterprise systems anticipated in his theory, when it was published in 1968. For this exercise, a reading of Bertalanffy's (1968) book "General Systems Theory: Foundations, Development, Applications" is recommended (see "References").
3. Read the paper "Exposition of a New Theory on the Measurement of Risk" written by Daniel Bernoulli (1738). Write an essay that summarizes the major new concepts Bernoulli introduced about risk and its measurement. Describe why Bernoulli's paper is foundational to the emergence of economic science and human choice in the presence of uncertain prospects. For this exercise, a reading of Bernoulli's (1738) paper is recommended.
4. Suppose you have a choice to receive \$10,000 with certainty or participate in a lottery with two possible outcomes. In one outcome, you might receive \$100,000 with probability p. In the other, you might receive nothing with probability $(1 - p)$. Survey your friends to discover their values of p that would lead them to play the lottery instead of accepting \$10,000 with certainty. What is your value of p? What axiom of expected utility theory does this problem represent?
5. Suppose you have two equally attractive outcomes x and y each with probability p and suppose a third outcome z with probability $(1-p)$ is introduced. If you decide that lottery A that produces outcome x with probability p and outcome z with probability $(1-p)$ is equally attractive to a lottery B that produces outcome y with probability p and outcome z with probability $(1-p)$, then what axiom of expected utility theory are you following?
6. Compute the Arrow–Pratt measure of absolute risk aversion for the following utility functions, where w denotes a person's wealth position and $0 \le w \le 1$.

$$\text{(A) } U(w) = \sqrt{w}, \quad \text{(B) } U(w) = w, \quad \text{(C) } U(w) = w - \frac{1}{2}w^2.$$

7. From Exercise 6, graph each utility function and its Arrow–Pratt measure. Identify which utility function behaves with an

Arrow–Pratt measure that is independent of wealth, increasing with increasing wealth and decreasing with increasing wealth.

8. Read the paper "Truth and Probability" (1926) by F. P. Ramsey, University of Cambridge. Summarize Ramsey's key arguments on the logical consistency of subjective utility and subjective probability as measures of value and chance. Compare Ramsey's thoughts on when a person will engage in a risky prospect with those written by Daniel Bernoulli (1738) and von Neumann and Morgenstern (1944) in their theory of rational choice. For this exercise, Ramsey's paper can be obtained from http://homepage.newschool.edu/~het/texts/ramsey/ramsess.pdf.

3

Foundations of Risk and Decision Theory

3.1 Introduction

This chapter introduces the mathematical foundations of risk and decision theory. Topics include elements of probability theory—the original formalism for studying the nature of risk. This is followed by an introduction to modern decision theory and how it aids human judgment in the presence of uncertainty. The chapter concludes with a discussion on how these topics apply to the analysis of risk in engineering systems.

3.2 Elements of Probability Theory

Whether it is a storm's intensity, an arrival time, or the success of a decision, the word "probable" or "likely" has long been part of our language. Most people have an appreciation for the impact of chance on the occurrence of an event. In the last 350 years, the theory of probability has evolved to explain the nature of chance and how it may be studied.

Probability theory is the formal study of events whose outcomes are uncertain. Its origins trace to seventeenth-century gambling problems. Games that involved playing cards, roulette wheels, and dice provided mathematicians a host of interesting problems. Solutions to many of these problems yielded the first principles of modern probability theory. Today, probability theory is of fundamental importance in science, engineering, and business.

Engineering risk management aims at identifying and managing events whose outcomes are uncertain. In particular, it focuses on events that, if they occur, have unwanted impacts or consequences to a project. The phrase *if they occur* means these events are probabilistic in nature. Thus, understanding them in the context of probability concepts is essential. This chapter presents an introduction to these concepts and illustrates how they apply to managing risks in engineering systems.

We begin this discussion with the traditional look at dice. If a six-sided die is tossed, there are clearly six possible outcomes for the number that appears on the upturned face. These outcomes can be listed as elements in a set $\{1,2,3,4,5,6\}$. The set of all possible outcomes of an experiment, such as tossing a six-sided die, is called the *sample space*, which we denote by Ω. The individual outcomes of Ω are called sample points, which we denote by ω.

An *event* is any subset of the sample space. An event is *simple* if it consists of exactly one outcome. Simple events are also referred to as *elementary* events or elementary outcomes. An event is *compound* if it consists of more than one outcome. For instance, let A be the event an odd number appears and B be the event an even number appears in a single toss of a die. These are compound events, which may be expressed by the sets $A = \{1,3,5\}$ and $B=\{2,4,6\}$. Event A occurs *if and only if* one of the outcomes in A occurs. The same is true for event B.

Events can be represented by sets. New events can be constructed from given events according to the rules of set theory. The following presents a brief review of set theory concepts.

Union. For any two events A and B of a sample space, the new event $A \cup B$ (which reads A union B) consists of all outcomes either in A or in B or in both A and B. The event $A \cup B$ occurs if either A or B occurs. To illustrate the union of two events, consider the following: if A is the event an odd number appears in the toss of a die and B is the event an even number appears, then the event $A \cup B$ is the set $\{1,2,3,4,5,6\}$, which is the sample space for this experiment.

Intersection. For any two events A and B of a sample space Ω, the new event $A \cap B$ (which reads A *intersection* B) consists of all outcomes that are in *both* A and B. The event $A \cap B$ occurs *only if both A and B occur.* To illustrate the intersection of two events, consider the following: if A is the event a six appears in the toss of a die, B is the event an odd number appears, and C is the event an even number appears, then the event $A \cap C$ is the simple event $\{6\}$; on the other hand, the event $A \cap B$ contains no outcomes. Such an event is called the *null event*. The null event is traditionally denoted by $\varnothing$. In general, if $A \cap B = \varnothing$, we say events A and B are *mutually exclusive* (*disjoint*). For notation convenience, the intersection of two events A and B is sometimes written as AB, instead of $A \cap B$.

Complement. The complement of event A, denoted by A^c, consists of all outcomes in the sample space Ω that are not in A. The event A^c occurs if and only if A does not occur. The following illustrates the complement of an event. If C is the event an even number appears in the toss of a die, then C^c is the event an odd number appears.

Subset. Event A is said to be a subset of event B if all the outcomes in A are also contained in B. This is written as $A \subset B$.

In the preceding discussion, the sample space for the toss of a die was given by $\Omega = \{1,2,3,4,5,6\}$. If we *assume* the die is fair, then any outcome in the sample space is as likely to appear as any other. Given this, it is reasonable to conclude the proportion of time each outcome is expected to occur is 1/6. Thus, the probability of each simple event in the sample space is

$$P(\{1\}) = P(\{2\}) = P(\{3\}) = P(\{4\}) = P(\{5\}) = P(\{6\}) = \frac{1}{6}$$

Similarly, suppose B is the event an odd number appears in a single toss of the die. This compound event is given by the set $B=\{1,3,5\}$. Since there are three ways event B can occur out of six possible, the probability of event B is $P(B)=3/6=1/2$. The following presents a view of probability known as the equally likely interpretation.

Equally likely interpretation. In this view, if a sample space Ω consists of a finite number of outcomes n, which are all equally likely to occur, then the probability of each simple event is $1/n$. If an event A consists of m of these n outcomes, then the probability of event A is

$$P(A) = \frac{m}{n} \tag{3.1}$$

In this interpretation, it is assumed the sample space consists of a *finite* number of outcomes and all outcomes are equally likely to occur. What if the sample space is finite but the outcomes are *not* equally likely? In these situations, probability might be measured in terms of how frequently a particular outcome occurs when the experiment is repeatedly performed under identical conditions. This leads to a view of probability known as the frequency interpretation.

Frequency interpretation. In this view, the probability of an event is the limiting proportion of time the event occurs in a set of n repetitions of the experiment. In particular, we write this as

$$P(A) = \lim_{n\to\infty} \frac{n(A)}{n}$$

where $n(A)$ is the number of times in n repetitions of the experiment the event A occurs. In this sense, $P(A)$ is the limiting frequency of event A. Probabilities measured by the frequency interpretation are referred to as *objective probabilities.* There are many circumstances where it is appropriate to work with objective probabilities. However, there are limitations with this interpretation of probability. It restricts events to those that can be subjected to repeated trials conducted under *identical conditions.* Furthermore, it is not clear how many trials of an experiment are needed to obtain an event's limiting frequency.

Axiomatic definition. In 1933, Russian mathematician A. N. Kolmogorov presented a definition of probability in terms of three axioms. These axioms define probability in a way that encompasses the *equally likely and frequency interpretations* of probability. It is known as the axiomatic definition of probability. Based on this definition, it is assumed for each event A, in the sample space Ω, there is a real number $P(A)$ that denotes the probability of A. In accordance with Kolmogorov's axioms, introduced in Chapter 1, probability is simply a numerical measure that satisfies the following:

Axiom 1 $0 \leq P(A) \leq 1$ for any event A in Ω

Axiom 2 $P(\Omega) = 1$

Axiom 3 For any sequence of mutually exclusive events $A_1, A_2, \ldots$ defined on Ω

$$P\left(\bigcup_{i=1}^{\infty} A_i\right) = \sum_{i=1}^{\infty} P(A_i)$$

For any *finite sequence* of mutually exclusive events $A_1, A_2, \ldots, A_n$ defined on Ω

$$P\left(\bigcup_{i=1}^{n} A_i\right) = \sum_{i=1}^{n} P(A_i)$$

The first axiom states the probability of any event is a nonnegative number in the interval 0–1. In Axiom 2, the sample space Ω is sometimes referred to as the *sure* or *certain event*; therefore, we have $P(\Omega)$ equal to 1. Axiom 3 states for any sequence of mutually exclusive events, the probability of at least one of these events occurring is the sum of their respective probabilities. In Axiom 3, this sequence may also be finite. From these axioms, five basic theorems of probability are derived.

Theorem 3.1

The probability event A occurs is one minus the probability it will not occur; that is,

$$P(A) = 1 - P(A^c)$$

Theorem 3.2

The probability associated with the null event $\varnothing$ is zero, that is,

$$P(\varnothing) = 0$$

Theorem 3.3

If events A_1 and A_2 are mutually exclusive, then

$$P(A_1 \cap A_2) \equiv P(A_1 A_2) = 0$$

Theorem 3.4

For any two events A_1 and A_2

$$P(A_1 \cup A_2) = P(A_1) + P(A_2) - P(A_1 \cap A_2)$$

Theorem 3.5

If event A_1 is a subset of event A_2, then

$$P(A_1) \leq P(A_2)$$

Measure of belief interpretation. From the axiomatic view, probability need only be a numerical measure satisfying the three axioms stated by Kolmogorov. Given this, it is possible for probability to reflect a "measure of belief" in an event's occurrence. For instance, an engineer might assign a probability of 0.70 to the event *the radar software for the Advanced Air Traffic Control System will not exceed 100K lines of developed source instructions*. We consider this event to be nonrepeatable. It is not practical, or possible, to build this system n times (and under identical conditions) to determine whether this probability is indeed 0.70. When an event such as this arises, its probability may be assigned. Probabilities based on personal judgment, or measure of belief, are known as *subjective probabilities*.

Subjective probabilities are most common in engineering system projects. Such probabilities are typically assigned by expert technical judgment. The engineer's probability assessment of 0.70 is a subjective probability. Ideally, subjective probabilities should be based on available evidence and previous experience with similar events. Subjective probabilities are suspect if premised on limited insights or no prior experience. Care is also needed in soliciting subjective probabilities. They must certainly be plausible *and* they must be *consistent* with Kolmogorov's axioms and the theorems of probability, which stem from these axioms. Consider the following:

> The XYZ Corporation has offers on two contracts A and B. Suppose the proposal team made the following subjective probability assignments. The chance of winning contract A is 40%, the chance of winning contract B is 20%, the chance of winning *contract A or contract B* is 60%, and the chance of winning *both contract A and contract B* is 10%. It turns out this set of probability assignments is *not* consistent with the axioms and theorems of probability! Why is this?* If the chance of winning contract B was changed to 30%, then this *set of probability assignments* would be consistent.

Kolmogorov's axioms, and the resulting theorems of probability, *do not suggest* how to assign probabilities to events. Instead, they provide a way to verify that probability assignments are consistent, whether these probabilities are objective or subjective.

Risk versus uncertainty. There is an important distinction between the terms *risk* and *uncertainty*. Risk is the chance of loss or injury. In a situation that

* The answer can be seen from Theorem 3.4.

includes favorable and unfavorable events, risk is the *probability an unfavorable event occurs*. Uncertainty is the *indefiniteness about the outcome of a situation*. We analyze uncertainty *for the purpose of measuring risk*. In systems engineering, the analysis might focus on measuring the risk of failing to achieve performance objectives, overrunning the budgeted cost, or delivering the system too late to meet user needs. Conducting the analysis often involves degrees of subjectivity. This includes defining the events of concern and, when necessary, subjectively specifying their occurrence probabilities. Given this, it is fair to ask whether it is meaningful to apply rigorous mathematical procedures to such analyses. In a speech before the 1955 Operations Research Society of America meeting, Charles J. Hitch addressed this question. He stated:

> Systems analyses provide a framework which permits the judgment of experts in many fields to be combined to yield results that transcend any individual judgment. The systems analyst may have to be content with better rather than optimal solutions; or with devising and costing sensible methods of hedging; or merely with discovering critical sensitivities. We tend to be worse, in an absolute sense, in applying analysis or scientific method to broad context problems; but unaided intuition in such problems is also much worse in the absolute sense. Let's not deprive ourselves of any useful tools, however short of perfection they may fail.
>
> **(Hitch, 1955)**

Conditional Probability and Bayes' Rule

In many circumstances, the probability of an event is conditioned on knowing another event has taken place. Such a probability is known as a *conditional probability*. *Conditional probabilities* incorporate information about the occurrence of another event. The conditional probability of event A given event B has occurred is denoted by $P(A \mid B)$. If a pair of dice is tossed, then the probability the sum of the toss is even is 1/2. This probability is known as a *marginal* or *unconditional probability*.

How would this unconditional probability change (i.e., be conditioned) if it was *known* the sum of the toss was a number less than 10? This is discussed in the following problem.

PROBLEM 3.1

A pair of dice is tossed, and the sum of the toss is a number less than 10. Given this, compute the probability this sum is an even number.

Solution

Suppose we define event A and event B as follows:

A: The sum of the toss is even.
B: The sum of the toss is a number less than 10.

TABLE 3.1

Outcomes Associated with Event B

$$\begin{bmatrix} \{(1,1)\} & \{(1,2)\} & \{(1,3)\} & \{(1,4)\} & \{(1,5)\} & \{(1,6)\} \\ \{(2,1)\} & \{(2,2)\} & \{(2,3)\} & \{(2,4)\} & \{(2,5)\} & \{(2,6)\} \\ \{(3,1)\} & \{(3,2)\} & \{(3,3)\} & \{(3,4)\} & \{(3,5)\} & \{(3,6)\} \\ \{(4,1)\} & \{(4,2)\} & \{(4,3)\} & \{(4,4)\} & \{(4,5)\} & \\ \{(5,1)\} & \{(5,2)\} & \{(5,3)\} & \{(5,4)\} & & \\ \{(6,1)\} & \{(6,2)\} & \{(6,3)\} & & & \end{bmatrix}$$

The sample space Ω contains 36 possible outcomes; however, in this case, we want the subset of Ω containing *only* those outcomes whose toss yielded a sum less than 10. This subset is shown in Table 3.1. It contains 30 outcomes. Within Table 3.1, only 14 outcomes are associated with the event *the sum of the toss is even given it is a number less than 10.*

$$\begin{Bmatrix} \{(1,1)\},\{(1,3)\},\{(1,5)\},\{(2,2)\},\{(2,4)\},\{(2,6)\},\{(3,1)\},\{(3,3)\},\{(3,5)\} \\ \{(4,2)\},\{(4,4)\},\{(5,1)\},\{(5,3)\},\{(6,2)\} \end{Bmatrix}$$

Therefore, the probability of this event is $P(A|B)=14/30$.

If A and B are events in the same sample space Ω, then $P(A \mid B)$ is the probability of event A within the subset of the sample space defined by event B. Formally, the *conditional probability of event A given event B has occurred,* where $P(B)>0$, is

$$P(A|B) = \frac{P(A \cap B)}{P(B)} \tag{3.2}$$

The *conditional probability of event B given event A has occurred,* where $P(A)>0$, is

$$P(B|A) = \frac{P(B \cap A)}{P(A)} \tag{3.3}$$

PROBLEM 3.2

A proposal team from XYZ Corporation has offers on two contracts A and B. The team made subjective probability assignments on the chances of winning these contracts. They assessed a 40% chance on the event winning contract A, a 50% chance on the event winning contract B, and a 30% chance on the event winning both contracts. Given this, what is the probability of

(A) Winning at least one of these contracts?

(B) Winning contract A and not winning contract B?

(C) Winning contract A if the proposal team has won at least one contract?

Solution

(A) Winning at least one contract means winning either contract A *or* contract B or both contracts. This event is represented by the set $A \cup B$. From Theorem 3.4,

$$P(A \cup B) = P(A) + P(B) - P(A \cap B)$$

therefore

$$P(A \cup B) = 0.40 + 0.50 - 0.30 = 0.60$$

(B) The event winning contract A and not winning contract B is represented by the set $A \cap B^c$. From the Venn diagram in Figure 3.1, observe that $P(A) = P((A \cap B^c) \cup (A \cap B))$.

Since the events $A \cap B^c$ and $A \cap B$ are mutually exclusive (disjoint), from Theorem 3.3 and Theorem 3.4, we have

$$P(A) = P(A \cap B^c) + P(A \cap B)$$

This is equivalent to $P(A \cap B^c) = P(A) - P(A \cap B)$; therefore,

$$P(A \cap B^c) = P(A) - P(A \cap B) = 0.40 - 0.30 = 0.10$$

(C) If the proposal team has won one of the contracts, the probability of winning contract A must be revised (or conditioned) on this information. This means we must compute $P(A \mid A \cup B)$. From Equation 3.2,

$$P(A \mid A \cup B) = \frac{P(A \cap (A \cup B))}{P(A \cup B)}$$

Since $P(A) = P(A \cap (A \cup B))$, we have

$$P(A \mid A \cup B) = \frac{P(A \cap (A \cup B))}{P(A \cup B)} = \frac{P(A)}{P(A \cup B)} = \frac{0.40}{0.60} = \frac{2}{3} \approx 0.67$$

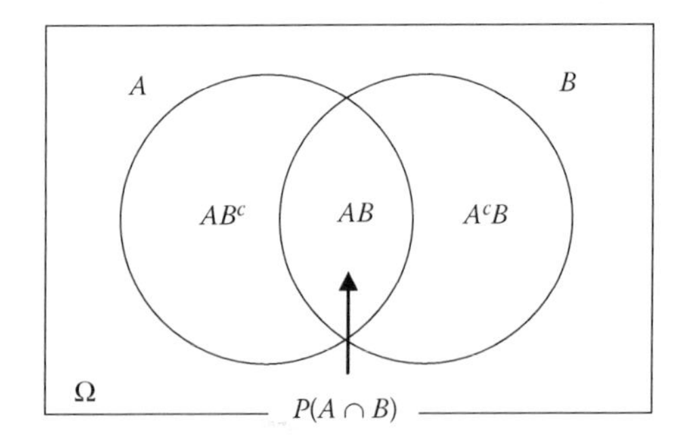

FIGURE 3.1
Venn diagram for $P(A) = P((A \cap B^C) \cup (A \cap B))$.

A consequence of conditional probability is obtained if we multiply Equations 3.2 and 3.3 by $P(B)$ and $P(A)$, respectively. This yields

$$P(A \cap B) = P(B)P(A|B) = P(A)P(B|A) \quad (3.4)$$

Equation 3.4 is known as the *multiplication rule*. The multiplication rule provides a way to express the probability of the intersection of two events in terms of their conditional probabilities. An illustration of this rule is presented in Problem 3.3.

PROBLEM 3.3

A box contains memory chips of which 3 are defective and 97 are nondefective. Two chips are drawn at random, one after the other, without replacement. Determine the probability

(A) Both chips drawn are defective.

(B) The first chip is defective and the second chip is nondefective.

Solution

(A) Let A and B denote the event the first and second chips drawn from the box are *defective*, respectively. From the multiplication rule, we have

$$P(A \cap B) = P(A)P(B|A)$$

$$= P(\text{first chip defective})\, P(\text{second chip defective}|\text{first chip defective})$$

$$= \frac{3}{100}\left(\frac{2}{99}\right) = \frac{6}{9900}$$

(B) To determine the probability the first chip drawn is defective and the second chip is *nondefective*, let C denote the event the second chip drawn is nondefective. Thus,

$$P(A \cap C) = P(AC) = P(A)P(C|A)$$

$$= P(\text{first chip defective})\, P(\text{second chip nondefective}|\text{first chip defective})$$

$$= \frac{3}{100}\left(\frac{97}{99}\right) = \frac{291}{9900}$$

In this example, the sampling was performed without replacement. Suppose the chips sampled were replaced; that is, the first chip selected was replaced before the second chip was selected. In that case, the probability of a defective chip being selected on the second drawing is independent of the outcome of the first chip drawn. Specifically,

P(second chip defective) = P(first chip defective) = 3/100

$$\text{so } P(A \cap B) = \frac{3}{100}\left(\frac{3}{100}\right) = \frac{9}{10000} \text{ and } P(A \cap C) = \frac{3}{100}\left(\frac{97}{100}\right) = \frac{291}{10000}$$

Independent Events

Two events A and B are said to be *independent* if and only if

$$P(A \cap B) = P(A)P(B) \tag{3.5}$$

and *dependent* otherwise. Events $A_1, A_2, \ldots, A_n$ are (mutually) *independent* if and only if the probability of the intersection of any subset of these n events is the product of their respective probabilities.

For instance, events $A_1, A_2,$ and A_3 are independent (or mutually independent) if the following equations are satisfied:

$$P(A_1 \cap A_2 \cap A_3) = P(A_1)P(A_2)P(A_3) \tag{3.5a}$$

$$P(A_1 \cap A_2) = P(A_1)P(A_2) \tag{3.5b}$$

$$P(A_1 \cap A_3) = P(A_1)P(A_3) \tag{3.5c}$$

$$P(A_2 \cap A_3) = P(A_2)P(A_3) \tag{3.5d}$$

It is possible to have three events $A_1, A_2,$ and A_3 for which Equations 3.5b through 3.5d hold but Equation 3.5a does not hold. Mutual independence implies pair-wise independence, in the sense that Equations 3.5b through 3.5d hold, but the converse is not true.

There is a close relationship between independent events and conditional probability. To see this, suppose events A and B are independent. This implies

$$P(AB) = P(A)P(B)$$

From this, Equations 3.2 and 3.3 become, respectively, $P(A \mid B) = P(A)$ and $P(B \mid A) = P(B)$. When two events are independent, the occurrence of one event has no impact on the probability the other event occurs. To illustrate independence, suppose a fair die is tossed. Let A be the event an odd number appears. Let B be the event one of these numbers $\{2,3,5,6\}$ appears. From this,

$$P(A) = 1/2 \quad \text{and} \quad P(B) = 2/3$$

Since $A \cap B$ is the event represented by the set $\{3,5\}$, we can readily state $P(A \cap B) = 1/3$. Therefore, $P(A \cap B) = P(AB) = P(A)P(B)$ and we conclude events A and B are independent.

Dependence can be illustrated by tossing two fair dice. Suppose A is the event the sum of the toss is odd and B is the event the sum of the toss is even. Here, $P(A \cap B) = 0$ and $P(A)$ and $P(B)$ are each 1/2. Since $P(A \cap B) \neq P(A)P(B)$, events A and B are dependent.

It is important not to confuse the meaning of independent events with mutually exclusive events. If events A and B are mutually exclusive, the event *A and B* is empty; that is, $A \cap B = \varnothing$. This implies $P(A \cap B) = P(\varnothing) = 0$. If events A and B are *independent* with $P(A) \neq 0$ and $P(B) \neq 0$ then A and B cannot be mutually exclusive since $P(A \cap B) = P(A)P(B) \neq 0$.

Random Variables

To illustrate the concept of a random variable, consider the set of all possible outcomes associated with tossing two fair six-sided dice. Suppose x represents the sum of the toss. Define X as a variable that takes on only values given by x. If the sum of the toss is 2, then $X = 2$; if the sum of the toss is 3, then $X = 3$. Numerical values of X are associated with events defined from the sample space Ω for this experiment, which is given in Table 3.2. In particular,

$X = 2$ is associated with this simple event $\{(1,1)\}$*

$X = 3$ is associated with these two simple events $\{(1,2)\}, \{(2,1)\}$

$X = 4$ is associated with these three simple events $\{(1,3)\}, \{(2,2)\}, \{(3,1)\}$

Here, X is called a random variable. Formally, a *random variable* is a real-valued function defined over a sample space. The sample space is the *domain* of a random variable. Traditionally, random variables are denoted by capital letters such as X.

Random variables can be characterized as discrete or continuous. A random variable is *discrete* if its set of possible values is finite or countably infinite. A random variable is *continuous* if its set of possible values is uncountable.

TABLE 3.2

Sample Space Associated with Tossing Two Six-Sided Dice

$$\begin{bmatrix} \{(1,1)\} & \{(1,2)\} & \{(1,3)\} & \{(1,4)\} & \{(1,5)\} & \{(1,6)\} \\ \{(2,1)\} & \{(2,2)\} & \{(2,3)\} & \{(2,4)\} & \{(2,5)\} & \{(2,6)\} \\ \{(3,1)\} & \{(3,2)\} & \{(3,3)\} & \{(3,4)\} & \{(3,5)\} & \{(3,6)\} \\ \{(4,1)\} & \{(4,2)\} & \{(4,3)\} & \{(4,4)\} & \{(4,5)\} & \{(4,6)\} \\ \{(5,1)\} & \{(5,2)\} & \{(5,3)\} & \{(5,4)\} & \{(5,5)\} & \{(5,6)\} \\ \{(6,1)\} & \{(6,2)\} & \{(6,3)\} & \{(6,4)\} & \{(6,5)\} & \{(6,6)\} \end{bmatrix}$$

* The outcomes from tossing two dice are recorded as (d_1, d_2), where d_1 and d_2 are the numbers appearing on the upturned faces of the first and second die, respectively. Therefore, in this discussion, $x = d_1 + d_2$.

Discrete Random Variables

Consider the set of all possible outcomes associated with tossing two fair six-sided dice. Suppose x represents the sum of the toss and X is a random variable that takes on only values given by x.

The sample space Ω for this experiment consists of the 36 outcomes in Table 3.2. The random variable X is discrete since the *only* possible values are $x = 2,3,4,5,6,\ldots,12$. The function that describes probabilities associated with the event $\{X = x\}$ for all *feasible values* of x is shown in Figure 3.2. This function is known as the *probability function* of X. Mathematically, the probability function of a discrete random variable X is defined as

$$p_X(x) = P(X = x) \tag{3.6}$$

The probability function is also referred to as the *probability mass function* or the *frequency function* of X. The probability function associates probabilities to events described by distinct (single) points of interest. Over all *feasible* (possible) values of x, probability functions satisfy the following conditions:

$$p_X(x) \geq 0$$

$$\sum p_X(x) = 1 \text{ over all } x$$

If x is *not a feasible* value of X, then

$$p_X(x) = P(X = x) = P(\emptyset) = 0$$

It is often of interest to determine probabilities associated with events of the form $\{X \leq x\}$. For instance, suppose we wanted the probability that the sum of the numbers resulting from the toss of two fair dice will not exceed 7. This is equivalent to computing $P(X \leq 7)$. In this instance, $P(X \leq 7) = P(\{X=2\} \cup \{X=3\} \cup \ldots \cup \{X=7\})$. Thus, X can take a value not exceeding 7 if and only if X takes on one of the values 2,3,…,7. Since

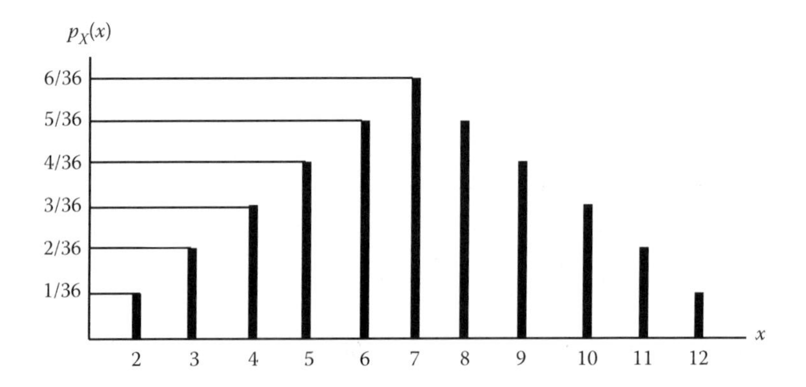

FIGURE 3.2
Probability function for the sum of two dice tossed.

events $\{X=2\},\{X=3\},\ldots,\{X=7\}$ are mutually exclusive, from Axiom 3 and Figure 3.2, we have

$$P(X \le 7) = P(X=2) + P(X=3) + \cdots + P(X=7) = \frac{21}{36}$$

The function that produces probabilities for events of the form $\{X \le x\}$ is known as the *cumulative distribution function* (CDF). Formally, if X is a discrete random variable, then its CDF is defined by

$$F_X(x) = P(X \le x) = \sum_{t \le x} p_X(t) \quad (-\infty < x < \infty) \tag{3.7}$$

In terms of the preceding discussion, we would write $P(X \le 7)$ as

$$\begin{aligned} F_X(7) = P(X \le 7) &= \sum_{t \le 7} p_X(t) = p_X(2) + p_X(3) + \cdots + p_X(7) \\ &= P(X=2) + P(X=3) + \cdots + P(X=7) = 21/36 \end{aligned}$$

The CDF for the random variable with probability function in Figure 3.2 is pictured in Figure 3.3. The CDF in Figure 3.3 is a step function—a characteristic of CDFs for *discrete random variables*. The height of each step along the CDF is the probability the value associated with that step occurs. In Figure 3.3, the probability that $X = 3$ is the height of the step between $X = 2$ and $X = 3$; that is, $P(X = 3) = 3/36 - 1/36 = 2/36$.

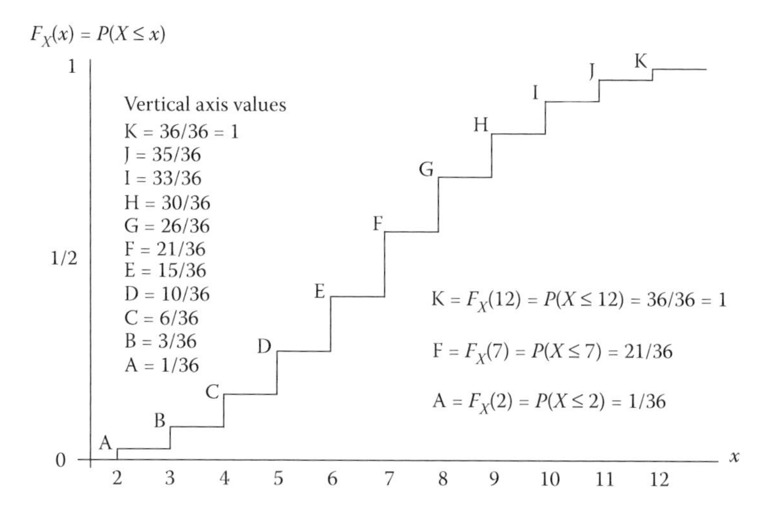

FIGURE 3.3
Cumulative distribution function for the sum of two dice tossed.

Continuous Random Variables

As mentioned earlier, a random variable is continuous if its set of possible values is uncountable. For instance, suppose T is a random variable representing the duration (in hours) of a device. If the possible values of T are given by $\{t : 0 \le t \le 2500\}$, then T is a *continuous random variable*.

In general, we say X is a *continuous random variable* if there exists a *nonnegative function* $f_X(x)$, defined on the real line, such that for any interval A

$$P(X \in A) = \int_A f_X(x)\,dx$$

The function $f_X(x)$ is called the *probability density function* (PDF) of X. Unlike the probability function for a discrete random variable, the PDF *does not* directly produce a probability; that is, $f_X(a)$ does not produce $p_X(a)$ as defined by Equation 3.6. Here, the probability X is contained in any subset of the real line is determined by integrating $f_X(x)$ over that subset. Since X *must* assume some value on the real line, it will always be true that

$$\int_{-\infty}^{\infty} f_X(x)\,dx \equiv P(X \in (-\infty, \infty)) = 1$$

In this case, the CDF of the random variable X is defined as

$$F_X(x) = P(X \le x) = P(X \in (-\infty, x]) = \int_{-\infty}^{x} f_X(t)\,dt \tag{3.8}$$

A useful way to view Equation 3.8 is shown by Figure 3.4. If we assume $f_X(x)$ is a PDF, then from calculus we can interpret the probabilities of the events $\{X \le a\}$ and $\{a \le X \le b\}$ as the areas of the indicated regions in Figure 3.4. When X is a continuous *random variable*, the probability $X = a$ is zero because

$$P(X = a) = P(a \le X \le a) = \int_a^a f_X(x)\,dx = 0 \tag{3.9}$$

From this, it is seen the inclusion or exclusion of an interval's endpoints does not affect the probability X falls in the interval. Thus, if a and b are any two real numbers then

$$P(a < X \le b) = P(a < X < b) = P(a \le X < b) = P(a \le X \le b) = F_X(b) - F_X(a) \tag{3.10}$$

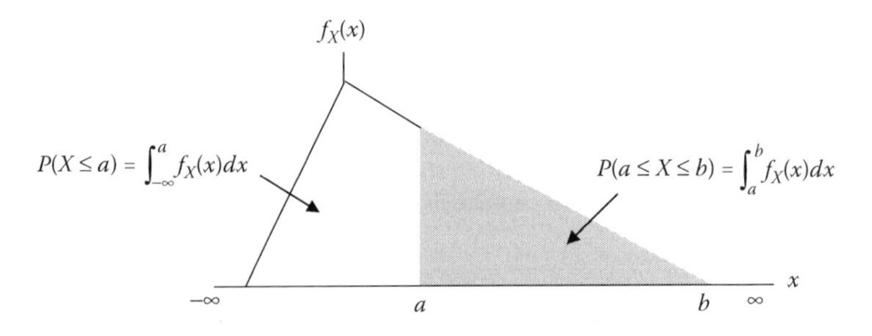

FIGURE 3.4
A probability density function.

when X is a *continuous random variable*.

In general, for *any* discrete or continuous random variable, the value of $F_X(x)$ at any x must be a number in the interval $0 \le F_X(x) \le 1$. The function $F_X(x)$ is always continuous from the right. It is nondecreasing as x increases; that is, if $x_1 < x_2$, then $F_X(x_1) \le F_X(x_2)$. Finally,

$$\lim_{x \to -\infty} F_X(x) = 0 \text{ and } \lim_{x \to \infty} F_X(x) = 1$$

Bayes' Rule

Suppose we have a collection of events A_i representing possible conjectures about a topic. Furthermore, suppose we have some initial probabilities associated with the "truth" of these conjectures. Bayes' rule* provides a way to update (or revise) initial probabilities when new information about these conjectures is evidenced. Bayes' rule is a consequence of conditional probability.

Suppose we partition a sample space Ω into a finite collection of three mutually exclusive events. In Figure 3.5, define these as A_1, A_2, and A_3, where $A_1 \cup A_2 \cup A_3 = \Omega$. Let B denote an arbitrary event contained in Ω. With this, we can write $B = (A_1 \cap B) \cup (A_2 \cap B) \cup (A_3 \cap B)$. Since events $(A_1 \cap B)$, $(A_2 \cap B)$, $(A_3 \cap B)$ are mutually exclusive, it follows from Axiom 3 that $P(B) = P(A_1 \cap B) + P(A_2 \cap B) + P(A_3 \cap B)$. From the multiplication rule given in Equation 3.4, $P(B)$ can be expressed in terms of conditional probability as

$$P(B) = P(A_1)P(B|A_1) + P(A_2)P(B|A_2) + P(A_3)P(B|A_3)$$

This equation is known as the *total probability law*. Its generalization is

$$P(B) = \sum_{i=1}^{n} P(A_i)P(B|A_i)$$

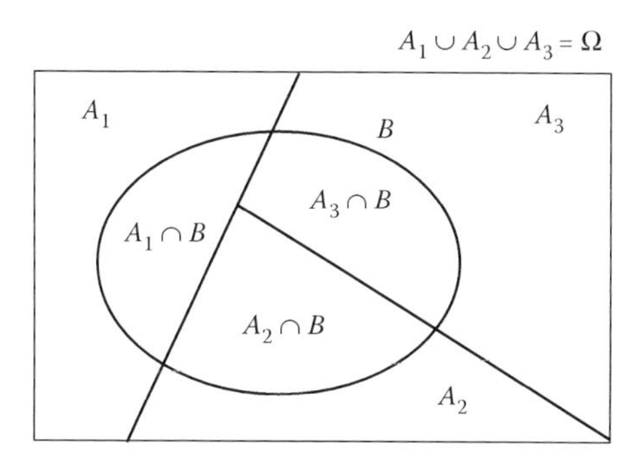

FIGURE 3.5
Partitioning Ω into three mutually exclusive sets.

* Named in honor of Thomas Bayes (1702–1761), an English minister and mathematician.

where $\Omega = \bigcup_{i=1}^{n} A_i$ and $A_i \cap A_j = \emptyset$ and $i \neq j$. The conditional probability for each event A_i given event B has occurred is

$$P(A_i|B) = \frac{P(A_i \cap B)}{P(B)} = \frac{P(A_i)P(B|A_i)}{P(B)}$$

When the total probability law is applied to this equation, we have

$$P(A_i|B) = \frac{P(A_i)P(B|A_i)}{\sum_{i=1}^{n} P(A_i)P(B|A_i)} \tag{3.11}$$

This equation is known as *Bayes' rule.*

PROBLEM 3.4

The ChipyTech Corporation has three divisions D_1, D_2, and D_3 that each manufactures a specific type of microprocessor chip. From the total annual output of chips produced by the corporation, D_1 manufactures 35%, D_2 manufactures 20%, and D_3 manufactures 45%. Data collected from the quality control group indicate 1% of the chips from D_1 are defective, 2% of the chips from D_2 are defective, and 3% of the chips from D_3 are defective. Suppose a chip was randomly selected from the total annual output produced and it was found to be defective. What is the probability it was manufactured by D_1? By D_2? By D_3?

Solution

Let A_i denote the *event* the selected chip was produced by division D_i $(i=1,2,3)$. Let B denote the event the selected chip is defective. To determine the probability the defective chip was manufactured by D_i, we must compute the conditional probability $P(A_i | B)$ for $i=1,2,3$. From the information provided, we have

$$P(A_1) = 0.35, \quad P(A_2) = 0.20, \quad \text{and} \quad P(A_3) = 0.45$$

$$P(B|A_1) = 0.01, \quad P(B|A_2) = 0.02, \quad P(B|A_3) = 0.03$$

The total probability law and Bayes' rule are used to determine $P(A_i|B)$ for each $i = 1, 2,$ and 3. With this, $P(B)$ can be written as

$$P(B) = P(A_1)P(B|A_1) + P(A_2)P(B|A_2) + P(A_3)P(B|A_3)$$

$$P(B) = 0.35(0.01) + 0.20(0.02) + 0.45(0.03) = 0.021$$

and from Bayes' rule, we can write

$$P(A_i| B) = \frac{P(A_i)P(B|A_i)}{\sum_{i=1}^{n} P(A_i)P(B|A_i)} = \frac{P(A_i)P(B|A_i)}{P(B)}$$

$$P(A_1|B) = \frac{P(A_1)P(B|A_1)}{P(B)} = \frac{0.35(0.01)}{0.021} = 0.167$$

$$P(A_2|B) = \frac{P(A_2)P(B|A_2)}{P(B)} = \frac{0.20(0.02)}{0.021} = 0.190$$

$$P(A_3|B) = \frac{P(A_3)P(B|A_3)}{P(B)} = \frac{0.45(0.03)}{0.021} = 0.643$$

Table 3.3 provides a comparison of $P(A_i)$ with $P(A_i|B)$ for each $i=1,2,3$. The probabilities given by $P(A_i)$ are the probabilities the selected chip will have been produced by division D_i before it is randomly selected and before it is known whether the chip is defective. Therefore, $P(A_i)$ are the *prior*, or *a priori* (before-the-fact) probabilities. The probabilities given by $P(A_i|B)$ are the probabilities the selected chip was produced by division D_i after it is known the selected chip is defective. Therefore, $P(A_i|B)$ are the *posterior*, or *a posteriori* (after-the-fact) probabilities. Bayes' rule provides a means for computing posterior probabilities from the known prior probabilities $P(A_i)$ and the conditional probabilities $P(B|A_i)$ for a particular situation or experiment.

Bayes' rule established a philosophy that became known as *Bayesian inference* and *Bayesian decision theory*. These areas play important roles in the application of probability theory to systems engineering problems. In the total probability law, we may think of A_i as representing possible states of nature to which an engineer assigns subjective probabilities. These subjective probabilities are the prior probabilities, which are often premised on personal judgments based on past experience. In general, Bayesian methods offer a powerful way to revise, or update, probability assessments as new information becomes available.

Bayesian Inference in Engineering Risk Management

This discussion introduces a technique known as *Bayesian inference*. Bayesian inference is a way to examine how an initial belief in the truth of a hypothesis H may change when evidence e relating to it is observed. This is done by a repeated application of Bayes' rule.

TABLE 3.3

Bayes' Probability Updating: Problem 3.4 Summary

i	$P(A_i)$	$P(A_i \mid B)$
1	0.35	0.167
2	0.20	0.190
3	0.45	0.643

Suppose an engineering firm was awarded a project to develop a software application. Suppose there are a number of challenges associated with this and among them are staffing the project, managing multiple development sites, and functional requirements that continue to evolve.

Given these challenges, suppose the project's management team believes they have a 50% chance of completing the software development in accordance with the customer's planned schedule. From this, how might management use Bayes' rule to monitor whether the *chance* of completing the project on schedule is increasing or decreasing?

As mentioned earlier, *Bayesian inference* is a procedure that takes evidence, observations, or indicators, as they emerge, and applies Bayes' rule to infer the truthfulness or falsity of a hypothesis, in terms of its probability. In this case, the hypothesis H is *Project XYZ will experience significant delays in completing its software development.*

Suppose at time t_1 the project's management comes to recognize that *Project XYZ has been unable to fully staff to the number of software engineers needed for this effort*. In Bayesian inference, we treat this as an observation or evidence that has some bearing on the truthfulness of H. This is illustrated in Figure 3.6. Here, H is the hypothesis "node" and e_1 is the evidence node contributing to the truthfulness of H.

Given the evidence-to-hypothesis relationship in Figure 3.6, we can form the following equations from Bayes' rule.

$$P(H|e_1) = \frac{P(H)P(e_1|H)}{P(H)P(e_1|H) + P(H^c)P(e_1|H^c)} \tag{3.12}$$

$$P(H|e_1) = \frac{P(H)P(e_1|H)}{P(H)P(e_1|H) + (1 - P(H))P(e_1|H^c)} \tag{3.13}$$

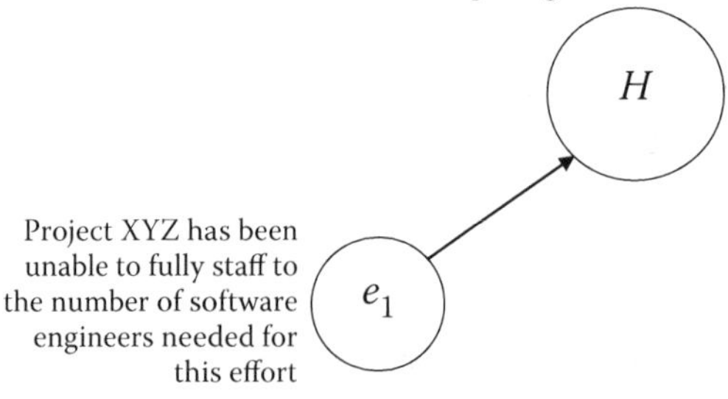

FIGURE 3.6
Evidence observed at time t_1.

Here, $P(H)$ is the team's initial or *prior* subjective (judgmental) probability that Project XYZ will be completed in accordance with the customer's planned schedule. Recall from the preceding discussion this was $P(H)=0.50$. The other terms in Equation 3.12 (or Equation 3.13) are defined as follows: $P(H|e_1)$ is the probability H is true given evidence e_1, the term $P(e_1|H)$ is the probability evidence e_1 would be observed given H is *true*, and the term $P(e_1|H^c)$ is the probability evidence e_1 would be observed given H is *not true*.

Suppose this team's experience with e_1 is that staffing shortfalls is a factor that contributes to delays in completing software development projects. Given this, suppose they judge $P(e_1|H)$ and $P(e_1|H^c)$ to be 0.60 and 0.25, respectively. From the evidence e_1 and the team's probability assessments related to e_1, we can compute a revised probability that Project XYZ will experience significant delays in completing its software development. This revised probability is given by the following equation.

$$P(H|e_1)=\frac{P(H)P(e_1|H)}{P(H)P(e_1|H)+(1-P(H))P(e_1|H^c)}$$

$$=\frac{(0.50)(0.60)}{(0.50)(0.60)+(1-0.50)(0.25)}=0.706 \qquad (3.14)$$

Notice the effect evidence e_1 has on increasing the probability that Project XYZ will experience a significant schedule delay. We have gone from the initial or *prior* probability of 50% to a *posterior* probability of just over 70%.

In the Bayesian inference community, this is sometimes called *updating*, that is, updating the "belief" in the truthfulness of a hypothesis in light of observations or evidence that adds new information to the initial or prior assessments.

Next, suppose the management team observed two more evidence nodes at, say, time t_2. Suppose these are in addition to the continued relevance of evidence node e_1. Suppose the nature of evidence nodes e_2 and e_3 are described in Figure 3.7. Now, what is the chance Project XYZ will experience a significant schedule delay given all the evidence collected in the set shown in Figure 3.7? Bayesian updating will again be used to answer this question.

Here, we will show how Bayesian updating is used to sequentially revise the *posterior* probability computed in Equation 3.14, to account for the observation of new evidence nodes e_2 and e_3. We begin by writing the following:

$$P(H|e_1 \cap e_2)\equiv P(H|e_1e_2) \qquad (3.15)$$

$$P(H|e_1e_2)=\frac{P(H|e_1)P(e_2|H)}{P(H|e_1)P(e_2|H)+(1-P(H|e_1))P(e_2|H^c)} \qquad (3.16)$$

$$P(H|e_1 \cap e_2 \cap e_3)\equiv P(H|e_1e_2e_3) \qquad (3.17)$$

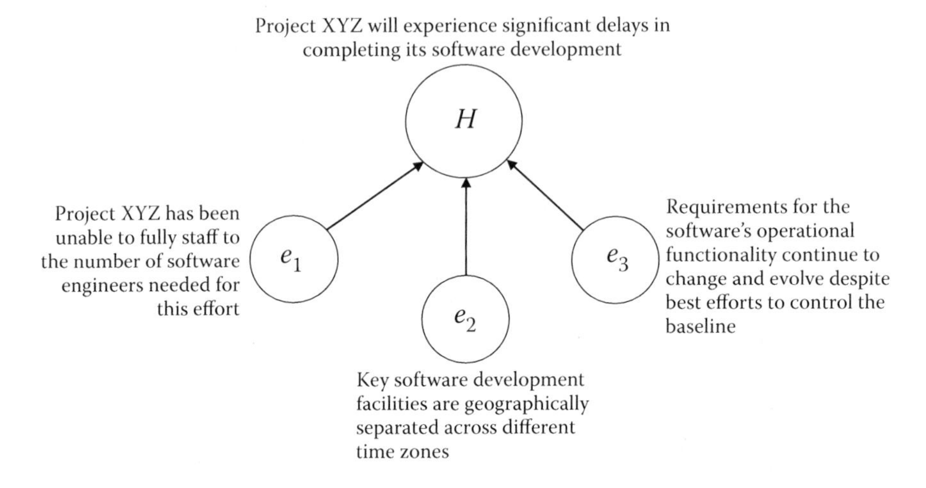

FIGURE 3.7
Evidence e_2 and e_3 observed at time t_2.

$$P(H|e_1e_2e_3) = \frac{P(H|e_1e_2)P(e_3|H)}{P(H|e_1e_2)P(e_3|H) + (1 - P((H|e_1e_2)|e_1))P(e_3|H^c)} \tag{3.18}$$

Suppose the management team made the following assessments:

$$P(e_2|H) = 0.90,\ P(e_2|H^c) = 0.45$$
$$P(e_3|H) = 0.95,\ P(e_3|H^c) = 0.10$$

Substituting them first into Equation 3.16 and then into Equation 3.18 yields the following:

$$P(H|e_1e_2) = 0.83 \quad \text{and} \quad P(H|e_1e_2e_3) = 0.98$$

Thus, given the influence of *all* the evidence observed to date, we can conclude hypothesis H is almost certain to occur. Figure 3.8 illustrates the findings from this analysis.

Writing a Risk Statement

Probability is a measure of the chance an *event* may or may not occur. Furthermore, all probabilities are conditional in the broadest sense that one can always write the following*:

$$P(A) = P(A|\Omega)$$

where A is an event (a subset) contained in the sample space Ω.

* This result derives from the fact that $P(\Omega|A)=1$.

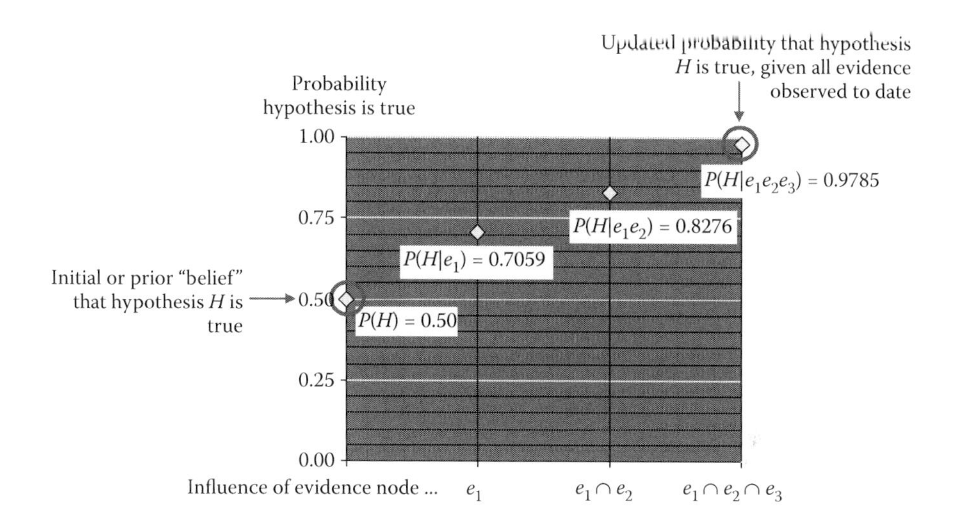

FIGURE 3.8
Bayesian updating: truthfulness of hypothesis H.

In a similar way, one may consider subjective or judgmental probabilities as conditional probabilities. The conditioning event (or events) may be experience with the occurrence of events known to have a bearing on the occurrence probability of the future event. Conditioning events can also manifest themselves as evidence, as discussed in the previous section on Bayesian inference.

Given these considerations, a best practice for expressing an identified risk is to write it in a form known as the *risk statement*. A risk statement aims to provide clarity and descriptive information about the identified risk so a reasoned and defensible assessment can be made on the risk's occurrence probability and its areas of impact (if the risk event occurs).

A protocol for writing a risk statement is the *Condition-If-Then* construct. This construct is a recognition that a risk event is, by its nature, a probabilistic event and one that, if it occurs, has unwanted consequences. An example of the *Condition-If-Then* construct is shown in Figure 3.9.

In Figure 3.9, the *Condition* reflects what is known today. It is the *root cause* of the identified risk event. Thus, the *Condition* is an event that has occurred, is presently occurring, or will occur with certainty. Risk events are future events that may occur because of the *Condition* present. The following is an illustration of this protocol.

Suppose we have the following two events. Define the *Condition* as event B and the *If* as event A (the risk event)

B = {*Current test plans are focused on the components of the subsystem and not on the subsystem as a whole*}

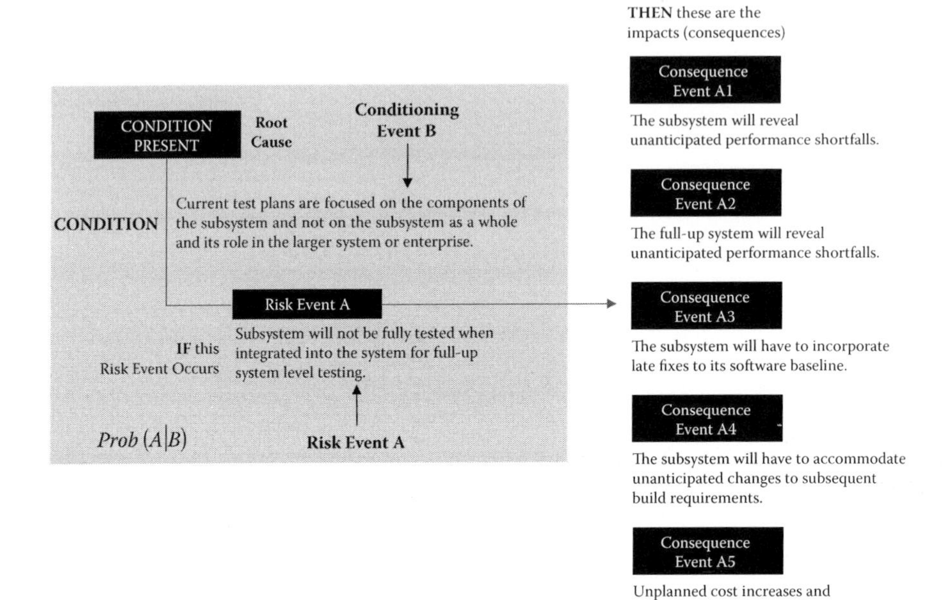

FIGURE 3.9
An illustration of the *Condition If-Then* construct (Garvey, 2008).

A = {*Subsystem will not be fully tested when integrated into the system for full-up system-level testing*}

The risk statement is the *Condition-If* part of the construct; specifically,

RISK STATEMENT: *{The subsystem will not be fully tested when integrated into the system for full-up system-level testing because current test plans are focused on the components of the subsystem and not on the subsystem as a whole}*

From this, we see the *Condition-If* part of the construct is equivalent to a probability event; formally, we can write

$$0 < P(A|B) = \alpha < 1$$

where α is the probability risk event A occurs given the conditioning event B (the root cause event) has occurred. As explained earlier, a risk event is equivalent to a probability event; it is the *Condition-If* part of the risk statement construct. The *Then* part of this construct contains additional information; that is, information on the risk's consequences.

In summary, a best practice formalism for writing a risk is to follow the *Condition-If-Then* construct. Here, the *Condition* is the same as described above (i.e., it is the root cause). The *If* is the associated risk event. The *Then* is the consequence, or set of consequences, that will impact the engineering system project if the risk event occurs.

3.3 The Value Function

Many decisions involve choosing the "best" or "most preferred" option among a set of competing options. In this section, we introduce the field of decision theory and discuss elements of this subject designed to identify not only the best option but an ordering of options from most-to-least preferred, as a function of how well each performs against evaluation criteria.

In engineering risk management, decision makers often need to order risks from most-to-least critical for a variety of purposes. A primary one is to decide where risk mitigation resources should be allocated. In this context, risks are analogous to options. Their criticality is a function of multiple evaluation criteria, such as a risk's impact on an engineering system's cost, schedule, or technical performance.

Modern decision theory has much of its theoretical basis in the classic work "Decisions with Multiple Objectives: Preferences and Value Tradeoffs" (Keeney and Raiffa, 1976). In this, Keeney and Raiffa present the foundations of preference theory and multiattribute value and utility function theory. This section explores these concepts and illustrates their application from an engineering risk management perspective. We begin with the value function.

A *value function* is a real-valued mathematical function defined over an evaluation criterion (or attribute) that represents an option's measure of "goodness" over the levels of the criterion. A measure of "goodness" reflects a decision maker's judged value in the performance of an option (or alternative) across the levels of a criterion (or attribute).

A value function is usually designed to vary from 0 to 1 (or 0 to 100) over the range of levels (or scores) for a criterion. In practice, the value function for a criterion's least preferred level (or score) takes the value 0. The value function for a criterion's most preferred level (or score) takes the value 1.

Figure 3.10 illustrates a buyer's value function for the criterion *Car Color.* Because this function is defined over a criterion, it is known as a *single-dimensional value function* (SDVF). The notation in Figure 3.10 operates as follows. The letter capital X denotes the criterion *Car Color.* The letter small x denotes the level (or score) for a specific option or alternative associated with criterion X. The notation $V_X(x)$ denotes the value of x. For example, for the criterion X = Car Color the option x = Green = 3 has a value of 2/3; that is, $V_{\text{Car Color}}(\text{Green}) = V_{\text{Car Color}}(3) = 2/3$.

In Figure 3.10, suppose a buyer has the following preferences for the criterion *Car Color.* A yellow car is the least preferred color, whereas a black car is the most preferred color. These colors receive a value of 0 and 1, respectively. Furthermore, the value function in Figure 3.10 shows the buyer's increasing value of color as the level of the criterion moves from the color yellow to the color black. Here, red is preferred to yellow, green is preferred to red, blue is preferred to green, and black is preferred to blue.

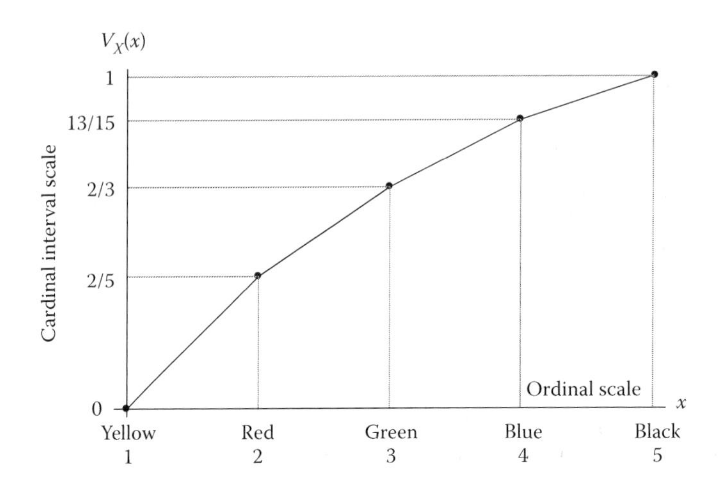

FIGURE 3.10
A value function for car color (Garvey, 2008).

In Figure 3.10, the values show not only an ordering of preferences but suppose the buyer's strength of preference for one color over another is also captured. Here, the smallest increment (change) in value occurs between blue and black. If we use this increment as a reference standard, then it can be shown, for this buyer, the value increment between yellow and red is three times the value increment between blue and black; the value increment between red and green is two times the value increment between blue and black; the value increment between green and blue is one and a half times the value increment between blue and black.

The expression "value increment" or "increment in value" refers to the degree the buyer, in this case, prefers the higher level (score) to the lower level (score) (Kirkwood, 1997). In Figure 3.10, the value increment between yellow and red is greater than that between blue and black. Increasing from yellow to red is more preferable to increasing from blue to black.

Since the buyer's value function in Figure 3.10 features a preference ordering and a strength of preference between the criterion's levels, we say this function is a *measurable value function*. In a measurable value function, the value difference between any two levels (or scores) within a criterion (or attribute) represents a decision-maker's strength of preference between the two levels (or scores). The vertical axis of a measurable value function is a *cardinal interval scale* measure of the strength of a decision-maker's preferences. For this reason, a measurable value function is also referred to as a *cardinal value function*. Refer to Kirkwood (1997) and Dyer and Sarin (1979) for an in-depth technical discussion on measurable value functions.

Figure 3.10 also illustrates how a value function can combine cardinal and ordinal features. In this case, the vertical axis is a cardinal interval scale, whereas the horizontal axis is an ordinal scale. In Figure 3.10, the values along the vertical axis have, to the decision-maker, meaningful preference differences between them. The horizontal axis, in Figure 3.10, is ordinal in the sense that red is preferred to yellow, green is preferred to red, blue is preferred to green, and black is preferred to blue. Along this axis, we have an ordering of preference only that is preserved. The distance between colors, along the horizontal axis, is indeterminate (not meaningful).

Measurement Scales

A measurement scale is a particular way of assigning numbers or labels to an attribute or measure. In measurement and decision theory, there are four commonly used measurement scales (Stevens, 1946). These are *nominal scale, ordinal scale, interval scale,* and *ratio scale.*

Nominal scale. A nominal scale is a measurement scale in which attributes are assigned a label (i.e., a name). It is only a qualitative scale. Nominal data can be counted, but no quantitative differences, or preference ordering of the attributes are implied in a nominal scale. From this, it follows that arithmetic operations are without meaning in a nominal scale. Figure 3.11 illustrates a nominal scale for a set of U.S. cities, labeled A, B, C, and D.

Ordinal scale. An ordinal scale is a measurement scale in which attributes are assigned a number that represents order or rank. For example, a person might rate the quality of different ice cream flavors according to the scale in Figure 3.12. Here, a scale of one to four is assigned to "Worst," "Good," "Very Good," and "Best," respectively. The numerical values indicate only relative order in the sequence. The distance between the numbers is arbitrary and has no meaning. One could have easily assigned the number "forty" to the attribute "Best" while still preserving the order of the sequence.

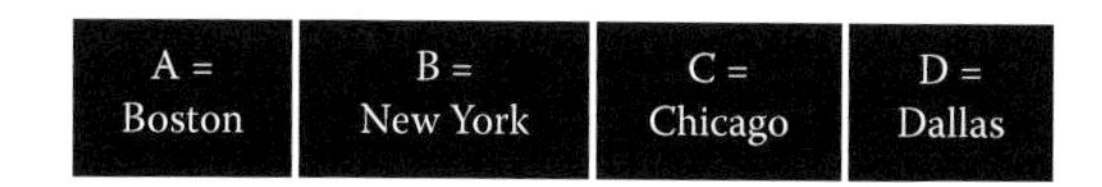

FIGURE 3.11
A nominal scale.

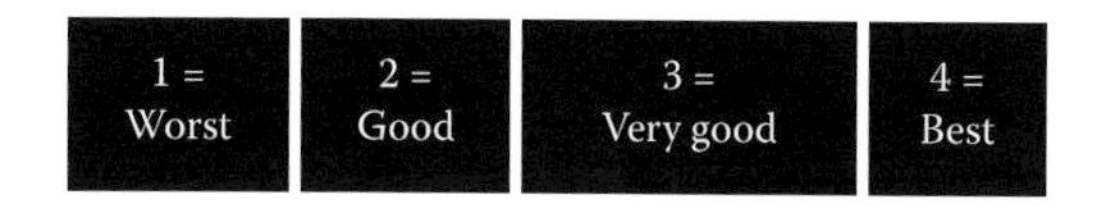

FIGURE 3.12
An ordinal scale.

In an ordinal scale, such as the one shown in Figure 3.12, it does not follow that "Good" is twice as valuable as "Worst" or "Best" is twice as valuable as "Good." We can only say that "Best" is more valued than "Very Good," "Very Good" is more valued than "Good," and "Good" is more valued than "Worst". In each case, we cannot say by how much they are more valued. Data along an ordinal scale is more insightful than that along a nominal scale because the ordinal scale provides information on preference order or rank. However, because the distance between values in the ordinal scale is arbitrary, arithmetic operations on ordinal data is impermissible.

Interval scale. An interval scale is a cardinal measurement scale in which attributes are assigned numbers such that differences between them have meaning. The zero point on an interval scale is chosen for convenience and does not necessarily represent the absence of the attribute being measured. Examples of interval scales are the Fahrenheit or Celsius temperature scales. The zero point in a temperature scale does not mean the absence of temperature. In particular, 0°C is assigned as the freezing point of water.

Because distances between numbers in an interval scale have meaning, addition and subtraction of interval scale numbers is permitted; however, because the zero point is arbitrary, multiplication and division of interval scale numbers are not permitted. For example, we can say that 75°F is 25°F hotter than 50°F; but we cannot say 75°F is 50% hotter than 50°F. *However, ratios of differences can be expressed meaningfully; for example, one difference can be one-half or twice or three times another.*

When working with *measurable value functions,* such differences are referred to as *preference differences.* As mentioned earlier, a measurable value function is one that is monotonic in preferences and value differences represent relative strength of preference. Thus, large value differences between options indicate the difference in preference between them is greater, say, than the difference in preference between other options. The numerical amount of this difference represents the relative amount of preference difference. The concept of value differences is considered a "primitive" in decision theory, that is, not a concept derived from other conditions.

Ratio scale. A ratio scale is a cardinal measurement scale with a "true zero." Here, attributes are assigned numbers such that (1) differences between the numbers reflect differences of the attribute, and (2) ratios between the numbers reflect ratios of the attribute. On a ratio scale, the zero point is a true zero in that it represents a complete absence of the characteristic being measured by the attribute. All arithmetic operations are permitted on numbers that fall along a ratio scale. Examples of ratio scales include measures such as distance, weight, money.

Constructed scale. A constructed scale is a measurement scale specific to the evaluation criterion being measured. Constructed scales are developed for a specific decision context. They are often defined when natural scales are not possible or are not practical to use. They are also used when natural scales exist but additional context is desired and hence are used to supplement natural scales with additional information or context for the decision-maker.

Table 3.4 illustrates a constructed scale by Kirkwood (1997). It shows a scale for the security impacts of a networking strategy for a collection of personal computers. Table 3.4 also shows a mapping from the ordinal scale to its equivalent cardinal interval scale. The constructed scale in Table 3.4 can be viewed by its single dimensional value function (SDVF). Figure 3.13 is the SDVF for the information in this table. Constructed scales are common in decision theory. Many examples are in Kirkwood (1997), Keeney (1992), Clemen (1996), and Garvey (2008).

Developing a Value Function

The value functions in Figure 3.10 and Figure 3.13 are known as piecewise linear single dimensional value functions. They are formed by individual line segments joined together at their "value points." Piecewise linear SDVFs are commonly developed in cases when only a few levels (or scores) define a criterion.

One way to develop a piecewise linear SDVF is the *value increment approach*, described by Kirkwood (1997). This requires increments of value be specified between a criterion's levels (or scores). The sum of these value increments is equal to 1. In Figure 3.10, the value increments from the lowest level to the highest level are, respectively,

$$\left\{\frac{2}{5}, \frac{4}{15}, \frac{3}{15}, \frac{2}{15}\right\} = \left\{\frac{6}{15}, \frac{4}{15}, \frac{3}{15}, \frac{2}{15}\right\} \tag{3.19}$$

TABLE 3.4

A Constructed Scale for Security Impact (Kirkwood, 1997)

Ordinal Scale	Definition	Cardinal Interval Scale
–2	The addition of the network causes a potentially serious decrease in system control and security for the use of data or software.	$V_X(-2) = 0$
–1	There is a noticeable but acceptable diminishing of system control and security.	$V_X(-1) = 0.50$
0	There is no detectable change in system control or security.	$V_X(0) = 0.83$
1	System control or security is enhanced by the addition of a network.	$V_X(1) = 1$

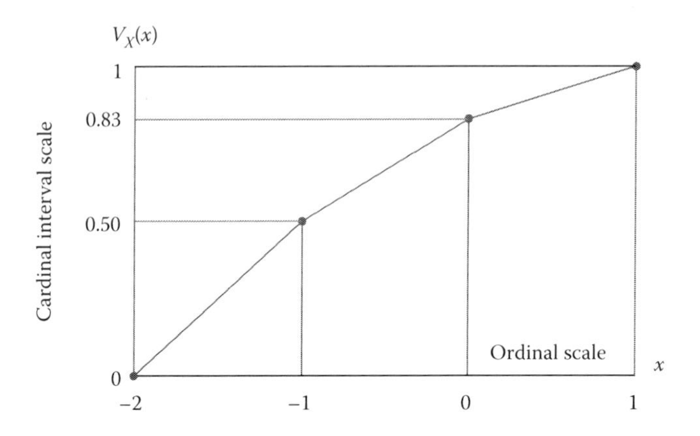

FIGURE 3.13
A SDVF for Table 3.4 (Kirkwood, 1997).

Their sum is equal to 1 and 2/15 is the smallest value increment. The value function in Figure 3.10 reveals, for this buyer, the smallest value increment for car color is between blue and black. A generalization of this is shown in Figure 3.14.

In Figure 3.14, the smallest value increment for criterion X is between levels (or scores) A4 and A5 and is denoted by Δ. Subsequent value increments are multiples of the smallest value increment; that is, $a\Delta$, $b\Delta$, and $c\Delta$, where a, b, and c are positive constants. It follows that

$$c\Delta + b\Delta + a\Delta + \Delta = 1 \text{ or equivalently } \Delta = \frac{1}{a+b+c+1} \tag{3.20}$$

With this, the following equations are true:

$$V_X(\text{A1}) = 0$$

$$V_X(\text{A2}) = V_X(\text{A1}) + c\Delta = c\Delta$$

$$V_X(\text{A3}) = V_X(\text{A2}) + b\Delta = c\Delta + b\Delta$$

$$V_X(\text{A4}) = V_X(\text{A3}) + a\Delta = c\Delta + b\Delta + a\Delta$$

$$V_X(\text{A5}) = V_X(\text{A4}) + \Delta = c\Delta + b\Delta + a\Delta + \Delta = 1$$

The value increment approach uses the property that ratios of differences between values can be expressed meaningfully along an interval scale; that is, one difference (or increment) between a pair of values can be a multiple of the difference (or increment) between another pair of values. Suppose we relate this approach to the value function for *Car Color*, in Figure 3.10. The value increments for the different levels (or scores) for car color, as multiples

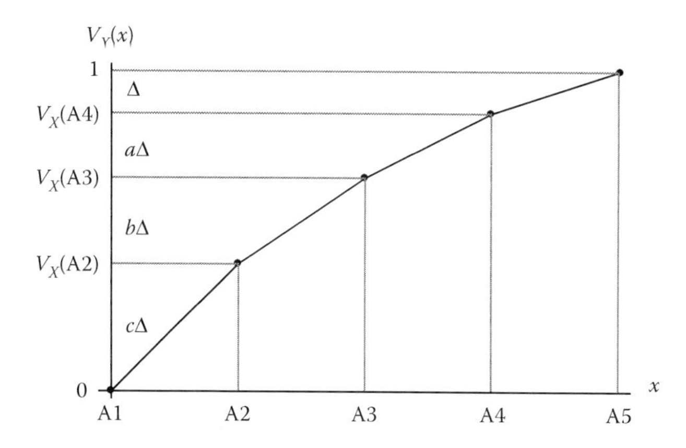

FIGURE 3.14
A piecewise linear SDVF.

of the smallest value increment Δ, are shown in Figure 3.15. From this, it follows that

$$V_X(\text{Yellow}) = 0$$

$$V_X(\text{Red}) = V_X(\text{Yellow}) + 3\Delta = 0 + 3\Delta = \frac{6}{15} = \frac{2}{5}$$

$$V_X(\text{Green}) = V_X(\text{Red}) + 2\Delta = \frac{2}{5} + \frac{4}{15} = \frac{2}{3}$$

$$V_X(\text{Blue}) = V_X(\text{Green}) + \frac{3}{2}\Delta = \frac{2}{3} + \frac{3}{2}\frac{2}{15} = \frac{13}{15}$$

$$V_X(\text{Black}) = V_X(\text{Blue}) + \Delta = \frac{13}{15} + \frac{2}{15} = 1$$

Another approach to specifying a SDVF is the direct assessment of value. This is sometimes referred to as *direct rating*. Here, the value function for a criterion's option (or alternative) with the least preferred level (or score) is assigned the value zero. The value function for a criterion's option with the most preferred level is assigned the value one.

Next, the intermediate options, or alternatives, are ranked such that their ranking reflects a preference ordering along the horizontal axis of the value function. With this, the values of these intermediate options (or alternatives) are directly assessed such that they fall between 0 and 1 along the vertical axis of the value function. The spacing (or distance) between the values of these intermediate options is intended to reflect the strength of preference of the expert (or team) making the assessments for one option over another.

Because values are directly assessed along a cardinal interval scale, it is important to check for consistency. Discussed later in this section, differences

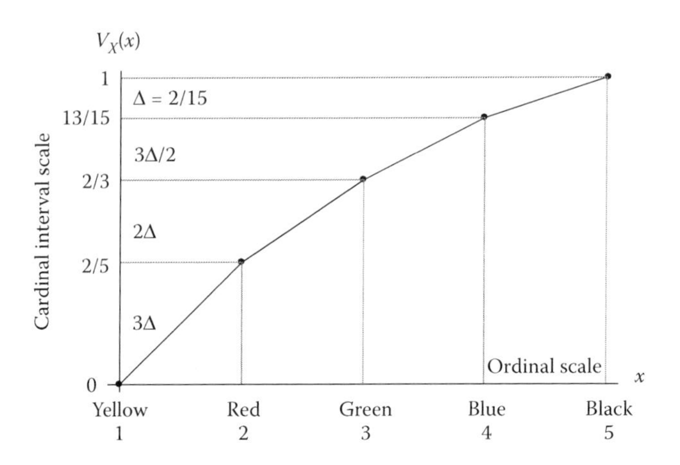

FIGURE 3.15
A value function for car color: A value increment view.

in values along an interval scale have meaning. For example, a value difference of 0.30 points between two options (or alternatives) should reflect an improvement in value that is exactly twice that measured by a difference of 0.15 points between two other options (or alternatives). When implementing a direct preference rating approach with a group of inviduals, check for bias and possible dominance of opinion by one or more participants.

Exponential Value Function

A special type of value function can represent increasing or decreasing values (preferences) for criteria characterized by a continuous range of levels (or scores). Figure 3.16 illustrates an *exponential value function* for the criterion *Probability of Intercept*.

In Figure 3.16, higher probabilities of a successful intercept are more valued than lower probabilities. The scores for this criterion vary continuously across the range of probability, that is, between 0 and 1 along the horizontal axis. The following definitions of the exponential value function are from Kirkwood (1997).

Definition 3.1

If values (preferences) are monotonically increasing over the levels (scores) for an evaluation criterion X, then the exponential value function is given by

$$V_X(x) = \begin{cases} \dfrac{1 - e^{-(x - x_{\min})/\rho}}{1 - e^{-(x_{\max} - x_{\min})/\rho}} & \text{if } \rho \neq \infty \\ \dfrac{x - x_{\min}}{x_{\max} - x_{\min}} & \text{if } \rho = \infty \end{cases} \tag{3.21}$$

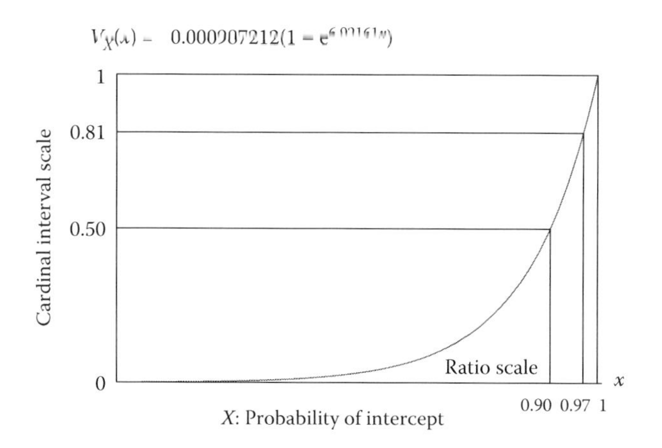

FIGURE 3.16
An exponential value function for probability of intercept.

Definition 3.2

If values (preferences) are monotonically decreasing over the levels (scores) for an evaluation criterion X, then the exponential value function is given by

$$V_X(x) = \begin{cases} \dfrac{1 - e^{-(x_{\max} - x)/\rho}}{1 - e^{-(x_{\max} - x_{\min})/\rho}} & \text{if } \rho \neq \infty \\ \dfrac{x_{\max} - x}{x_{\max} - x_{\min}} & \text{if } \rho = \infty \end{cases}. \tag{3.22}$$

A family of exponential value functions is shown in Figure 3.17. The left-most picture reflects exponential value functions for monotonically increasing preferences over the criterion X. The right-most picture reflects exponential value functions for monotonically decreasing preferences over the criterion X.

The shape of the exponential value function is governed by the parameter ρ, referred to as the *exponential constant* (Kirkwood, 1997). One procedure for determining ρ relies on identifying the midvalue associated with the range of levels (or scores) for the evaluation criterion of interest.

Definition 3.3

The midvalue of a criterion X over a range of possible levels (scores) for X is defined to be the level (score) x such that the difference in *value* between the

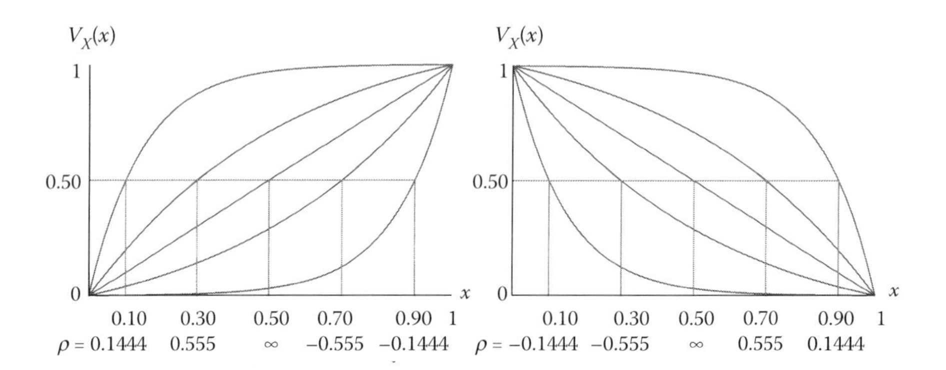

FIGURE 3.17
Families of exponential value functions.

lowest level (score) x_{min} and the midvalue x_{mid} is the same as the difference in *value* between x_{mid} and the highest level (score) x_{max} (Kirkwood, 1997).

From this definition, it follows that the SDVF for the midvalue of X will always equal 0.50; that is, $V_X(x_{mid})=0.50$. If x_{min}, x_{max}, and x_{mid} are known, then Equation 3.21 or Equation 3.22 can be numerically solved for ρ. For example, in Figure 3.16 suppose the midvalue x_{mid} for the criterion *Probability of Intercept* was assessed to be 0.90. Since this criterion is characterized by increasing preferences, Equation 3.21 is the appropriate form of the exponential value function. To determine ρ, in this case, we need to solve the following:

$$V_X(0.90) = 0.50 = \frac{1-e^{-(0.90-0)/\rho}}{1-e^{-(1-0)/\rho}} = \frac{1-e^{-(0.90)/\rho}}{1-e^{-(1)/\rho}} \tag{3.23}$$

Solving Equation 3.23 numerically yields $\rho = -0.1444475$. Today, a number of software applications are available to solve Equation 3.23, such as Microsoft's Excel Goal Seek or Solver routines. The following illustrates an exponential value function for monotonically decreasing preferences.

Suppose the criterion *Repair Time* for a mechanical device ranges from 10 hours to 30 hours. Suppose the midvalue for this criterion was assessed at 23 hours. We need to determine the exponential constant ρ for this exponential value function, as depicted in Figure 3.18.

Since values (preferences) are monotonically decreasing over the levels (scores) for the criterion X, *Repair Time*, the exponential value function is given by Equation 3.22, that is,

$$V_X(x) = \frac{1-e^{-(x_{max}-x)/\rho}}{1-e^{-(x_{max}-x_{min})/\rho}}$$

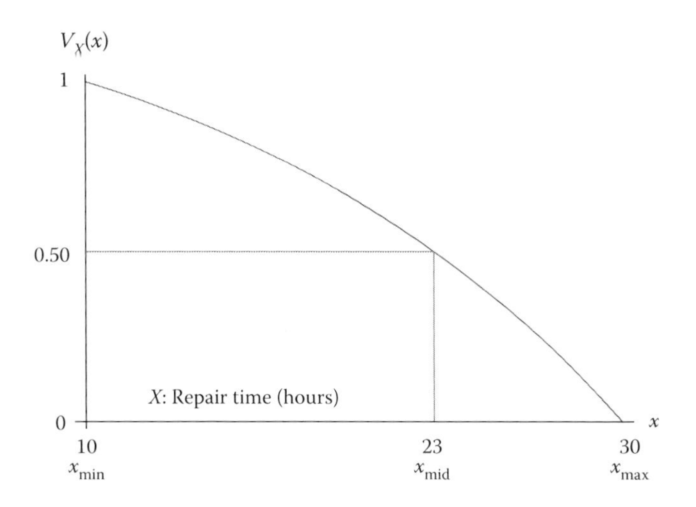

FIGURE 3.18
An exponential value function for repair time.

which, for this case, is

$$V_X(x) = \frac{1 - e^{-(30-x)/\rho}}{1 - e^{-(30-10)/\rho}} = \frac{1 - e^{-(30-x)/\rho}}{1 - e^{-(20)/\rho}}$$

Since $x_{mid} = 23$, we have

$$V_X(23) = 0.50 = \frac{1 - e^{-(30-23)/\rho}}{1 - e^{-(20)/\rho}} = \frac{1 - e^{-7/\rho}}{1 - e^{-20/\rho}} \quad (3.24)$$

Solving Equation 3.24 numerically yields $\rho = 15.6415$.

In this discussion, the exponential value function's exponential constant ρ was solved by setting $V_X(x_{mid})$ equal to 0.50 and numerically solving for ρ; that is, for a given criterion X its score (level) x is assessed such that it represents the midvalue of the value function $V_X(x)$.

In practice, one is not restricted to solving ρ based on the midvalue. An evaluator may prefer to assess the level (score) x associated with a value function's value of, say, $V_X(x_{0.25}) = 0.25$ or $V_X(x_{0.75}) = 0.75$ instead of $V_X(x_{mid}) = 0.50$. In these cases, a procedure similar to the one described here can be used to determine ρ.

The Value of Probability

In Figure 3.16, an exponential value function is defined over a criterion that represented the probability of an event. In general, value functions can be defined for criteria that represent event probabilities whose outcomes are uncertain. Increasing probabilities that an event will occur with preferred outcomes might be modeled by increasing values of a value function. The exponential value function provides a way to shape relationships between value and probability. Figure 3.19 illustrates a family of these relationships. The general equation for this family is as follows:

$$V_X(x) = \begin{cases} (1-e^{-x/\rho})/(1-e^{-1/\rho}) & \text{if } \rho \neq \infty \\ x & \text{if } \rho = \infty \end{cases} \quad (3.25)$$

Equation 3.25 is a particular case of Equation 3.21. Table 3.5 provides the exponential constants for each value function from left to right in Figure 3.19. These constants ensure that the value function defined by Equation 3.25 is equal to 0.50 for the given probability shown in Table 3.5.

The Additive Value Function

Deciding the "best" alternative from a number of competing alternatives is often based on their performance across multiple criteria. Given a set of n value functions defined over n criteria, what is an alternative's overall performance across these criteria? First, some definitions.

Definition 3.4

A criterion Y is preferentially independent of another criterion X if preferences for particular outcomes of Y do not depend on the level (or score) of criterion X.

Informally, preference independence is present if a decision-maker's preference ranking for one criterion does not depend on fixed levels (scores) of

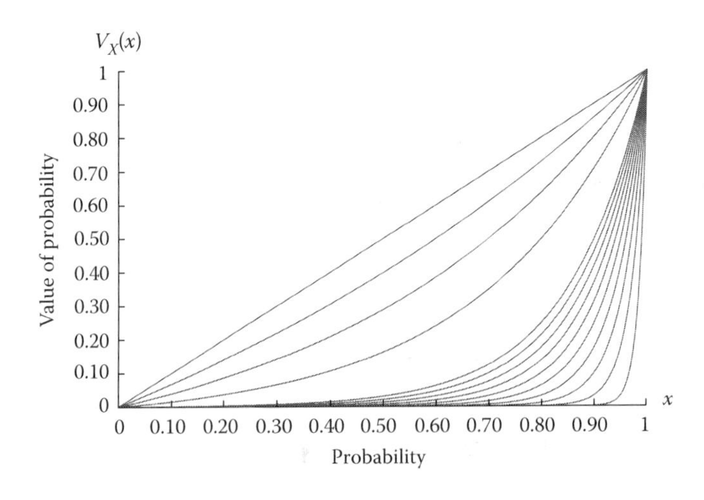

FIGURE 3.19
A family of value functions for the criterion probability.

TABLE 3.5

Exponential Constants for the Value Functions in Figure 3.19

Probability	Exponential Constant
0.50	$\rho=\infty$
0.60	$\rho=-1.2163$
0.70	$\rho=-0.555225$
0.80	$\rho=-0.304759$
0.90	$\rho=-0.144475$
0.91	$\rho=-0.129928$
0.92	$\rho=-0.115444$
0.93	$\rho=-0.100996$
0.94	$\rho=-0.0865629$
0.95	$\rho=-0.0721349$
0.96	$\rho=-0.0577078$
0.97	$\rho=-0.0432809$
0.98	$\rho=-0.0288539$
0.99	$\rho=-0.014427$

other criteria in the decision space. Consider a buyer's selection preference for a new car. If criterion Y is price and criterion X is color, then Y is preference independent of X if the buyer prefers a lower price to a higher price regardless of the car's color.

Definition 3.5

If an evaluator's preference for ith criterion (in a set of n criteria) remains the same regardless of the level (or score) of the other criteria, then the ith criterion is preference independent of the other criteria. If each criterion is preference independent of the other criteria, then the entire set of criteria is called *mutually preference independent*.

Definition 3.6

A value function $V_Y(y)$ is an additive value function if there exists n SDVFs $V_{X_1}(x_1), V_{X_2}(x_2), V_{X_3}(x_3),\ldots, V_{X_n}(x_n)$ satisfying

$$V_Y(y) = w_1 V_{X_1}(x_1) + w_2 V_{X_2}(x_2) + w_3 V_{X_3}(x_3) + \cdots + w_n V_{X_n}(x_n)$$

where w_i for $i=1,\ldots,n$ are nonnegative weights whose values range between 0 and 1 and where $w_1 + w_2 + w_3 + \cdots + w_n = 1$.

Theorem 3.6

If a set of criteria is mutually preferentially independent, then the evaluator's preferences can be represented by an additive value function.

A proof of Theorem 3.6 is beyond the scope of this text. The reader is referred to Keeney and Raiffa (1976) for its proof.

Given (1) the SDVFs $V_{X_1}(x_1), V_{X_2}(x_2), V_{X_3}(x_3), \ldots, V_{X_n}(x_n)$ each range in value between 0 and 1 and (2) the weights each range in value between 0 and 1 and sum to unity, it follows that $V_Y(y)$ will range between 0 and 1. Thus, the higher the value of $V_Y(y)$ the more preferred is the alternative; similarly, the lower the value of $V_Y(y)$ the less preferred is the alternative.

CASE 3.1

Consider the following case. Suppose a buyer needs to identify which of five car options is best across three evaluation criteria *Car Color, Miles per Gallon (MPG)*, and *Price*. Furthermore, suppose the criterion *MPG* is twice as important as the criterion *Car Color* and *Car Color* is half as important as *Price*. Suppose the buyer made the value assessments in Figure 3.20 for each criterion. Assume these criteria are mutually preferentially independent.

Solution

The three criteria are assumed to be mutually preferentially independent. It follows that the additive value function can be used to generate an overall score for the performance of each car across the three evaluation criteria. In this case, the additive value function is

$$V_Y(y) = w_1 V_{X_1}(x_1) + w_2 V_{X_2}(x_2) + w_3 V_{X_3}(x_3) \tag{3.26}$$

where $V_{X_1}(x_1)$, $V_{X_2}(x_2)$, and $V_{X_3}(x_3)$ are the value functions for *Car Color, MPG,* and *Price*, respectively; and, w_i for $i=1,2,3$ are nonnegative weights (importance weights) whose values range between 0 and 1 and where $w_1 + w_2 + w_3 = 1$.

Weight Determination

Since the criterion *MPG* was given to be twice as important as the criterion *Car Color* and *Car Color* was given to be half as important as *Price*, the weights in Equation 3.26 are determined as follows: let w_1 denote the weight for *Car Color*, w_2 denote the weight for *MPG*, and w_3 denote the weight for *Price*. From this, $w_2 = 2w_1$ and $w_1 = w_3/2$. Thus,

$$w_2 = 2\left(\frac{1}{2}w_3\right) = w_3$$

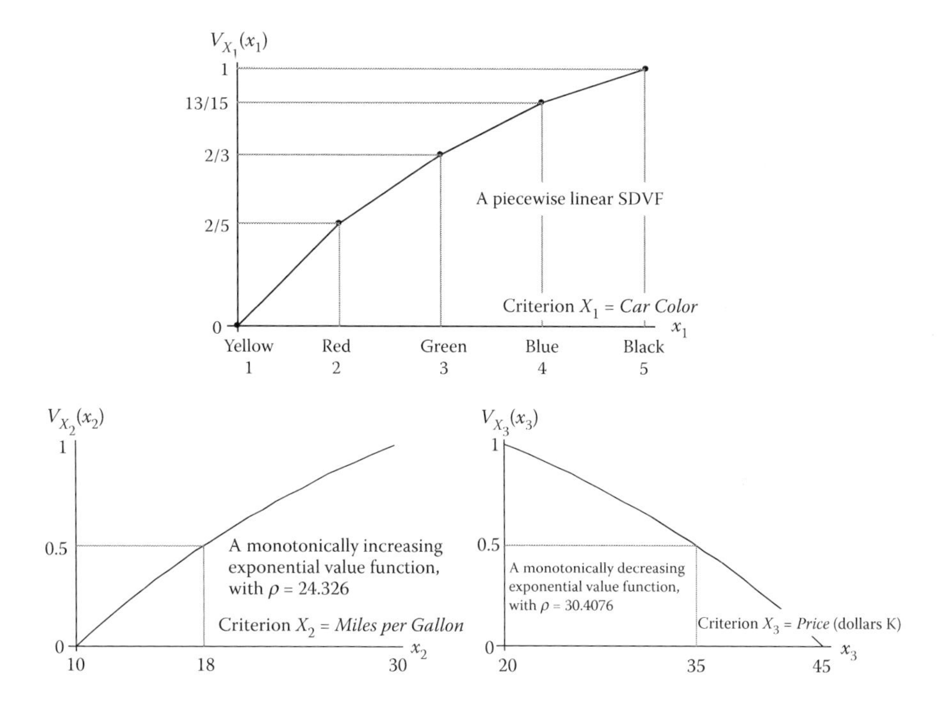

FIGURE 3.20
SDVFs for Case 3.1.

Since $w_1 + w_2 + w_3 = 1$, it follows that

$$\frac{1}{2}w_3 + w_3 + w_3 = 1 \Rightarrow \frac{5}{2}w_3 = 1 \Rightarrow w_3 = \frac{2}{5} \Rightarrow w_2 = \frac{2}{5} \Rightarrow w_1 = \frac{1}{2}w_3 = \frac{1}{5}$$

From this, Equation 3.26 can be written as

$$V_Y(y) = \frac{1}{5}V_{X_1}(x_1) + \frac{2}{5}V_{X_2}(x_2) + \frac{2}{5}V_{X_3}(x_3) \tag{3.27}$$

Determine Performance Matrix

Next, suppose the buyer collected data on the five car options according to their performance against each of the three criteria *Car Color, MPG,* and *Price*. In Table 3.6, the left half of the matrix shows the data on each car option across these criteria. The right half of the matrix shows the equivalent numerical scores from the value functions shown in Figure 3.20. The scores in the last column of Table 3.6 come from applying the value functions in Figure 3.20 to Equation 3.27. In Equation 3.27, the term $V_{X_1}(x_1)$ is given by the top function in

TABLE 3.6

A Performance Matrix of the Buyer's Car Options

	Criterion Level (Score)				Equivalent Value Scores			Overall Value Score
	Color	MPG	Price		Color	MPG	Price	
Car 1	4	15	30	Car 1	0.87	0.33	0.69	0.58
Car 2	1	22	25	Car 2	0.00	0.69	0.86	0.62
Car 3	5	18	38	Car 3	1.00	0.50	0.37	0.55
Car 4	3	12	42	Car 4	0.67	0.14	0.17	0.26
Car 5	2	28	21	Car 5	0.40	0.93	0.97	0.84

Figure 3.20; the terms $V_{X_2}(x_2)$ and $V_{X_3}(x_3)$ are shown at the bottom of Figure 3.20. They are defined as

$$V_{X_2}(x_2) = \frac{1 - e^{-(x_2 - 10)/24.326}}{1 - e^{-(30-10)/24.326}}$$

$$V_{X_3}(x_3) = \frac{1 - e^{-(45 - x_3)/30.4076}}{1 - e^{-(45-20)/30.4076}}$$

For example, from Equation 3.27, Car 1 has the following overall value score.

$$V_Y(y) = \frac{1}{5}V_{X_1}(4) + \frac{2}{5}V_{X_2}(15) + \frac{2}{5}V_{X_3}(30)$$

$$V_Y(y) = \frac{1}{5}\frac{13}{15} + \frac{2}{5}(0.33) + \frac{2}{5}(0.69) = 0.58$$

The overall value scores of the remaining car options are similarly computed. From this, Car 5 is the "best" choice followed by Car 2, Car 1, Car 3, and Car 4.

Sensitivity Analysis

Following an initial ranking of alternatives, a sensitivity analysis is often performed to identify what drives the ranking results. Frequently, this analysis is designed based on the sensitivity of rankings to changes in the importance weights of the evaluation criteria, which are *Car Color, MPG,* and *Price* in Case 3.1.

Recall that the additive value function is a weighted average of the individual SDVFs of each of the evaluation criteria. Here, the weights are nonnegative and sum to 1. Because of this, as one weight varies the other weights must also change such that their sum remains equal to 1. An algebraic rule for automatically tracking the change in the other weights as one weight varies is described by Kirkwood (1997). Consider the case of a three-term additive value model given by Equation 3.28.

$$V_Y(y) = w_1 V_{X_1}(x_1) + w_2 V_{X_2}(x_2) + w_3 V_{X_3}(x_3) \quad (3.28)$$

The weights in Equation 3.28 are nonnegative and have the property that they sum to 1. One procedure for varying the weights, for the purpose of a sensitivity analysis, is the *ratio method*. The ratio method operates as follows. Suppose w_2 is selected as the weight to vary. Let $w_{0,1}$ and $w_{0,3}$ denote the original set of weights for w_1 and w_3 established for the initial ranking. Then, formulas for w_1 and w_3 as a function of w_2 are, respectively,

$$w_1 = (1 - w_2)\left(\frac{w_{0,1}}{w_{0,1} + w_{0,3}}\right) \quad 0 \le w_2 \le 1 \quad (3.29)$$

$$w_3 = (1 - w_2)\left(\frac{w_{0,3}}{w_{0,1} + w_{0,3}}\right) \quad 0 \le w_2 \le 1 \quad (3.30)$$

So, w_1 and w_3 will automatically change as w_2 varies. This change will be such that $w_1 + w_2 + w_3 = 1$. This formulation also keeps the values for w_1 and w_3 in the same ratio as the ratio of their original weight values; that is, it can be shown that

$$\frac{w_3}{w_1} = \frac{w_{0,3}}{w_{0,1}}$$

In Case 3.1, $w_1 = 1/5, w_2 = 2/5$, and $w_3 = 2/5$. These were the original weights established for the initial ranking. Observe the ratio of w_3 to w_1 equals 2. This is the ratio preserved by Equations 3.29 and 3.30, where, for the sensitivity analysis, we set $w_{0,1} = 1/5$ and $w_{0,3} = 2/5$. Thus, for a sensitivity analysis on the weights in Case 3.1, we have the following:

$$w_1 = (1 - w_2)\left(\frac{1}{3}\right) \quad 0 \le w_2 \le 1 \quad (3.31)$$

$$w_3 = (1 - w_2)\left(\frac{2}{3}\right) \quad 0 \le w_2 \le 1 \quad (3.32)$$

In Case 3.1, $w_2 = 0.2$ instead of its original value of 2/5. From Equations 3.31 and 3.32, it follows that

$$w_1 = (1 - 0.2)\left(\frac{1}{3}\right) = \frac{4}{15} \quad \text{and} \quad w_3 = (1 - 0.2)\left(\frac{2}{3}\right) = \frac{8}{15}$$

Equation 3.26 then becomes

$$V_Y(y) = \frac{4}{15}V_{X_1}(x_1) + \frac{2}{10}V_{X_2}(x_2) + \frac{8}{15}V_{X_3}(x_3)$$

From Table 3.6, for Car 1 we have

$$V_Y(y) = \frac{4}{15}(0.87) + \frac{2}{10}(0.33) + \frac{8}{15}(0.69) = 0.67$$

Similar calculations can be done for the rest of the cars in the set of alternatives. Table 3.7 summarizes the results of these calculations, as the weight for MPG w_2 varies from zero to one in increments of 0.1. A graph of the results

TABLE 3.7

Case 3.1: Sensitivity Analysis on MPG

		Sensitivity Analysis MPG Weight Variation										
	Original Case	**0**	**0.10**	**0.20**	**0.30**	**0.40**	**0.50**	**0.60**	**0.70**	**0.80**	**0.90**	**1**
Car 1	0.58	0.75	0.71	0.67	0.63	0.58	0.54	0.50	0.46	0.42	0.37	0.33
Car 2	0.62	0.57	0.59	0.60	0.61	0.62	0.63	0.65	0.66	0.67	0.68	0.69
Car 3	0.55	0.58	0.57	0.56	0.55	0.55	0.54	0.53	0.52	0.52	0.51	0.50
Car 4	0.26	0.34	0.32	0.30	0.28	0.26	0.24	0.22	0.20	0.18	0.16	0.14
Car 5	0.84	0.78	0.80	0.81	0.83	0.84	0.86	0.87	0.89	0.90	0.92	0.93

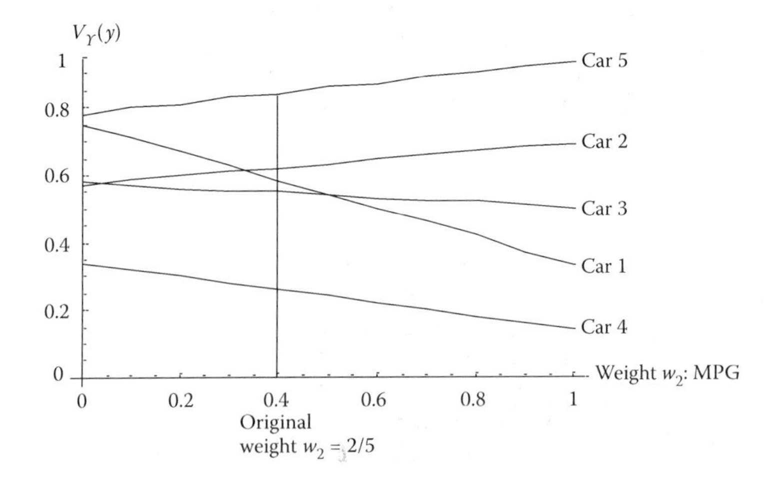

FIGURE 3.21
Case 3.1: Sensitivity analysis on MPG.

in Table 3.7 is shown in Figure 3.21. Car 5 is the clear winner and dominates the overall value score. Car 4 dominates the "loss column" falling below all other car option scores.

3.4 Risk and Utility Functions

In Section 3.3, risk was not considered in the analysis that identified the best alternative from a set of competing alternatives. Modeling decision problems by value functions is appropriate when there is *certainty* in how an alternative rates across the levels of criteria. What if there is *uncertainty* in the ratings of these alternatives? How might decisions be made in this circumstance? Addressing this involves a preference function called a *utility function.*

In general, value functions model decision-maker preferences when outcomes of an evaluation criterion are known. Utility functions model decision-maker preferences when these outcomes are uncertain. A utility function is a value function* that assigns a probability distribution to all possible outcomes of a criterion that represents an alternative's measure of "goodness" over the levels of that criterion. In this way, utility functions capture a decision-maker's probabilistic preference or risk attitude for the outcome of each alternative in a set of competing choices.

Discussed in Chapter 2, modern decision theory began with axioms stated by von Neumann and Morgenstern (1944) in their seminal work on a theory of games and rational choice. These axioms are known as the axioms of choice, the axioms of expected utility theory, or the preference axioms. The axioms of von Neumann and Morgenstern recognize the desirability of one alternative over others must consider a decision-maker's preferences for the outcome of each alternative and the probabilities these outcomes are realized. This form of utility is known as the von Neumann–Morgenstern (vNM) utility.

3.4.1 vNM Utility Theory

Risk is the chance an unwanted event occurs, as mentioned in Chapter 2. Taking a risk is a choice to gamble on an event whose outcome is uncertain. Risk is the probability an unfavorable outcome is realized. However, a favorable or unfavorable outcome is a personal determination—one governed by a person's attitude toward risk and their concept of value or worth. Thus, the study of risk is the study of chance and the study of choice. vNM utility theory is a formalism designed to capture the dualism of chance and choice.

* A utility function is a value function, but a value function *is not necessarily* a utility function (Keeney and Raiffa, 1976).

To understand vNM utility theory, we first introduce the concepts of lotteries and risk attitudes.

Lotteries and Risk Attitudes

Risk can be characterized in the way a person evaluates uncertain outcomes. Uncertain outcomes can be portrayed as a lottery. A *lottery* is an event whose outcome is determined by chance. It is sometimes called a *gamble* or a *risky prospect*. A lottery X may be written as

$$X \equiv L_X = (x_1, p_1; x_2, p_2; x_3, p_3) = \begin{cases} x_1 \text{ with probability } p_1 \\ x_2 \text{ with probability } p_2 \\ x_3 \text{ with probability } p_3 \end{cases} \tag{3.33}$$

where $L_X \equiv X$ is a random event whose outcome is x_i with probability $p_i\,(i=1,2,3)$ and $p_1+p_2+p_3=1$. Equation 3.33 can be written as a vector of probabilities. For instance,

$$L_X = (x_1, p_1; x_2, p_2; x_3, p_3) \equiv (p_1, p_2, p_3),\ 0 \le p_1, p_2, p_3 \le 1 \tag{3.34}$$

with $p_1+p_2+p_3=1$. The lottery given by Equation 3.34 is known as a *simple lottery*. A lottery whose outcome is a set of simple lotteries is known as a *compound lottery*. A compound lottery has a probability distribution on a set of simple lotteries. For instance, L'_Y is a compound lottery.

$$L'_Y = (q_1, L_1; q_2, L_2) \quad 0 \le q_1, q_2 \le 1 \tag{3.35}$$

where the lottery L'_Y awards lottery L_1 with probability q_1 and lottery L_2 with probability q_2 with $q_1+q_2=1$. A compound lottery can be reduced to a simple lottery, as illustrated by the lotteries L'_Y and L_Y in Figure 3.22.

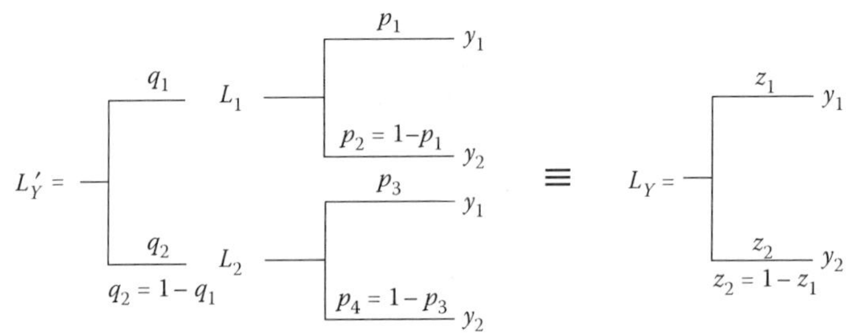

$L'_Y =$; q_1 ; L_1 ; p_1 ; y_1 ; $p_2 = 1 - p_1$; y_2 ; q_2 ; L_2 ; $q_2 = 1 - q_1$; p_3 ; y_1 ; $p_4 = 1 - p_3$; y_2 ; $\equiv$; $L_Y =$; z_1 ; y_1 ; z_2 ; y_2 ; $z_2 = 1 - z_1$

FIGURE 3.22
A compound lottery reduced to a simple lottery.

Using probability rules, the compound lottery L'_Y can be reduced to L_Y, that is,

$$L'_Y = L_Y = (z_1, z_2) = (q_1 p_1 + (1-q_1)p_3, q_1(1-p_1) + (1-q_1)(1-p_3)) \quad 0 \le z_1, z_2 \le 1$$

with $z_1 + z_2 = 1$. Thus, the probabilities of realizing outcomes y_1 and y_2, respectively, are

$$z_1 = q_1 p_1 + (1 - q_1)p_3$$
$$z_2 = q_1(1 - p_1) + (1 - q_1)(1 - p_3)$$

From this, observe that L'_Y and $q_1 L_1 + (1-q_1)L_2$ have the same vector of probabilities; that is,

$$\begin{aligned} q_1 L_1 + (1 - q_1)L_2 &= q_1(p_1, 1 - p_1) + (1 - q_1)(p_3, 1 - p_3) \\ &= (q_1 p_1 + (1 - q_1)p_3, q_1(1 - p_1) + (1 - q_1)(1 - p_3)) \end{aligned}$$

Preferences to engage in lotteries are indicated by the following conventions. The notation $L_1 \succ L_2$ indicates a person has a strict preference for lottery L_1 over lottery L_2; $L_1 \sim L_2$ indicates a person is indifferent between lottery L_1 and lottery L_2; $L_1 \succeq L_2$ indicates a person has a weak preference for lottery L_1 over lottery L_2 (i.e., L_1 is not worse than L_2).

In vNM utility theory, lotteries are used to order preferences over uncertain outcomes. In this theory, persons are rational if their decisions to engage in risky prospects follow the vNM *preference axioms*. Introduced in Chapter 2, these axioms (in preference notation) are as follows:

Completeness Axiom

Either $L_1 \succ L_2$ or $L_2 \succ L_1$ or $L_1 \sim L_2$

Transitivity Axiom

If $L_1 \succeq L_2$ and $L_2 \succeq L_3$, then $L_1 \succeq L_3$

Continuity Axiom

If $L_1 \succeq L_2$ and $L_2 \succeq L_3$, then there exists a probability q such that $L_2 \sim qL_1 + (1-q)L_3$

Independence (Substitution) Axiom

If $L_1 \succ L_2$, then $qL_1 + (1-q)L_3 \succ qL_2 + (1-q)L_3$

Monotonicity Axiom

If $L_1 \succ L_2$ and $q_1 > q_2$, then $q_1 L_1 + (1-q_1)L_2 \succ q_2 L_1 + (1-q_2)L_2$. The monotonicity axiom stems from the preceding axioms.

A fundamental result from these axioms is the vNM *expected utility theorem*. This theorem proved that if a person's preference ordering over lotteries (or risky prospects) follows the vNM axioms, then a real-valued order-preserving cardinal utility function exists on an interval scale.

Theorem 3.7: The Expected Utility Theorem

If preferences over any two lotteries L_1 and L_2 satisfy the vNM axioms, then there exists an interval scale real number $U(x_i)$ associated to each outcome x_i such that, for $i = 1, 2, 3, \ldots, n$

$$L_1(p_1, p_2, \ldots, p_n) \succeq L_2(q_1, q_2, \ldots, q_n) \text{ if and only if}$$

$$p_1U(x_1) + p_2U(x_2) + \ldots + p_nU(x_n) \geq q_1U(x_1) + q_2U(x_2) + \cdots + q_nU(x_n) \quad 0 \leq p_i, q_i \leq 1.$$

The expected utility theorem proved that preference order-preserving, real-valued, cardinal utility functions exist on interval scales. In this context, utility is a cardinal measure of strength of preference for outcomes under uncertainty. Because of their interval scale properties, utility functions are unique up to positive linear transformations. This means a person's preference order for engaging in risky prospects is preserved up to $aU(x) + b \, (a > 0)$ and ratios of differences in utility are invariant. Thus, the origin and unit of measurement of a utility scale can be set arbitrarily. Traditionally, this scale is set such that the utility of the least preferred outcome is 0 utils and the utility of the most preferred outcome is 1 util.

Definition 3.7

The *expected utility* of lottery L_X with possible outcomes $\{x_1, x_2, x_3, \ldots, x_n\}$ is

$$E(U(L_X)) = E(U(X)) = p_1U(x_1) + p_2U(x_2) + p_3U(x_3) + \cdots + p_nU(x_n) \quad (3.36)$$

where p_i is the probability L_X produces x_i (for $i = 1, \ldots, n$) and $U(x_i)$ is the cardinal utility associated to each outcome x_i.

Definition 3.8

The *expected value* of lottery L_X with possible outcomes $\{x_1, x_2, x_3, \ldots, x_n\}$ is

$$E(L_X) \equiv E(X) = p_1x_1 + p_2x_2 + p_3x_3 + \cdots + p_nx_n \quad (3.37)$$

where p_i is the probability L_X produces x_i (for $i = 1, \ldots, n$).

From the expected utility theorem, a decision rule for choosing between risky prospects is to select the one that offers the maximum expected utility (MEU). Person's who invoke this rule are sometimes called *expected utility maximizers*. However, the decision to participate in a lottery rests with a person's willingness to take risks. This can be characterized by a concept known as certainty equivalent.

Definition 3.9

A *certainty equivalent* of lottery X is an amount x_{CE} such that the decision-maker is indifferent between X and the amount x_{CE} for certain.

For example, what amount of dollars would you be willing to receive with certainty that makes you indifferent between that amount and engaging in lottery X, given as follows:

$$L_X \equiv X = \begin{cases} \text{Win}\ \$500 \text{ with probability } 0.60 \\ \text{Lose}\ \$150 \text{ with probability } 0.40 \end{cases}$$

The expected value of this lottery is $E(L_X) = E(X) = 0.60(\$500) + 0.40(-\$150) =$ \$240. If you would be indifferent between receiving \$200 with certainty and engaging in the lottery, then we say your certainty equivalent for this lottery is $x_{CE} = 200$. In this example, we say this person is *risk averse*. He is willing to accept, with certainty, an amount of dollars less than the expected amount that might be received if the decision was made to participate in the lottery. A person is considered to be a risk-taker or *risk seeking* if their certainty equivalent is greater than the expected value of the lottery. A person is considered *risk neutral* if their certainty equivalent is equal to the expected value of the lottery.

There is a mathematical relationship between certainty equivalent, expected value, and risk attitude. People with increasing preferences whose risk attitude is *risk averse* will always have a certainty equivalent less than the expected value of an outcome. People with decreasing preferences whose risk attitude is *risk averse* will always have a certainty equivalent greater than the expected value of an outcome.

People with increasing preferences whose risk attitude is *risk seeking* will always have a certainty equivalent greater than the expected value of an outcome. People with decreasing preferences whose risk attitude is *risk seeking* will always have a certainty equivalent less than the expected value of an outcome.

People whose risk attitude is *risk neutral* will always have a certainty equivalent equal to the expected value of an outcome. This is true regardless of whether a person has increasing or decreasing preferences.

3.4.2 Utility Functions

There is a class of mathematical functions that exhibit these behaviors with respect to risk averse, risk seeking, and risk neutral attitudes. They are referred to as *utility functions*. A family of such functions is shown in Figure 3.23.

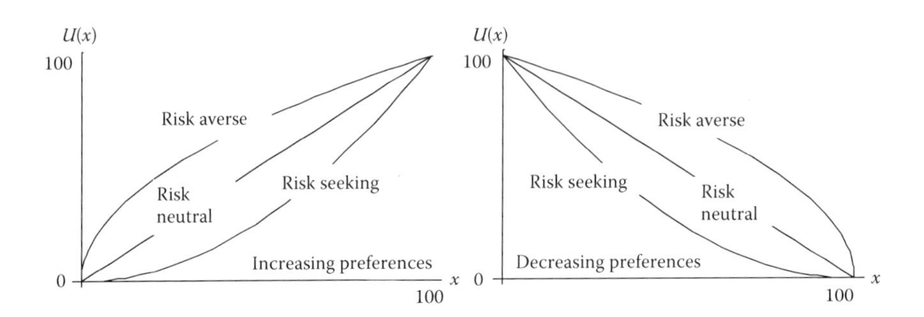

FIGURE 3.23
Utility functions capture risk attitudes.

A *utility* is a measure of worth, satisfaction, or preference an outcome has for an individual. It is a dimensionless number that is sometimes referred to as "util." A *utility function* is a real-valued mathematical function that relates uncertain outcomes along the horizontal axis to measures of worth or "utils" along the vertical axis. A utility function defined over a criterion, or attribute, is known as a *single-dimensional utility function.*

As discussed earlier, the vertical axis of a cardinal utility function is an interval scale and usually runs between 0 and 1 or between 0 to 100 utils (as shown in Figure 3.23). With this convention, the utility of the least preferred outcome is assigned 0 utils and the utility of the most preferred outcome is assigned 100 utils. Higher preferred outcomes have higher "utils" than lower preferred outcomes.

Utility functions generally take one of the shapes shown in Figure 3.23. They are concave, linear, or convex. A concave utility function appears "hill-like" and is always associated with a risk averse person. Concave functions have the property that they lie above a chord (a line segment) connecting any two points on the curve. A linear function is always associated with a risk neutral person. A convex function appears "bowl-like" and is always associated with a risk-seeking person. Convex functions lie below a chord connecting any two points on the curve.

Utility functions, as representations of a person's risk attitude, exhibit a number of relationships between the expected value of a lottery, the expected utility of a lottery, and the utility of a lottery's expected value. These measures can be related to the concept of certainty equivalent.

The relationship between $E(X)$ and $E(U(X))$ of a lottery X can be seen by looking at the utility function in Figure 3.24. Figure 3.24 shows this relationship for a monotonically increasing risk averse utility function. Similar relationships can be developed for other utility function shapes, such as those in Figure 3.23. In Figure 3.24, the equation of the chord $\gamma(x)$ below $U(x)$ is given by Equation 3.38.

$$\gamma(x) = \frac{U(b) - U(a)}{b - a}(x - b) + U(b) \tag{3.38}$$

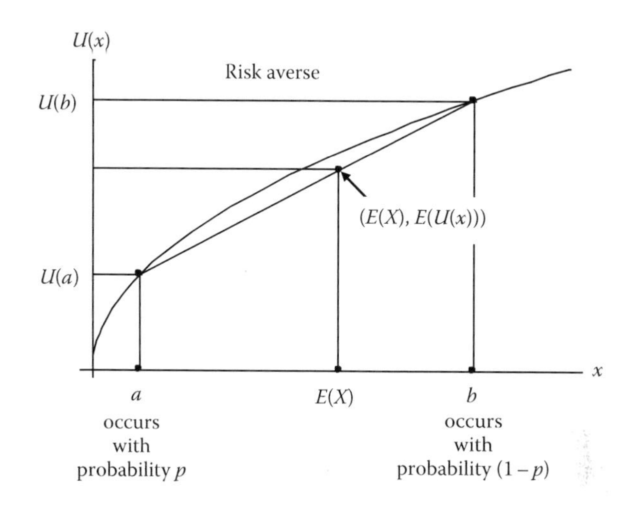

FIGURE 3.24
Relationship between $E(X)$ and $E(U(x))$.

Here, we have $E(X) = pa + (1 - p)b$. If we set $x = E(X)$ in Equation 3.38, then it can be shown that

$$\gamma(E(X)) = pU(a) + (1 - p)U(b)$$

From Definition 3.7, $pU(a) + (1 - p)U(b) = E(U(x))$, therefore,

$$\gamma(E(X)) = E(U(x)) \tag{3.39}$$

The certainty equivalent x_{CE} is that value on the horizontal axis where a person is indifferent between the lottery and receiving x_{CE} with certainty. From this, it follows that the utility of x_{CE} must equal the expected utility of the lottery, that is,

$$U(x_{\mathrm{CE}}) = E(U(x)) \tag{3.40}$$

or, equivalently

$$U(x_{\mathrm{CE}}) = \gamma(E(X)) \tag{3.41}$$

as shown in Figure 3.25. From Equation 3.40, when a single-dimensional utility function $U(x)$ has been specified, the certainty equivalent x_{CE} can be determined as

$$x_{\mathrm{CE}} = U^{-1}(E(U(x))) \tag{3.42}$$

where $U^{-1}(x)$ is the inverse of $U(x)$.

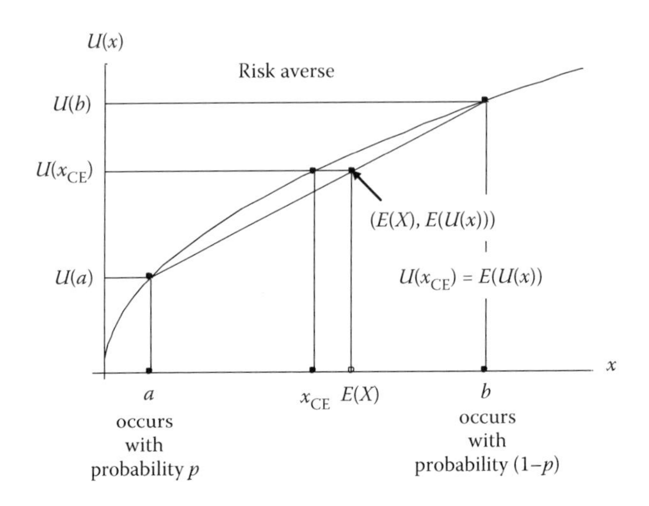

FIGURE 3.25
Relationship between $E(X), E(U(x))$, and x_{CE}.

PROBLEM 3.5

Consider the lottery L_X given as follows:

$$L_X \equiv X = \begin{cases} \text{Win \$80K with probability } 0.60 \\ \text{Win \$10K with probability } 0.40 \end{cases}$$

Determine the certainty equivalent x_{CE} if a person's utility function is given by $U(x) = 10\sqrt{x}$, where x is in dollars thousand ($K).

Solution

Figure 3.26 shows $U(x) = 10\sqrt{x}$ is a monotonically increasing, risk averse, preference function. To determine x_{CE} for this lottery, we first compute its expected value, that is,

$$E(L_X) = E(X) = p_1 x_1 + p_2 x_2 = 0.60(\$80\text{K}) + 0.40(\$10\text{K}) = \$52\text{K}$$

Since $U(x)$ is a monotonically increasing risk averse utility function, it follows that x_{CE} must be less than $E(X)$. From Equation 3.38, the chord $\gamma(x)$ between $10 \le x \le 80$ is given by

$$\gamma(x) = \frac{U(80) - U(10)}{80 - 10}(x - 80) + U(80) \Rightarrow \gamma(x) = 0.8257(x - 80) + 89.4$$

From Equation 3.39, $\gamma(E(X)) = pU(a) + (1 - p)U(b) = E(U(x))$; thus,

$$\gamma(52) = 0.8257(52 - 80) + 89.4 = 66.3 = E(U(x))$$

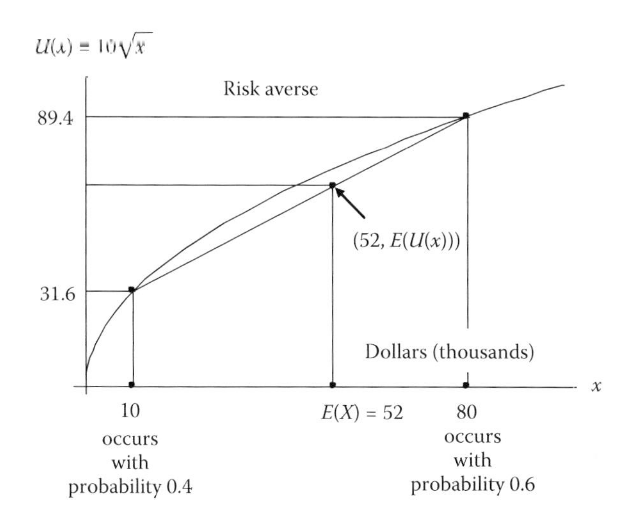

FIGURE 3.26
Problem 3.5 x_{CE} derivation.

The expected utility of L_X could also have been determined from Definition 3.7, that is,

$$E(U(x)) = p_1U(10) + p_2U(80) = 0.40(31.6) + 0.60(89.4) = 66.3$$

From Equation 3.40, we have $U(x_{CE}) = E(U(x)) = 66.3$. If $U(x) = 10\sqrt{x}$, then $U^{-1}(x) = (x/10)^2$. Therefore, the certainty equivalent for this lottery is

$$x_{CE} = U^{-1}E(U(x)) = (66.3/10)^2 = 43.93 \approx 44$$

These results are summarized in Figure 3.27.

In Figure 3.27, notice there is a dot above the coordinate (52, 66.3). This point denotes the utility of the expected value, that is, $U(E(X))$. A property of concave functions (risk averse utility functions) is $U(E(X)) > E(U(x))$. Likewise, a property of convex functions (risk-seeking utility functions) is $U(E(X)) < E(U(x))$. Risk neutral utility functions have the property $U(E(X)) = E(U(x))$. Figure 3.28 illustrates these relationships for monotonically increasing preferences.

Suppose a utility function is scaled such that its vertical axis ranges from 0 to 1. Suppose a lottery has two outcomes. Outcome a occurs with probability p. Outcome b occurs with probability $(1-p)$. If preferences are monotonically increasing (i.e., more is better than less) such that $U(a) = 0$ and $U(b) = 1$, then it can be shown that $E(U(x)) = 1 - p$. Similarly, if the utility function is monotonically decreasing, (i.e., less is better) such that $U(a) = 1$ and $U(b) = 0$, then it can be shown that $E(U(x)) = p$.

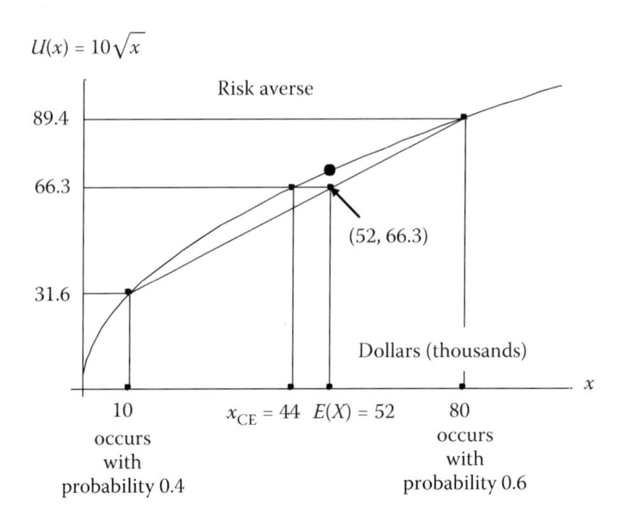

FIGURE 3.27
Problem 3.5 summary.

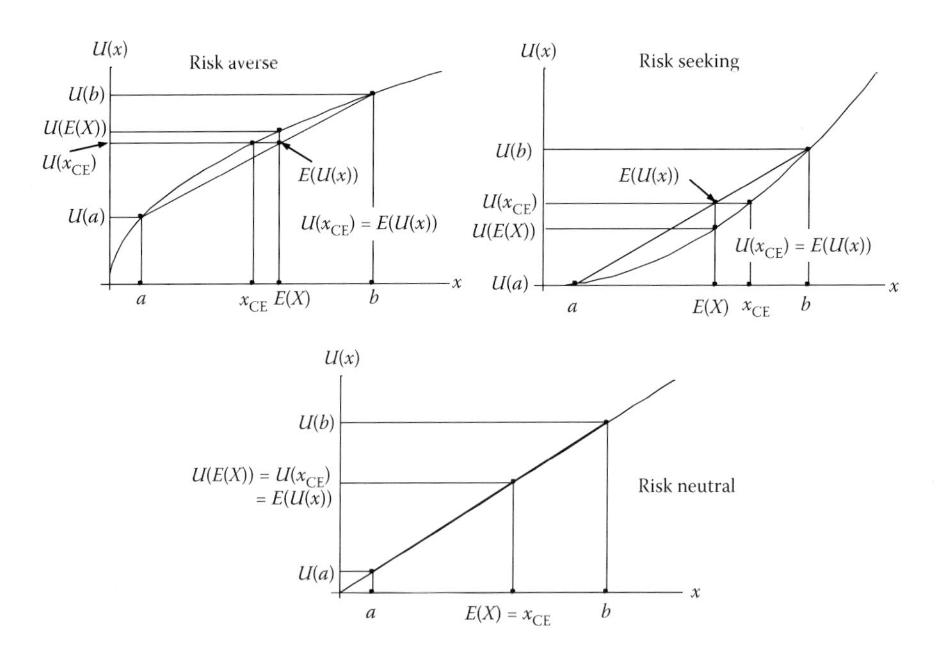

FIGURE 3.28
A family of utility functions for monotonically increasing preferences.

The Exponential Utility Function

A special type of utility function known as the exponential utility function (Kirkwood, 1997) can represent a broad class of utility function shapes or risk attitudes. Similar in form to the exponential value function, the following defines the exponential utility function.

Definition 3.10

If utilities are monotonically increasing over the levels (scores) for an evaluation criterion X, then the exponential utility function is given by

$$U(x) = \begin{cases} \dfrac{1-e^{-(x-x_{min})/\rho}}{1-e^{-(x_{max}-x_{min})/\rho}} & \text{if } \rho \neq \infty \\ \dfrac{x-x_{min}}{x_{max}-x_{min}} & \text{if } \rho = \infty \end{cases} \tag{3.43}$$

Definition 3.11

If utilities are monotonically decreasing over the levels (scores) for an evaluation criterion X, then the exponential utility function is given by

$$U(x) = \begin{cases} \dfrac{1-e^{-(x_{max}-x)/\rho}}{1-e^{-(x_{max}-x_{min})/\rho}} & \text{if } \rho \neq \infty \\ \dfrac{x_{max}-x}{x_{max}-x_{min}} & \text{if } \rho = \infty \end{cases} \tag{3.44}$$

The function $U(x)$ is scaled such that it ranges from 0 to 1. In particular, for monotonically increasing preferences, $U(x_{min}) = 0$ and $U(x_{max}) = 1$. The opposite holds for monotonically decreasing preferences, that is, $U(x_{min}) = 1$ and $U(x_{max}) = 0$.

A family of exponential utility functions is shown in Figure 3.29. The left-most picture reflects exponential utility functions for monotonically increasing preferences ("more is better") over the criterion X. The right-most picture reflects exponential utility functions for monotonically decreasing preferences ("less is better") over the criterion X.

In Equations 3.43 and 3.44, the constant ρ is called the *risk tolerance*. The risk tolerance ρ reflects the risk attitude of a person's utility or preferences for a particular outcome. Positive values of ρ reflect a risk averse utility function. Negative values of ρ reflect a risk-seeking utility function. A ρ value of "infinity" reflects a risk neutral utility function.

Mentioned previously, an exponential utility function can be specified to represent many shapes that reflect a person's risk attitude. The shape is governed by ρ, whose magnitude reflects the degree a person is risk averse or risk seeking. If a person is neither risk averse nor risk seeking, then the

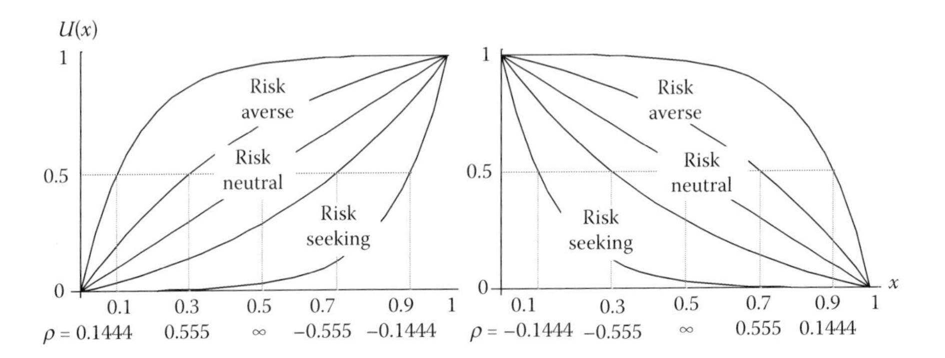

FIGURE 3.29
Families of exponential utility functions.

exponential utility function is a straight line. The following further discusses the exponential utility function and ways to determine its shape when either ρ or the certainty equivalent x_{CE} is given.

Consider an investment with the following two outcomes. Earn 10 million dollars ($M) with probability $p = 1/3$ or earn $20M with probability $(1-p) = 2/3$. Suppose $U(10) = 0$, $U(20) = 1$, and the certainty equivalent for this lottery was set at $ 13M. What is the value of ρ? To answer this question, first determine the expected earnings from this investment. From Definition 3.8, the expected earnings is

$$E(X) = (1/3)(\$10\text{M}) + (2/3)(\$20\text{M}) = \$16.67\text{M}$$

Since the certainty equivalent was set at $13M, we know this investor is risk averse, because $x_{CE} < E(X)$. We also have monotonically increasing preferences since earning more is better than earning less. So, the utility function should look something like one of the upper curves in the left side of Figure 3.29. In this case,

$$U(x) = \frac{1-e^{-(x-10)/\rho}}{1-e^{-(20-10)/\rho}} = \frac{1-e^{-(x-10)/\rho}}{1-e^{-(10)/\rho}} \tag{3.45}$$

From Definition 3.7, $E(U(x)) = 1 - p = 2/3$. From Equation 3.40, we have

$$U(x_{CE}) = E(U(x)) = 2/3 \tag{3.46}$$

Since x_{CE} was given to be equal to $13M, from Equation 3.45 it follows that

$$U(x_{CE}) = U(13) = \frac{1-e^{-(13-10)/\rho}}{1-e^{-(20-10)/\rho}} = \frac{1-e^{-(3)/\rho}}{1-e^{-(10)/\rho}} = 2/3 \tag{3.47}$$

Equation 3.47 is then solved numerically for ρ, which yields ρ=2.89139. This was determined from the *Mathematica®* routine **FindRoot** $[(1-\text{Exp}[-3/\rho])/(1-\text{Exp}[-10/\rho])==2/3,\{\rho,1\}$ A graph of this exponential utility function is shown in Figure 3.30.

From this discussion, when specifying an exponential utility function it is necessary to identify the value associated with the certainty equivalent. Once the certainty equivalent has been specified, the shape of the exponential utility function, which reflects the risk attitude of the individual, can be completely determined.

Figure 3.31 shows a family of exponential utility functions for various certainty equivalents, as they vary around the basic data in Figure 3.30. Notice the sharpness in risk averseness as the certainty equivalent moves away to the left of $E(X)$, in this case.

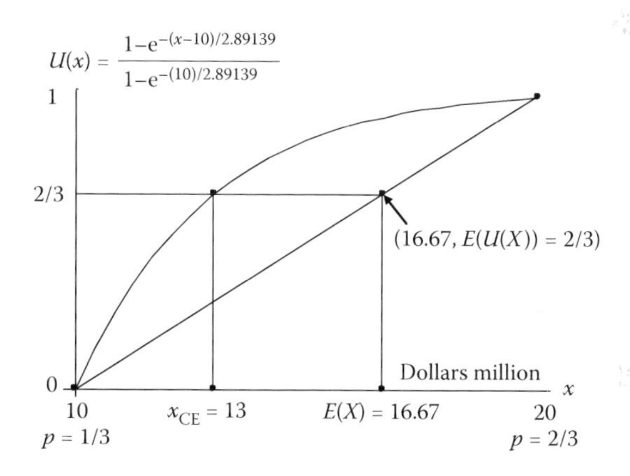

FIGURE 3.30
An exponential utility function.

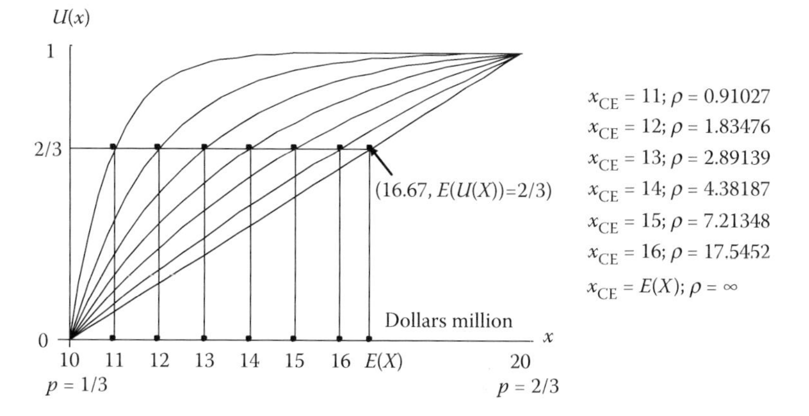

FIGURE 3.31
Families of risk averse exponential utility functions.

Suppose we are given the utility function in Figure 3.30 and we want to know the certainty equivalent. In this case, we have a value for ρ and we want the value for x_{CE} given ρ. To determine x_{CE}, it is necessary to apply Equation 3.42 to the exponential utility function. We need two formulas for this. One is the formula for the inverse of the exponential utility function. The other is the expected utility of the exponential utility function. These formulas are provided in the following theorems. It is left to the reader to derive these results.

Theorem 3.8

If $U(x)$ is a monotonically increasing exponential utility function, then the inverse function, expected utility, and certainty equivalent are as follows.

(a) *Inverse Function*

$$U^{-1}(x) = \begin{cases} x_{min} - \rho \ln(1 - x/k) & \text{if } \rho \neq \infty \\ x(x_{max} - x_{min}) + x_{min} & \text{if } \rho = \infty \end{cases} \tag{3.48}$$

where $k = 1/(1 - e^{-(x_{max} - x_{min})/\rho})$.

(b) *Expected Utility*

$$E(U(x)) = \begin{cases} k(1 - e^{x_{min}/\rho} E(e^{-x/\rho})) & \text{if } \rho \neq \infty \\ \dfrac{E(X) - x_{min}}{x_{max} - x_{min}} & \text{if } \rho = \infty \end{cases} \tag{3.49}$$

(c) *Certainty Equivalent*

$$x_{CE} = \begin{cases} -\rho \ln E(e^{-x/\rho}) & \text{if } \rho \neq \infty \\ E(X) & \text{if } \rho = \infty \end{cases} \tag{3.50}$$

Theorem 3.9

If $U(x)$ is a monotonically decreasing exponential utility function, then the inverse function, expected utility, and certainty equivalent are as follows.

(a) *Inverse Function*

$$U^{-1}(x) = \begin{cases} x_{max} + \rho \ln(1 - x/k) & \text{if } \rho \neq \infty \\ x_{max} - x(x_{max} - x_{min}) & \text{if } \rho = \infty \end{cases} \tag{3.51}$$

where $k = 1/(1 - e^{-(x_{max} - x_{min})/\rho})$.

(b) *Expected Utility*

$$E(U(x)) = \begin{cases} k(1-\mathrm{e}^{-x_{\max}/\rho}E(\mathrm{e}^{x/\rho})) & \text{if } \rho \neq \infty \\ \dfrac{x_{\max} - E(X)}{x_{\max} - x_{\min}} & \text{if } \rho = \infty \end{cases} \tag{3.52}$$

(c) *Certainty Equivalent*

$$x_{\text{CE}} = \begin{cases} \rho \ln E(\mathrm{e}^{x/\rho}) & \text{if } \rho \neq \infty \\ E(X) & \text{if } \rho = \infty \end{cases} \tag{3.53}$$

PROBLEM 3.6

Consider the utility function in Figure 3.30. Show that the certainty equivalent for this utility function is \$13M.

Solution

The utility function in Figure 3.30 is given by

$$U(x) = \frac{1-\mathrm{e}^{-(x-10)/2.89139}}{1-\mathrm{e}^{-(10)/2.89139}} = 1.0325(1-\mathrm{e}^{-(x-10)/2.89139}) \tag{3.54}$$

where $\rho = 2.89139$. Since this function is monotonically increasing, its certainty equivalent x_{CE} is given by Equation 3.50. Applying that equation, with reference to Figure 3.30, we have

$$x_{\text{CE}} = -\rho \ln E(\mathrm{e}^{-x/\rho}) = (-2.89139)\ln((1/3)\mathrm{e}^{-10/2.89139} + (2/3)\mathrm{e}^{-20/2.89139}) = 13$$

So, the certainty equivalent of the exponential value function is \$13M, as shown in Figure 3.30.

Thus far, we have worked with lotteries that represent uncertain events having a discrete number of chance outcomes. When the outcomes of a lottery are defined by a continuous probability density function, then

$$E(X) = \int_a^b x f_X(x)\,dx \tag{3.55}$$

$$E(U(x)) = \int_a^b U(x) f_X(x)\,dx. \tag{3.56}$$

Furthermore, the certainty equivalent x_{CE} becomes the solution to

$$U(x_{\text{CE}}) = E(U(x)) = \int_a^b U(x) f_X(x)\,dx. \tag{3.57}$$

PROBLEM 3.7

Consider the utility function in Figure 3.26. This function is given by $U(x) = 10\sqrt{x}$, where x is in dollars thousand (K). Determine $E(X)$ and the certainty equivalent x_{CE} if the lottery denoted by X is given by the uniform probability density function in Figure 3.32.

Solution

The equation for $f_X(x)$ in Figure 3.32 is

$$f_X(x) = \frac{1}{70} \quad 10 \le x \le 80$$

From Equation 3.55, we have

$$E(X) = \int_a^b x f_X(x)\,dx = \int_{10}^{80} x(1/70)\,dx = 45$$

Thus, the expected value of the lottery X is \$45K. From Equation 3.57, we have

$$E(U(x)) = \int_a^b U(x) f_X(x)\,dx = \int_{10}^{80} 10\sqrt{x}(1/70)\,dx = 65.135$$

From Equations 3.40 and 3.57, we have $U(x_{CE}) = E(U(x)) = 65.135$. Since $U(x) = 10\sqrt{x}$, it follows that $U(x_{CE}) = 10\sqrt{x_{CE}} = 65.135$. Solving this equation for x_{CE} yields $x_{CE} = 42.425$. Thus, the certainty equivalent is \$42.43K when rounded.

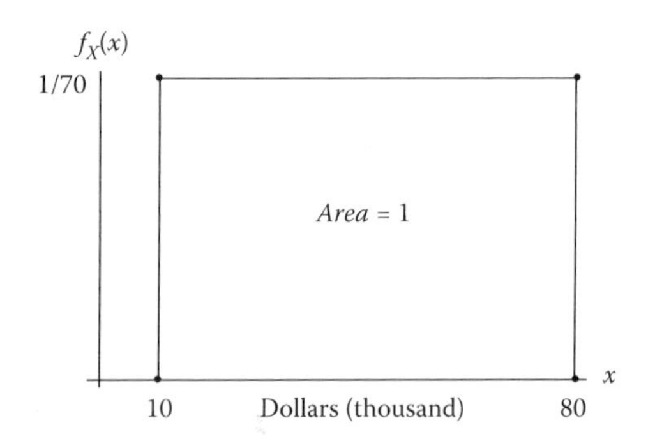

FIGURE 3.32
A uniform probability density function.

3.5 Multiattribute Utility—The Power Additive Utility Function

Thus far, utility has been discussed from a single-dimensional perspective. The preceding focused on a single-dimensional utility function, its properties, and its characteristics. This section examines multiattribute utility. This is concerned with specifying a utility function over multiple criteria or multiple attributes that characterize an option or alternative. The question is one of ranking these options as a function of how well they perform across a set of criteria or attributes.

Multiattribute utility functions come in several general forms. For purposes of this discussion, we focus on one form known as the *power-additive utility function*. Kirkwood (1997) provides an extensive discussion on this utility function, from which this material derives. The reader is directed to Keeney and Raiffa (1976), Clemen (1996), and von Winterfeldt and Edwards (1986) for other general forms of the multiattribute utility function.

When accounting for the risk attitude of a decision-maker, one can convert "values" from a value function into utilities. Doing this requires a function that takes values from a multiattribute value function and maps them into a corresponding set of utilities. The power-additive utility function is a multiattribute utility function that performs this mapping. The power-additive utility function covers a wide span of possible risk attitudes.

3.5.1 The Power-Additive Utility Function

The power-additive utility function is a multiattribute utility function similar in form to the exponential value function and the exponential utility function, which have been previously discussed. The following defines the power-additive utility function.

Definition 3.12

If utilities are *monotonically increasing* over the values of the additive value function $V_Y(y)$, then the power-additive utility function is given by

$$U(v) = \begin{cases} K(1 - e^{-(V_Y(y)/\rho_m)}) & \text{if } \rho_m \neq \infty \\ V_Y(y) & \text{if } \rho_m = \infty \end{cases} \tag{3.58}$$

where $K = 1/(1 - e^{-1/\rho_m})$ and $v = V_Y(y)$ is the additive value function given in Definition 3.6.

Definition 3.13

If utilities are *monotonically decreasing* over the values of the additive value function $V_Y(y)$, then the power-additive utility function is given by

$$U(v) = \begin{cases} K(1 - e^{-((1-V_Y(y))/\rho_m)}) & \text{if } \rho_m \neq \infty \\ 1 - V_Y(y) & \text{if } \rho_m = \infty \end{cases} \quad (3.59)$$

where $K = 1/(1 - e^{-1/\rho_m})$ and $v = V_Y(y)$ is the additive value function given in Definition 3.6.

As mentioned above, the value function $V_Y(y)$ is an additive value function; that is, there exists n SDVFs $V_{X_1}(x_1), V_{X_2}(x_2), V_{X_3}(x_3), \ldots, V_{X_n}(x_n)$ satisfying

$$V_Y(y) = w_1 V_{X_1}(x_1) + w_2 V_{X_2}(x_2) + w_3 V_{X_3}(x_3) + \cdots + w_n V_{X_n}(x_n)$$

where w_i for $i = 1, \ldots, n$ are nonnegative weights (importance weights) whose values range between 0 and 1 and where $w_1 + w_2 + w_3 + \cdots + w_n = 1$.

Given the conventions that (1) the SDVFs $V_{X_1}(x_1), V_{X_2}(x_2), V_{X_3}(x_3), \ldots, V_{X_n}(x_n)$ each range in value between 0 and 1 and (2) the weights each range in value between 0 and 1 and sum to unity, it follows that $V_Y(y)$ will range between 0 and 1. From this, the power-additive utility function will also range between 0 and 1.

In Definitions 3.12 and 3.13, we assume conditions for an additive value function hold, as well as an independence condition known as *utility independence*. *Utility independence* is a stronger form of independence than preferential independence. From Clemen (1996), an attribute X_1 is *utility independent* of attribute X_2 if preferences for uncertain choices involving different levels of X_1 are independent of the value of X_2.

3.5.2 Applying the Power-Additive Utility Function

The shape of the power-additive utility function is governed by a parameter known as *multiattribute risk tolerance* ρ_m. Figure 3.33 presents families of power-additive utility functions for various ρ_m and for increasing or decreasing preferences. Multiattribute risk averse utility functions have positive values for ρ_m. Multiattribute risk-seeking utility functions have negative values for ρ_m. The multiattribute risk neutral case occurs when ρ_m approaches infinity. Here, we have a straight line; this is where the expected value of the value function $V_Y(y)$ can be used to rank alternatives.

One approach to selecting ρ_m is to have the decision-maker review Figure 3.33 and select ρ_m that most reflects his risk attitude. An extremely risk averse decision-maker, where monotonically increasing preferences apply, might select ρ_m in the interval $0.05 \leq \rho_m \leq 0.15$. A less risk averse decision-maker,

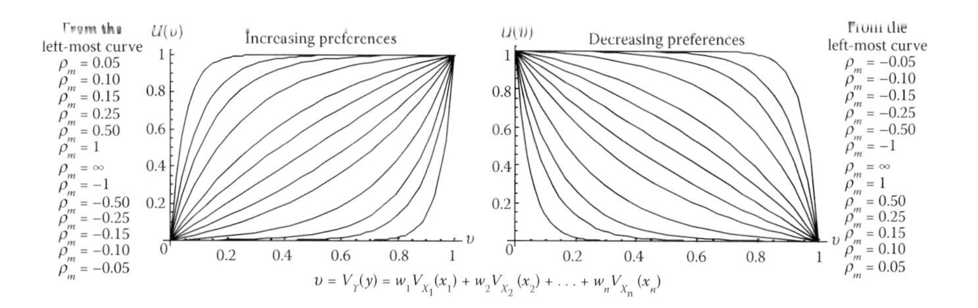

FIGURE 3.33
Families of power-additive utility functions.

where monotonically increasing preferences apply, might select ρ_m in the interval $0.15 < \rho_m \leq 1$. As ρ_m becomes increasingly large (approaches infinity), the decision-maker is increasingly risk neutral and the power-additive utility function approaches a straight line. As we shall see, when this occurs the expected value of the value function $V_Y(y)$ can be used to rank alternatives.

The information presented in Figure 3.33 provides a visual "look-up" procedure for selecting ρ_m. Alternative approaches involve the use of lotteries similar to those previously discussed. For a discussion on the use of lotteries to derive ρ_m, see Kirkwood (1997).

Theorem 3.10

If utilities are *monotonically increasing* over the values of the additive value function $V_Y(y)$ with the power-additive utility function given below:

$$U(v) = \begin{cases} K(1 - e^{-(V_Y(y)/\rho_m)}) & \text{if } \rho_m \neq \infty \\ V_Y(y) & \text{if } \rho_m = \infty \end{cases}$$

where $K = 1/(1 - e^{-1/\rho_m})$ and $v = V_Y(y)$ (given in Definition 3.6) then

$$E(U(v)) = \begin{cases} K(1 - E(e^{-(V_Y(y)/\rho_m)})) & \text{if } \rho_m \neq \infty \\ E(V_Y(y)) & \text{if } \rho_m = \infty \end{cases}$$

Theorem 3.11

If utilities are *monotonically decreasing* over the values of the additive value function $V_Y(y)$ with the power-additive utility function given below:

$$U(v) = \begin{cases} K(1 - e^{-((1 - V_Y(y))/\rho_m)}) & \text{if } \rho_m \neq \infty \\ 1 - V_Y(y) & \text{if } \rho_m = \infty \end{cases}$$

where $K = 1/(1-e^{-1/\rho_m})$ and $v = V_Y(y)$ (given in Definition 3.6) then

$$E(U(v)) = \begin{cases} K(1-E(e^{-((1-V_Y(y))/\rho_m)})) & \text{if } \rho_m \neq \infty \\ 1-E(V_Y(y)) & \text{if } \rho_m = \infty \end{cases}$$

Theorems 3.10 and 3.11 provide the way to compute the expected utilities of the power-additive utility function. Expected utilities provide measures with which to rank uncertain alternatives, from most-to-least preferred. The following presents a set of formulas needed to compute these expected utilities, when uncertainties are expressed as either discrete or continuous probability distributions.

First, we look at Theorem 3.10. Here, utilities are monotonically increasing over the values of the additive value function. From Theorem 3.10,

$$E(U(v)) = \begin{cases} K(1-E(e^{-(V_Y(y)/\rho_m)})) & \text{if } \rho_m \neq \infty \\ E(V_Y(y)) & \text{if } \rho_m = \infty \end{cases}$$

For the case where $\rho \neq \infty$, the term $E(e^{-(V_Y(y)/\rho_m)})$ can be written as follows:

$$E(e^{-(V_Y(y)/\rho_m)}) = E(e^{-(w_1 V_{X_1}(x_1) + w_2 V_{X_2}(x_2) + \ldots + w_n V_{X_n}(x_n))/\rho_m})$$

$$E(e^{-(V_Y(y)/\rho_m)}) = E(e^{-(w_1 V_{X_1}(x_1))/\rho_m}) E(e^{-(w_2 V_{X_2}(x_2))/\rho_m}) \ldots E(e^{-(w_n V_{X_n}(x_n))/\rho_m})$$

where the X_i's are independent random variables and where

$$E(e^{-(w_i V_{X_i}(x_i))/\rho_m}) = \begin{cases} \sum_{x_i} p_{X_i}(x_i) e^{-(w_i V_{X_i}(x_i))/\rho_m} & \text{if } X_i \text{ is discrete} \\ \int_{-\infty}^{\infty} e^{-(w_i V_{X_i}(x_i))/\rho_m} f_{X_i}(x_i)\,dx_i & \text{if } X_i \text{ is continuous} \end{cases} \quad (3.60)$$

In this equation, $p_{X_i}(x_i)$ is the probability the uncertain outcome X_i takes the score x_i if X_i is a discrete random variable. In Equation 3.60, $f_{X_i}(x_i)$ is the probability density function of X_i if X_i is a continuous random variable. In Theorem 3.10, when $\rho = \infty$ the term $E(V_Y(y))$ can be written as follows:

$$E(V_Y(y)) = w_1 E(V_{X_1}(x_1)) + w_2 E(V_{X_2}(x_2)) + \ldots + w_n E(V_{X_n}(x_n)) \quad (3.61)$$

where

$$E(V_{X_i}(x_i)) = \begin{cases} \sum_{x_i} p_{X_i}(x_i) V_{X_i}(x_i) & \text{if } X_i \text{ is discrete} \\ \int_{-\infty}^{\infty} V_{X_i}(x_i) f_{X_i}(x_i)\,dx_i & \text{if } X_i \text{ is continuous} \end{cases} \quad (3.62)$$

Next, we look at Theorem 3.11. Here, utilities are monotonically decreasing over the values of the additive value function. From Theorem 3.11,

$$E(U(v)) = \begin{cases} K(1 - E(\mathrm{e}^{-((1-V_Y(y))/\rho_m)})) & \text{if } \rho_m \neq \infty \\ 1 - E(V_Y(y)) & \text{if } \rho_m = \infty \end{cases}$$

For the case where $\rho \neq \infty$, the term $E(\mathrm{e}^{-((1-V_Y(y))/\rho_m)})$ can be written as

$$E(\mathrm{e}^{-((1-V_Y(y))/\rho_m)}) = E(\mathrm{e}^{-(1-(w_1 V_{X_1}(x_1) + w_2 V_{X_2}(x_2) + \cdots + w_n V_{X_n}(x_n)))/\rho_m})$$

$$E(e^{-((1-V_Y(y))/\rho_m)}) = E(e^{(-1+(w_1 V_{X_1}(x_1) + w_2 V_{X_2}(x_2) + \ldots + w_n V_{X_n}(x_n)))/\rho_m})$$

$$E(\mathrm{e}^{-((1-V_Y(y))/\rho_m)}) = \mathrm{e}^{-1/\rho_m} E(\mathrm{e}^{w_1(V_{X_1}(x_1))/\rho_m}) E(\mathrm{e}^{w_1(V_{X_2}(x_2))/\rho_m}) \ldots E(\mathrm{e}^{w_n(V_{X_n}(x_n))/\rho_m})$$

where the X_i's are independent random variables and where

$$E(\mathrm{e}^{(w_i V_{X_i}(x_i))/\rho_m}) = \begin{cases} \sum_{x_i} p_{X_i}(x_i)\mathrm{e}^{(w_i V_{X_i}(x_i))/\rho_m} & \text{if } X_i \text{ is discrete} \\ \int_{-\infty}^{\infty} \mathrm{e}^{(w_i V_{X_i}(x_i))/\rho_m} f_{X_i}(x_i)\,dx_i & \text{if } X_i \text{ is continuous} \end{cases} \quad (3.63)$$

In Equation 3.63, $p_{X_i}(x_i)$ is the probability the uncertain outcome X_i takes the score x_i if X_i is a discrete random variable. In Equation 3.63, $f_{X_i}(x_i)$ is the probability density function of X_i if X_i is a continuous random variable. In Theorem 3.11, when $\rho = \infty$ the term $1 - E(V_Y(y))$ can be written as

$$1 - E(V_Y(y)) = 1 - (w_1 E(V_{X_1}(x_1)) + w_2 E(V_{X_2}(x_2)) + \ldots + w_n E(V_{X_n}(x_n))) \quad (3.64)$$

where

$$E(V_{X_i}(x_i)) = \begin{cases} \sum_{x_i} p_{X_i}(x_i) V_{X_i}(x_i) & \text{if } X_i \text{ is discrete} \\ \int_{-\infty}^{\infty} V_{X_i}(x_i) f_{X_i}(x_i)\,dx_i & \text{if } X_i \text{ is continuous} \end{cases} \quad (3.65)$$

3.6 Applications to Engineering Risk Management

The preceding sections covered many topics. An introduction to probability theory and value and utility function theory was presented. This section offers two case discussions that illustrate the application of these topics in engineering risk management.

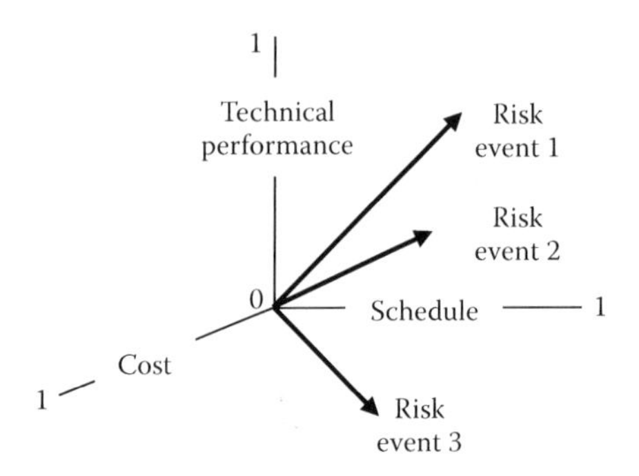

FIGURE 3.34
Typical risk impact dimensions.

The following case discussion shows how value functions may be used in engineering risk management. In general, value functions provide ways to measure the relative merit of an alternative in a set of competing alternatives. As shown in Section 3.3, merit can be measured as a function of multiple criteria. In engineering risk management, value functions can be created to measure the relative criticality of each risk in a set of competing risks. Risk criticality can also be measured as a function of multiple evaluation criteria. As shown in Figure 3.34, these criteria typically include a risk's impact on a system's cost, schedule, and technical performance. Additional areas of impact may also be defined. Risks with higher measures of criticality are candidates for greater management attention and resources than those with lesser measures of criticality.

3.6.1 Value Theory to Measure Risk

CASE 3.2

Consider a satellite communication system that interfaces to a number of networked subsystems. Suppose a data management architecture is being newly designed for the communication system as whole, one where the interfacing subsystem databases must support information exchanges. However, due to schedule pressures suppose the new database for the communication system will not be fully tested for compatibility with the existing subsystem databases when version 1.0 of the data management architecture is released.

Suppose the project team identified the following risk event: *Inadequate synchronization of the communication system's new database with the existing subsystem databases.* Suppose we have the risk statement given in Figure 3.35.

Here, two events are described. They are the conditioning event and the risk event. In Figure 3.35, the *Condition* is event B and the *If* is event A; that is,

B = {*The new database for the communication system will not be fully tested for compatibility with the existing subsystem databases when version 1.0 of the data management architecture is released*}

A = {*Inadequate synchronization of the communication system's new database with the existing subsystem databases*}

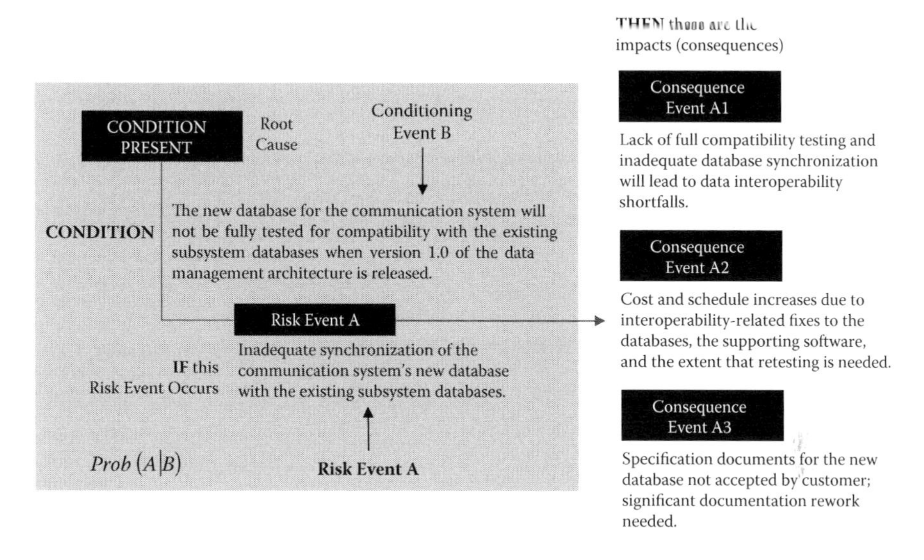

FIGURE 3.35
Risk statement: Lack of database synchronization.

From this, we can form the risk statement, as given below.

RISK STATEMENT: {*Inadequate synchronization of the communication system's new database with the existing subsystem databases because the new database for the communication system will not be fully tested for compatibility with the existing subsystem databases when version 1.0 of the data management architecture is released*}

Probability Assessment

Recall that a risk event is equivalent to a probability event; formally, $0 < P(A|B) = \alpha < 1$, where α is the probability risk event A occurs given the conditioning event B has occurred. Suppose the engineering team used the table in Figure 3.36 as a guide for their assessment of the chance risk event A occurs.

In support of their assessment, suppose they determined inadequate synchronization between the exchange of data across the system's subsystems is almost sure to occur if lack of full compatibility testing persists. Referring to the table in Figure 3.36, suppose the engineering team assigned a probability of 0.95 to risk event A; that is, $P(A|B) = \alpha = 0.95$. Furthermore, suppose they assumed a linear relationship between this probability and its value along the family of value functions in Figure 3.36; thus, $V(P(A|B)) = V(\alpha) = \alpha = 0.95$ in this case.

Impact Assessment

As shown in Figure 3.34, risk events can have multiconsequential impacts to an engineering system—especially systems in development. For this case discussion, suppose risks, if realized, affect the communication system's *Cost*, *Schedule*, *Technical Performance*, and program-related technical efforts (i.e., *Programmatics*). Suppose each risk's impact is represented by a

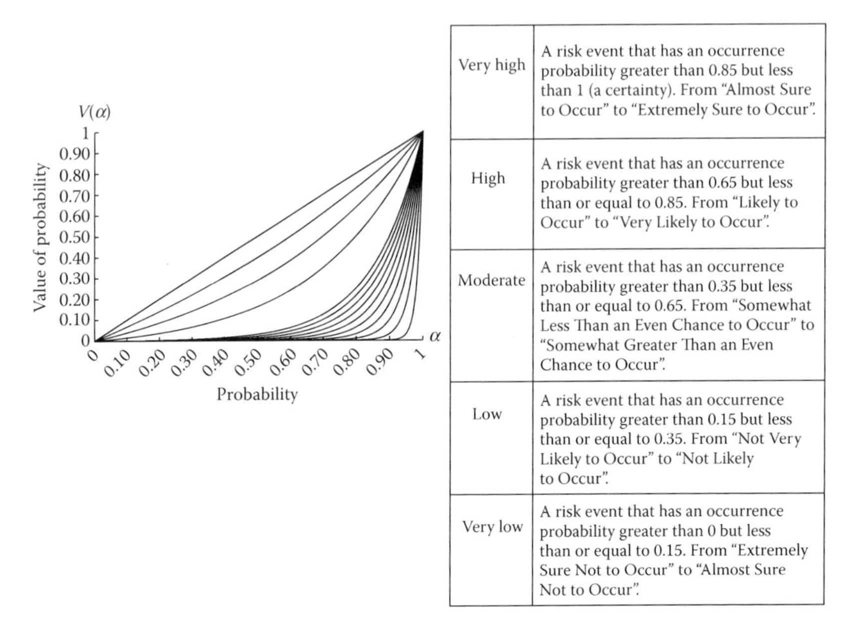

Very high	A risk event that has an occurrence probability greater than 0.85 but less than 1 (a certainty). From "Almost Sure to Occur" to "Extremely Sure to Occur".
High	A risk event that has an occurrence probability greater than 0.65 but less than or equal to 0.85. From "Likely to Occur" to "Very Likely to Occur".
Moderate	A risk event that has an occurrence probability greater than 0.35 but less than or equal to 0.65. From "Somewhat Less Than an Even Chance to Occur" to "Somewhat Greater Than an Even Chance to Occur".
Low	A risk event that has an occurrence probability greater than 0.15 but less than or equal to 0.35. From "Not Very Likely to Occur" to "Not Likely to Occur".
Very low	A risk event that has an occurrence probability greater than 0 but less than or equal to 0.15. From "Extremely Sure Not to Occur" to "Almost Sure Not to Occur".

FIGURE 3.36
Probability table and families of value functions.

value function, where increasing values of this function are associated with increasing levels of consequence.

Following the approach described in Section 3.3, a value function can be specified for each impact area. Depending on the area, these value functions might be piecewise linear or vary continuously across levels (or scores). Furthermore, some impact areas do not have a natural or common unit of measure. In these cases, a constructed scale may be needed.

Consider a risk event's impact on the technical performance of the communication system. Technical performance is a difficult impact dimension to express in a common unit. This is because technical performance can be measured in many ways, such as the number of millions of instructions per second or the weight of an end item. It is difficult, then, to specify for an engineering system a value function for *technical performance* along a common measurement scale. A constructed scale is often appropriate, which we will illustrate in this case.

Figure 3.37 illustrates a piecewise linear value function designed along a constructed scale for the *technical performance* impact area. Suppose this function was designed by the communication system's project team. There are many ways to define such a constructed scale and its associated value function. Table 3.8 provides one set of linguistic definitions for the levels (scores) corresponding to the value function in Figure 3.37.

In Figure 3.37, the anchor points 0 and 1 along the vertical axis are assigned by the team and set at level 1 and level 5, respectively, along the horizontal

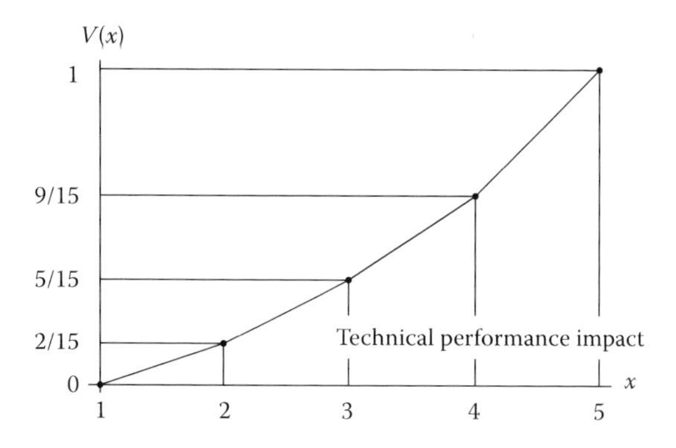

FIGURE 3.37
A value function for technical performance impact.

TABLE 3.8

A Constructed Scale for Technical Performance Impact

Ordinal Scale Level (Score)	Definition: Technical Performance Impact
5	A risk event that, if it occurs, impacts the system's operational capabilities (or the engineering of these capabilities) to the extent that critical technical performance (or system capability) shortfalls result.
4	A risk event that, if it occurs, impacts the system's operational capabilities (or the engineering of these capabilities) to the extent that technical performance (or system capability) is marginally below minimum acceptable levels.
3	A risk event that, if it occurs, impacts the system's operational capabilities (or the engineering or these capabilities) to the extent that technical performance (or system capability) falls well-below stated objectives but remains enough above minimum acceptable levels.
2	A risk event that, if it occurs, impacts the system's operational capabilities (or the engineering of these capabilities) to the extent that technical performance (or system capability) falls below stated objectives but well-above minimum acceptable levels.
1	A risk event that, if it occurs, impacts the system's operational capabilities (or the engineering of these capabilities) in a way that results in a negligible effect on overall performance (or achieving capability objectives for a build/block/increment), but regular monitoring for change is strongly recommended.

axis. Suppose it was decided the smallest increment Δ in value occurs between a level 1 and level 2 technical performance impact. If we use Δ as the reference standard, it can be seen the team decided the following: (1) the value increment between a level 2 and level 3 technical performance impact is one and a half times the smallest value increment Δ; (2) the value increment between a level 3 and level 4 technical performance impact is two times the smallest value increment Δ; and (3) the value increment between a level 4 and level 5 technical performance impact is three times the smallest value increment Δ.

From a risk management perspective, the value function in Figure 3.37 can be interpreted as follows. It reflects management's monotonically increasing preferences for risk events that, if they occur, score at increasingly higher levels along the technical performance impact scale. The higher a risk event scores along the value function in Figure 3.37, the greater its technical performance impact.

Illustrated in Figure 3.34, a risk event, can impact not only the technical performance of a system but also its cost and schedule. An unmitigated risk may negatively impact the cost of a system, in terms of increased dollars beyond the budget to address problems caused by the risk. In addition, there may be adverse schedule impacts in terms of missed milestones or schedule slippages beyond what was planned.

To address these concerns, suppose the communication system's project team designed the two value functions in Figure 3.38. These value functions capture a risk event's impacts on the system's cost and schedule, shown as single-dimensional monotonically increasing exponential value functions. In designing these value functions, suppose the project team decided a 5% increase in cost and a 3-month increase in schedule to be the midvalues for the cost and schedule value functions, respectively. From Definition 3.1, the general form of the value functions in Figure 3.38 is as follows:

$$V(x) = \frac{1 - e^{-(x - x_{min})/\rho}}{1 - e^{-(x_{max} - x_{min})/\rho}}$$

The specific equations for the value functions in Figure 3.38 are as follows:

$$\text{Cost Impact: } V_1(x) = \frac{1 - e^{-x/8.2}}{1 - e^{-20/8.2}} = 1.096(1 - e^{-x/8.2}) \tag{3.66}$$

$$\text{Schedule Impact: } V_2(x) = \frac{1 - e^{-x/4.44}}{1 - e^{-18/4.44}} = 1.018(1 - e^{-x/4.44}) \tag{3.67}$$

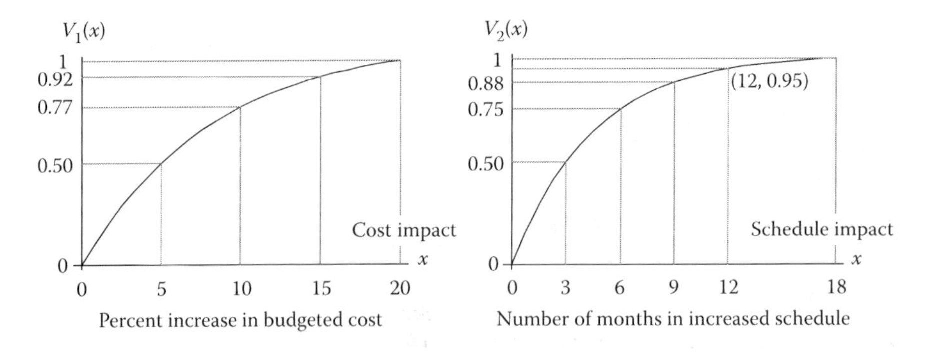

FIGURE 3.38
Illustrative value functions for cost and schedule impacts.

The value functions in Figure 3.38 can also be expressed as constructed scales, as shown in Tables 3.9 and 3.10.

Figure 3.34 illustrated three dimensions of an engineering system commonly impacted by risk. In practice, other dimensions can be impacted by risk. For example, a risk's programmatic impact is often as serious a concern as its impacts on a system's technical performance, cost, or schedule.

In this regard, programmatic impacts might refer to specific work products or activities necessary to advance the program along its milestones or its life cycle. Examples of technical work products include system architecture documents, system design documents, the system's engineering management plan, concepts of operation, and the system's logistics plan. Examples of programmatic work products include the system's integrated master schedule, its life cycle cost estimate, its risk management plan, and various acquisition or contract-related documents and plans.

TABLE 3.9

A Constructed Scale for Cost Impact

Ordinal Scale Level (Score)	Definition: Cost Impact
5	A risk event that, if it occurs, will cause more than a 15% increase but less than or equal to a 20% increase in the program's budget.
4	A risk event that, if it occurs, will cause more than a 10% increase but less than or equal to a 15% increase in the program's budget.
3	A risk event that, if it occurs, will cause more than a 5% increase but less than or equal to a 10% increase in the program's budget.
2	A risk event that, if it occurs, will cause more than a 2% but less than or equal to a 5% increase in the program's budget.
1	A risk event that, if it occurs, will cause less than a 2% increase in the program's schedule budget.

TABLE 3.10

A Constructed Scale for Schedule Impact

Ordinal Scale Level (Score)	Definition: Schedule Impact
5	A risk event that, if it occurs, will cause more than a 12 month increase in the program's schedule.
4	A risk event that, if it occurs, will cause more than a 9 month but less than or equal to a 12 month increase in the program's schedule.
3	A risk event that, if it occurs, will cause more than a 6 month but less than or equal to a 9 month increase in the program's schedule.
2	A risk event that, if it occurs, will cause more than a 3 month but less than or equal to a 6 month increase in the program's schedule.
1	A risk event that, if it occurs, will cause less than a 3 month increase in the program's schedule.

Figure 3.39 illustrates a value function that could be used to express a risk event's programmatic impacts. Table 3.11 shows a constructed scale associated to this value function. Such a scale would be developed in a manner similar to the preceding discussions. In Figure 3.39, the anchor points 0 and 1 along the vertical axis are assigned by the team and set at level 1 and level 5, respectively, along the horizontal axis. Suppose it was decided the smallest increment Δ in value occurs between a level 1 and level 2 programmatic impact. If we use Δ as the reference standard, it can be seen the team decided

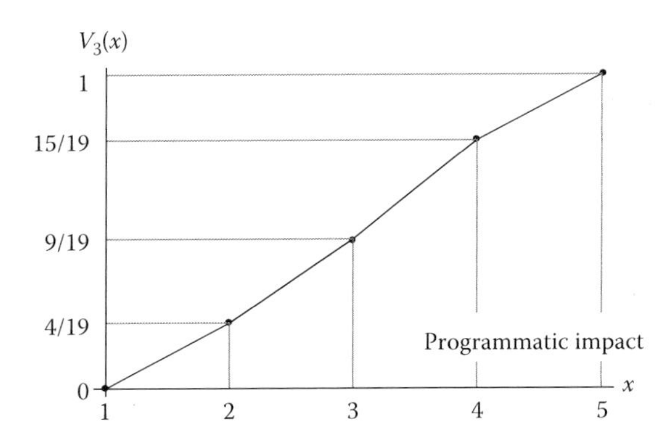

FIGURE 3.39
Illustrative value function for programmatic impact.

TABLE 3.11

A Constructed Scale for Programmatic Impacts

Ordinal Scale Level (Score)	Definition: Programmatic Impacts
5	A risk event that, if it occurs, impacts programmatic efforts to the extent that one or more critical objectives for technical or programmatic work products (or activities) will not be completed.
4	A risk event that, if it occurs, impacts programmatic efforts to the extent that one or more stated objectives for technical or programmatic work products (or activities) is marginally below minimum acceptable levels.
3	A risk event that, if it occurs, impacts programmatic efforts to the extent that one or more stated objectives for technical or programmatic work products (or activities) falls well-below goals but remains enough above minimum acceptable levels.
2	A risk event that, if it occurs, impacts programmatic efforts to the extent that one or more stated objectives for technical or programmatic work products (or activities) falls below goals but well-above minimum acceptable levels.
1	A risk event that, if it occurs, has little to no impact on programmatic efforts. Program advancing objectives for technical or programmatic work products (or activities) for a build/block/increment will be met, but regular monitoring for change is strongly recommended.

the following: (1) the value increment between a level 2 and level 3 programmatic impact is one and a quarter times the smallest value increment Δ; (2) the value increment between a level 3 and level 4 programmatic impact is one and a half times the smallest value increment Δ; and (3) the value increment between a level 4 and level 5 programmatic impact is the same as the value increment between a level 1 and level 2 programmatic impact. Figure 3.40 summarizes the four value functions associated with this case discussion.* We will return to these later in measuring a risk's criticality.

Measuring Overall Impact

One way to measure a risk event's overall impact is by the additive value function. If we assume preferential independence conditions, then, from Definition 3.6, we have

$$V(A) = w_1V_1(x_1) + w_2V_2(x_2) + w_3V_3(x_3) + w_4V_4(x_4) \tag{3.68}$$

where $V(A)$ is the overall impact of risk event A. The terms $V_1(x_1), V_2(x_2), V_3(x_3)$, and $V_4(x_4)$ are SDVFs for *Cost*, *Schedule*, *Technical Performance*, and *Programmatics* given in Figure 3.40. The parameters w_1, w_2, w_3, and w_4 are nonnegative weights whose values range between 0 and 1, and where $\Sigma_i w_i = 1$.

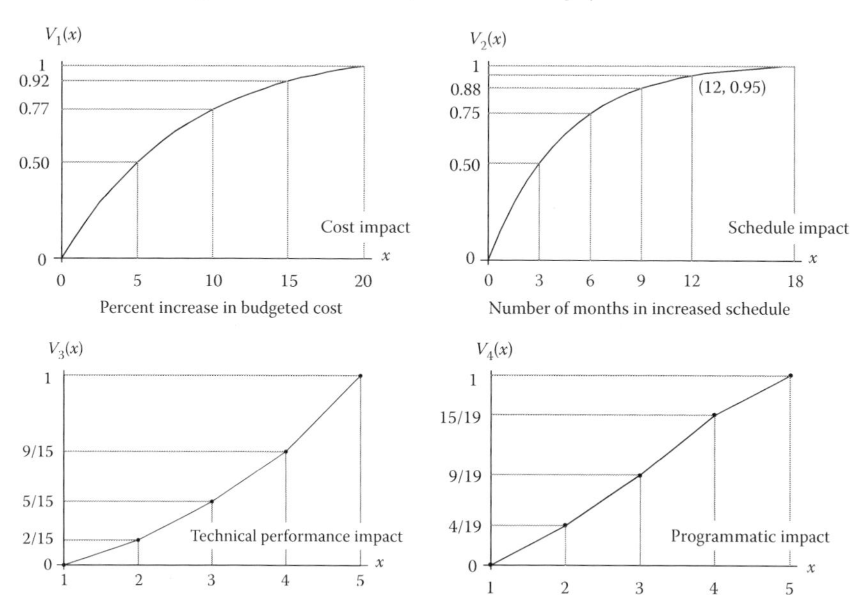

FIGURE 3.40
Case 3.2 value functions.

* The value functions in Figure 3.40 are illustrative. In practice, a project team must define their own criteria and specific value functions in ways that truly captures the impact areas of concern to the project and its management.

From the consequence events listed in Figure 3.35, suppose the project team assessed the impacts risk event A would have, *if it occurred* is shown in Table 3.12. A key feature in this table is the basis of assessment (BOA). A BOA is a written justification of the reason for a chosen rating. The BOA must be written such that it (1) clearly and concisely justifies the assessor's rationale and (2) enables this justification to be objectively reviewed by subject matter peers.

For this case, suppose the project team decided on the following weight assessments. Technical performance w_3 is twice as important as cost w_1; cost w_1 is twice as important as schedule; cost w_1 is twice as important as programmatics w_4. From this, we have the following:

$$w_3 = 2w_1;\ w_1 = 2w_2;\ w_1 = 2w_4$$

TABLE 3.12

Risk Event A Impact Assessments

Risk Event	*Inadequate synchronization of the communication system's new database with the existing subsystem databases because the new database for the communication system will not be fully tested for compatibility with the existing subsystem databases when version 1.0 of the data management architecture is released.*
	The consequence event descriptions in Figure 3.35 provide a starting point for the basis of assessments below. They support the team's judgments and supporting arguments for articulating the risk event's consequences, if it occurs, on the project's cost, schedule, programmatics, and the system's technical performance.
Cost Impact Level 4	This risk event, if it occurs, is estimated by the engineering team to cause a 12% increase in the project's current budget. The estimate is based on a careful assessment of the ripple effects across the project's cost categories for interoperability-related fixes to the databases, the supporting software, and the extent that re-testing is needed.
	Value Function Value: From Equation 3.66, $V_1(12) = 1.096(1 - e^{-12/8.2}) = 0.842$.
Schedule Impact Level 2	This risk event, if it occurs, is estimated by the engineering team to cause a 4-month increase in the project's current schedule. The estimate is based on a careful assessment of the ripple effects across the project's integrated master schedule for interoperability-related fixes to the databases, the supporting software, and the extent that re-testing is needed.
	Value Function Value: From Equation 3.67, $V_2(4) = 1.0181(1 - e^{-4/4.44}) = 0.604$.
Technical Performance Impact Level 4	This risk event, if it occurs, is assessed by the engineering team as one that will impact the system's operational capabilities to the extent that technical performance is marginally below minimum acceptable levels, depending on the location and extent of interoperability shortfalls.
	Value Function Value: From Figure 3.40, $V_3(4) = 9/15 = 0.60$.
Programmatic Impact Level 4	This risk event, if it occurs, is assessed by the engineering team as one that will impact programmatic efforts to the extent that one or more stated objectives for technical or programmatic work products (e.g., various specifications or activities) is marginally below minimum acceptable levels.
	Value Function Value: From Figure 3.40, $V_4(4) = 15/19 = 0.79$.

Since $w_1 + w_2 + w_3 + w_4 = 1$, it follows that

$$w_1 + \frac{1}{2}w_1 + 2w_1 + \frac{1}{2}w_1 = 1$$

thus, $w_1 = \frac{1}{4}$, $w_2 = \frac{1}{8}$, $w_3 = \frac{1}{2}$, and $w_4 = \frac{1}{8}$. From this, it follows that

$$V(A) = \frac{1}{4}V_1(x_1) + \frac{1}{8}V_2(x_2) + \frac{1}{2}V_3(x_3) + \frac{1}{8}V_4(x_4) \tag{3.69}$$

From Table 3.12, we have the following:

$$V(A) = \frac{1}{4}(0.842) + \frac{1}{8}(0.604) + \frac{1}{2}(0.60) + \frac{1}{8}(0.79) = 0.685 \tag{3.70}$$

Figure 3.41 shows a probability-impact plot of risk event A. Overall, this event might be of moderately high concern to the project team.

The preceding analysis approach can be extended to multiple risk events. Figure 3.42 shows a probability-impact plot of 25 risk events. Suppose each event was analyzed by the same process just discussed. In Figure 3.42, risk events 3, 5, and 8 appear to be ahead of the others in terms of occurrence probability and impact to the project. What about risk events 24, 4, and 1? How critical are they relative to risk events 3, 5, and 8? The following discusses a way to measure the *relative criticality* of each risk in a set of competing risks.

Relative Risk Criticality

An approach to rank order risk events from most-to-least critical is to form an additive value function that produces a risk score for each risk event. Suppose the risk score of risk event E is given by the following definition.

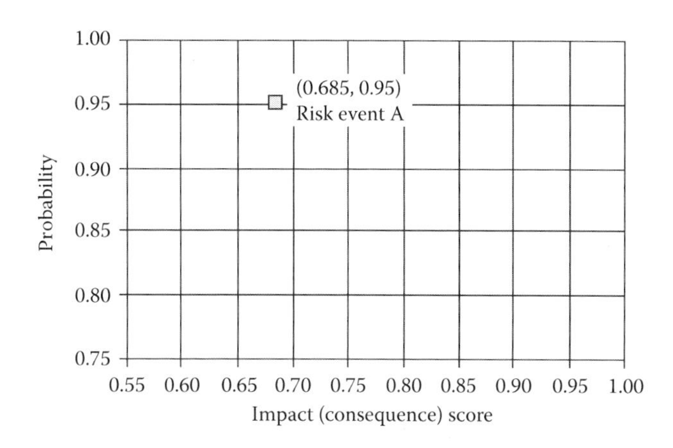

FIGURE 3.41
Case 3.2: A plot of risk event A.

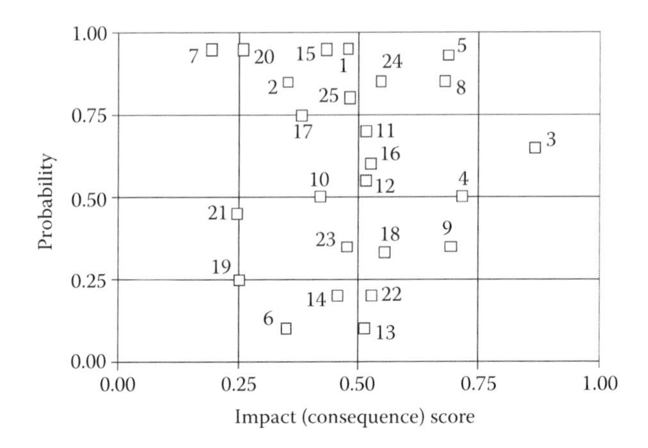

FIGURE 3.42
A scatter plot of 25 risk events.

Definition 3.14: Risk Event Risk Score

The *risk score* of risk event E is given by the additive value function

$$Risk\,Score(E) = RS(E) = u_1 V_{\text{Probability}}(E) + u_2 V_{\text{Impact}}(E) \tag{3.71}$$

subject to the considerations in Theorem 3.6. In Equation 3.71, the first term is a value function for the risk event's occurrence probability. The second term is a value function for the severity of the risk event's impact, if it occurs. The coefficients u_1, and u_2, are nonnegative weights such that $0 \le u_1 \le 1, 0 \le u_2 \le 1$, and $u_1 + u_2 = 1$. In Equation 3.71, these value functions can be designed to produce measures along either a 0 to 1 or a 0 to 100 cardinal interval scale, as discussed in Section 3.3.

Suppose Table 3.13 presents the data for each risk event plotted in Figure 3.42. From left to right, column one is the risk event number labeled in Figure 3.42. Column two is the assessment of each risk event's occurrence probability. For convenience, suppose we have assumed a linear relationship between probability and its value function. Column three is each risk event's overall impact score, computed by an application of Equation 3.69. Column four is each risk event's risk score, computed by Equations 3.71 and 3.72. In Equation 3.72, a risk event's overall impact score is twice as important as its assessed occurrence probability. With this,

$$RS(Ei) = \frac{1}{3} V_{\text{Probability}}(Ei) + \frac{2}{3} V_{\text{Impact}}(Ei) \tag{3.72}$$

where $i = 1, 2, 3, \ldots, 25$. Table 3.14 presents a relative risk ranking based on the value of each risk event's risk score. As mentioned above, the higher a risk

TABLE 3.13

Values and Scores for Risks in Figure 3.42

Risk Event #	Probability Direct Assessment	Impact Score	Risk Score
1	0.95	0.477	0.635
2	0.85	0.353	0.519
3	0.65	0.867	0.795
4	0.50	0.718	0.645
5	0.93	0.688	0.769
6	0.10	0.349	0.266
7	0.95	0.194	0.446
8	0.85	0.681	0.737
9	0.35	0.695	0.580
10	0.50	0.420	0.447
11	0.70	0.516	0.578
12	0.55	0.517	0.528
13	0.10	0.515	0.376
14	0.20	0.455	0.370
15	0.95	0.432	0.605
16	0.60	0.525	0.550
17	0.75	0.382	0.505
18	0.33	0.555	0.480
19	0.25	0.254	0.252
20	0.95	0.260	0.490
21	0.45	0.248	0.315
22	0.20	0.530	0.420
23	0.35	0.475	0.434
24	0.85	0.545	0.646
25	0.80	0.481	0.587

event's risk score the higher its rank position relative to the set of identified risks. Thus, risk event 3 is in first rank position. It has the highest risk score in the set shown in Table 3.13. Risk event 5 is in second rank position. It has the second highest risk score in the set shown in Table 3.13, and so forth.

In Table 3.14, observe the top five ranked risks are risk events 3, 5, 8, 24, and 4. This suggests the project team should focus further scrutiny on these five events to further confirm they indeed merit these top rank positions. This includes a further look at the basis of assessments behind the value function inputs chosen to characterize each risk event, which is used by the risk score equation to generate the above rankings. Last, it is best to treat any risk ranking as *suggestive* of a risk prioritization. Prioritization decisions with respect to where risk mitigation resources should be applied can be guided by this analysis but not solely directed by it. They are analytical filters that serve as aids to managerial decision-making. This completes Case 3.2.

TABLE 3.14

A Relative Ranking of the Risks in Figure 3.42

Rank	Risk Event #	Risk Score
1	3	0.795
2	5	0.769
3	8	0.737
4	24	0.646
5	4	0.645
6	1	0.635
7	15	0.605
8	25	0.587
9	9	0.580
10	11	0.578
11	16	0.550
12	12	0.528
13	2	0.519
14	17	0.505
15	20	0.490
16	18	0.480
17	10	0.447
18	7	0.446
19	23	0.434
20	22	0.420
21	13	0.376
22	14	0.370
23	21	0.315
24	6	0.266
25	19	0.252

3.6.2 Utility Theory to Compare Designs

CASE 3.3

Consider the following case. A new and highly sophisticated armored ground transport vehicle is currently being designed. There are three design alternatives undergoing engineering tests and performance trade studies. Suppose a set of evaluation criteria to evaluate these designs has been defined by the program's decision-makers. These criteria are *Operational Days*, *Maintenance/Service Time*, and *Cost* and are denoted by X_1, X_2, X_3, respectively. The criterion *Operational Days* refers to the number of days the vehicle can operate without maintenance or servicing. The criterion *Maintenance/Service Time* refers to the number of labor hours needed to service the vehicle to keep it operationally on duty. The criterion *Cost* refers to each vehicle's estimated recurring unit cost in dollars million.

Suppose a program's decision-makers assessed the criterion X_1 as twice as important as criterion X_2 and X_3 as twice as important as criterion X_2.

After careful deliberations, suppose the program's decision-makers defined a set of exponential value functions for each of the three criteria. These functions are shown in Figure 3.43. The equations for these value functions are given below.

$$V_{X_1}(x_1) = 1.30902(1 - e^{0.0320808(45 - x_1)})$$

$$V_{X_2}(x_2) = 1.38583(1 - e^{0.0639325(x_2 - 30)})$$

$$V_{X_3}(x_3) = -0.784058(1 - e^{-0.164433(x_3 - 8)})$$

Suppose the decision-makers also reviewed the graphs in Figure 3.33 and determined their multiattribute risk tolerance is represented by the curve with $\rho_m = 0.25$. So their preference structure reflects a monotonically increasing risk averse attitude over increasing values of the value function.

Suppose each design alternative is undergoing various engineering analyses, cost estimates, and simulations to assess their potential performance on the criteria in Figure 3.43. The results predicted from these analyses are

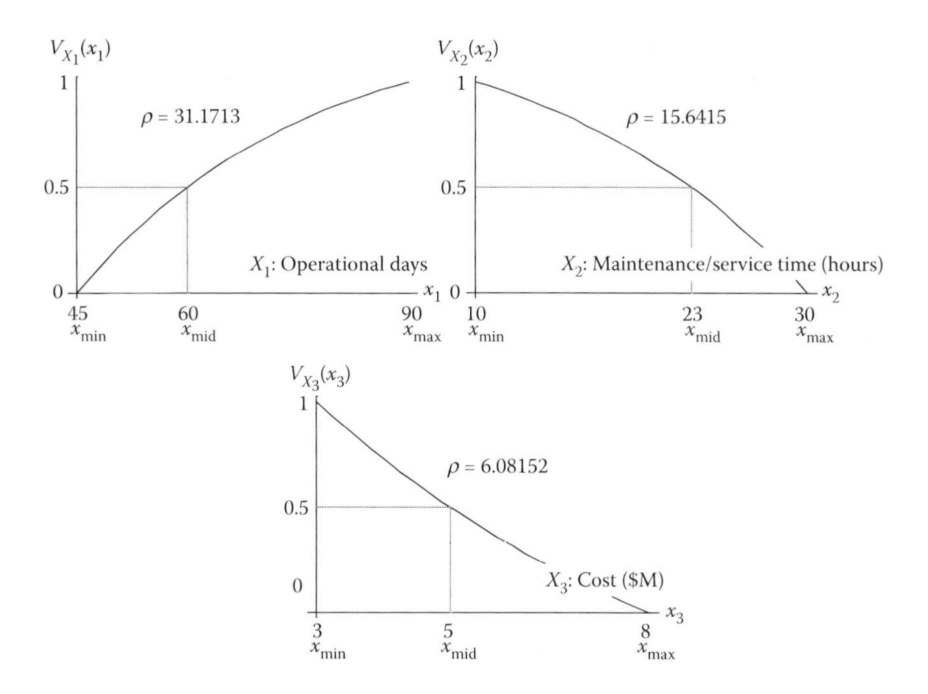

FIGURE 3.43
Exponential value functions for Case 3.3.

TABLE 3.15

Case 3.3: Design Alternative Performance Measures

Design Alternative	Criterion X_1	Criterion X_2	Criterion X_3
Alternative A	72 – 79 $X_1 \sim$ Unif (72, 79)	15 – 23 $X_2 \sim$ Unif (15, 23)	5.5 – 7 $X_3 \sim$ Unif (5.5, 7)
Alternative B	85 – 88 $X_1 \sim$ Unif (85, 88)	23 – 27 $X_2 \sim$ Unif (23, 27)	5 – 6.5 $X_3 \sim$ Unif (5, 6.5)
Alternative C	80 – 85 $X_1 \sim$ Unif (80, 85)	24 – 28 $X_2 \sim$Unif (24, 28)	4 – 5 $X_3 \sim$ Unif (4, 5)

Note: X_1 = Operational Days; X_2 = Maintenance/Service Hours; X_3 = Cost ($M)

summarized in Table 3.15. Suppose the uncertainties in the outcomes for each criterion are captured by a uniform probability density function—specified for each criterion within a given alternative.

From this information and the data in Table 3.15, determine which design alternative is performing "best," where best is measured as the alternative having the highest expected utility, in terms of the value of each design choice. In this case discussion, assume that conditions for an additive value function hold, as well as utility independence.

To determine which design alternative is performing "best," we will compute the expected utility of the value of each alternative, as well as compute each alternative's expected value. The alternative with the highest expected utility for value will be considered the "best" among the three design choices.

Since the decision-makers determined their multiattribute risk tolerance is represented by the exponential utility curve with $\rho_m = 0.25$, their preference structure reflects a monotonically increasing risk averse attitude over increasing values of the value function. Thus, Theorem 3.10 applies. This theorem will determine the expected utility for the value of each design alternative.

Applying Theorem 3.10: Analysis Setup

Since, in this case, $\rho_m = 0.25$ we have from Theorem 3.10

$$E(U(v)) = K(1 - E(e^{-(V_Y(y)/\rho_m)})) \tag{3.73}$$

where $K = 1/(1 - e^{-1/\rho_m})$ and $v = V_Y(y)$ (given in Definition 3.6). Given the parameters in this case, Equation 3.73 becomes

$$E(U(v)) = 1.01865736(1 - E(e^{-4V_Y(y)})) \tag{3.74}$$

where

$$v = V_Y(y) = \frac{2}{5}V_{X_1}(x_1) + \frac{1}{5}V_{X_2}(x_2) + \frac{2}{5}V_{X_3}(x_3) \tag{3.75}$$

and

$$V_{X_1}(x_1) = 1.30902(1 - e^{0.0320808(45 - x_1)}) \tag{3.76}$$

$$V_{X_2}(x_2) = 1.38583(1 - e^{0.0639325(x_2 - 30)}) \tag{3.77}$$

$$V_{X_3}(x_3) = -0.784058(1 - e^{-0.164433(x_3 - 8)}) \tag{3.78}$$

Next, we will look at the term $E(e^{-4V_Y(y)})$ in Equation 3.74. Here,

$$E(e^{-4V_Y(y)}) = E(e^{-4\left(\frac{2}{5}V_{X_1}(x_1) + \frac{1}{5}V_{X_2}(x_2) + \frac{2}{5}V_{X_3}(x_3)\right)}) \tag{3.79}$$

If we assume X_1, X_2, and X_3 are independent random variables, then

$$E(e^{-4V_Y(y)}) = E(e^{\frac{-8}{5}V_{X_1}(x_1)})E(e^{\frac{-4}{5}V_{X_2}(x_2)})E(e^{\frac{-8}{5}V_{X_3}(x_3)}) \tag{3.80}$$

where

$$E(e^{\frac{-8}{5}V_{X_1}(x_1)}) = \int_{-\infty}^{\infty} e^{\frac{-8}{5}V_{X_1}(x_1)} f_{X_1}(x_1)\, dx_1 \tag{3.81}$$

$$E(e^{\frac{-4}{5}V_{X_2}(x_2)}) = \int_{-\infty}^{\infty} e^{\frac{-4}{5}V_{X_2}(x_2)} f_{X_2}(x_2)\, dx_2 \tag{3.82}$$

$$E(e^{\frac{-8}{5}V_{X_3}(x_3)}) = \int_{-\infty}^{\infty} e^{\frac{-8}{5}V_{X_3}(x_3)} f_{X_3}(x_3)\, dx_3 \tag{3.83}$$

and $f_{X_i}(x_i)$ is the probability density function for X_i which, in this case discussion, is given to be a uniform distribution for each X_i.

Computation Illustration: Computing E(U(v)) and E(v) for Design Alternative A

First, compute the value of Equations 3.81 through 3.83 given the parameters in Table 3.15 for Design Alternative A. These computations are given below.

$$E(e^{\frac{-8}{5}V_{X_1}(x_1)}) = \int_{72}^{79} e^{\frac{-8}{5}(1.30902(1-e^{0.0320808(45-x_1)}))} \frac{1}{79-72}\, dx_1 = 0.271391$$

$$E(e^{\frac{-4}{5}V_{X_2}(x_2)}) = \int_{15}^{23} e^{\frac{-4}{5}(1.38583(1-e^{0.0639325(x_2-30)}))} \frac{1}{23-15}\, dx_2 = 0.57663$$

$$E(e^{\frac{-8}{5}V_{X_3}(x_3)}) = \int_{5.5}^{7} e^{\frac{-8}{5}(-0.784058(1-e^{-0.164433(x_3-8)}))} \frac{1}{7-5.5}\, dx_3 = 0.660046$$

Entering these values into Equation 3.80, we have

$$E(e^{-4V_Y(y)}) = E(e^{\frac{-8}{5}V_{X_1}(x_1)})E(e^{\frac{-4}{5}V_{X_2}(x_2)})E(e^{\frac{-8}{5}V_{X_3}(x_3)})$$

$$= (0.2713921)(0.57663)(0.660046) = 0.103292$$

Substituting this value for $E(e^{-4V_Y(y)})$ into Equation 3.74, we have

$$E(U(v)) = 1.01865736(1 - 0.103292) = 0.913438 \sim 0.91$$

Next, we proceed to compute the expected value $E(v)$ for this design alternative. Here, we need to determine $E(v)$ where

$$E(v) = E(V_Y(y)) = E(\frac{2}{5}V_{X_1}(x_1) + \frac{1}{5}V_{X_2}(x_2) + \frac{2}{5}V_{X_3}(x_3))$$

$$= \frac{2}{5}E(V_{X_1}(x_1)) + \frac{1}{5}E(V_{X_2}(x_2)) + \frac{2}{5}E(V_{X_3}(x_3)) \tag{3.84}$$

The terms in the Equation 3.84 are determined as follows:

$$E(V_{X_1}(x_1)) = \int_{-\infty}^{\infty} V_{X_1}(x_1) f_{X_1}(x_1)\,dx_1$$

$$= \int_{72}^{79} 1.30902(1 - e^{0.0320808(45-x_1)})\frac{1}{79-72}dx_1 = 0.815941$$

$$E(V_{X_2}(x_2)) = \int_{-\infty}^{\infty} V_{X_2}(x_2) f_{X_2}(x_2)\,dx_2$$

$$= \int_{15}^{23} 1.38583(1 - e^{0.0639325(x_2-30)})\frac{1}{23-15}dx_2 = 0.692384$$

$$E(V_{X_3}(x_3)) = \int_{-\infty}^{\infty} V_{X_3}(x_3) f_{X_3}(x_3)\,dx_3$$

$$= \int_{5.5}^{7} -0.784058(1 - e^{-0.164433(x_3-8)})\frac{1}{7-5.5}dx_3 = 0.264084$$

Substituting these into Equation 3.84, we have

$$E(v) = E(V_Y(y)) = E(\frac{2}{5}V_{X_1}(x_1) + \frac{1}{5}V_{X_2}(x_2) + \frac{2}{5}V_{X_3}(x_3))$$

$$= \frac{2}{5}E(V_{X_1}(x_1)) + \frac{1}{5}E(V_{X_2}(x_2)) + \frac{2}{5}E(V_{X_3}(x_3))$$

TABLE 3.16

Case 3.3: Summary Computations

Design Alternative	Criterion X_1	Criterion X_2	Criterion X_3	Expected Value $E(v)$	Expected Utility $E(U(v))$
Alternative A	72 – 79 $X_1 \sim$ Unif (72, 79)	15 – 23 $X_2 \sim$ Unif (15, 23)	5.5 – 7 $X_3 \sim$ Unif (5.5, 7)	0.57	0.91
Alternative B	85 – 88 $X_1 \sim$ Unif (85, 88)	23 – 27 $X_2 \sim$ Unif (23, 27)	5 – 6.5 $X_3 \sim$ Unif (5, 6.5)	0.60	0.93
Alternative C	80 – 85 $X_1 \sim$ Unif (80, 85)	24 – 28 $X_2 \sim$ Unif (24, 28)	4 – 5 $X_3 \sim$ Unif (4, 5)	0.67	0.95

Note: X_1 = Operational Days; X_2 = Maintenance/Service Hours; X_3 = Cost ($M).

$$= \frac{2}{5}(0.815941) + \frac{1}{5}(0.692384) + \frac{2}{5}(0.264084) \tag{3.85}$$

$$= 0.5704868 \sim 0.57$$

This concludes the computation illustration for Design Alternative A. The same types of computations are performed for the two other design alternatives. The results of these computations are summarized in Table 3.16. Design Alternative C is the "best" option in the set. This completes Case 3.3.

Questions and Exercises

1. State the interpretation of probability implied by the following:
   (A) The probability a tail appears on the toss of a fair coin is 1/2.
   (B) After recording the outcomes of 50 tosses of a fair coin, the probability a tail appears is 0.54.
   (C) It is with certainty the coin is fair!
   (D) The probability is 60% that the stock market will close 500 points above yesterday's closing count.
   (E) The design team believes there is less than a 5% chance the new microchip will require more than 12,000 gates.
2. A sack contains 20 marbles exactly alike in size but different in color. Suppose the sack contains 5 blue marbles, 3 green marbles, 7 red marbles, 2 yellow marbles, and 3 black marbles. Picking a single

marble from the sack and then replacing it, what is the probability of choosing the following:

(A) Blue marble? (B) Green marble? (C) Red marble?

(D) Yellow marble? (E) Black marble? (F) Nonblue marble

(G) Red or non-red marble?

3. If a fair coin is tossed, what is the probability of not obtaining a head? What is the probability of the event: (a head or not a head)?
4. Suppose A is an event (a subset) contained in the sample space Ω. Are the following probability statements true or false and why?

   (A) $P(A \cup A^C) = 1$ (B) $P(A|\Omega) = P(A)$
5. Suppose two 4-sided polygons are randomly tossed. Assuming the tetrahedrons are weighted fair, determine the set of all possible outcomes Ω. Assume each face is numbered 1, 2, 3, and 4. Let the sets A, B, C, and D represent the following events:

   A: The sum of the toss is even.

   B: The sum of the toss is odd.

   C: The sum of the toss is a number less than 6.

   D: The toss yielded the same number on each upturned face.

   (A) Find $P(A), P(B), P(C), P(A \cap B), P(A \cup B), P(B \cup C)$, and $P(B \cap C \cap D)$

   (B) Verify $P((A \cup B)^C) = P(A^C \cap B^C)$
6. The XYZ Corporation has offers on two contracts A and B. Suppose the proposal team made the following subjective probability assessments: the chance of winning contract A is 40%, the chance of winning contract B is 20%, the chance of winning contract A or contract B is 60%, the chance of winning both contracts is 10%.

   (A) Explain why the above set of probability assignments is *inconsistent* with the axioms of probability.

   (B) What must $P(B)$ equal such that it and the set of other assigned probabilities specified above are consistent with these axioms.
7. Suppose a coin is balanced such that tails appears three times more frequently than heads. Show the probability of obtaining a tail with such a coin is 3/4. What would you expect this probability to be if the coin was fair (equally balanced)?
8. Suppose the sample space of an experiment is given by $\Omega = A \cup B$. Compute $P(A \cap B)$ if $P(A) = 0.25$ and $P(B) = 0.80$.
9. If A and B are disjoint subsets of Ω show that

   (A) $P(A^c \cup B^c) = 1$ (B) $P(A^c \cap B^c) = 1 - [P(A) + P(B)]$
10. Two missiles are launched. Suppose there is a 75% chance missile A hits the target and a 90% chance missile B hits the target. If the

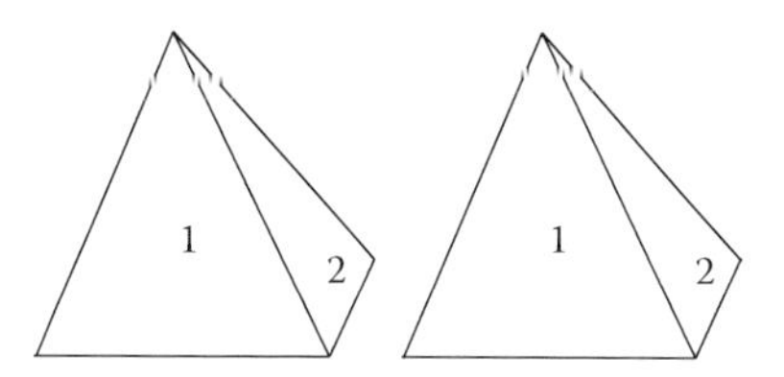

FIGURE 3.44
Two four-sided polygons for Exercise 5.

probability missile A hits the target is *independent* of the probability missile B hits the target, determine the probability missile A or missile B hits the target. Find the probability needed for missile A such that if the probability of missile B hitting the target remains at 90%, the probability missile A or missile B hits the target is 0.99.

11. Suppose A and B are independent events. Show that
    (A) The events A^c and B^c are independent.
    (B) The events A and B^c are independent.
    (C) The events A^c and B are independent.
12. Suppose A and B are independent events with $P(A) = 0.25$ and $P(B) = 0.55$. Determine the probability
    (A) At least one event occurs.
    (B) Event B occurs but event A does not occur.
13. Suppose A and B are independent events with $P(A) = r$ and the probability that "at least A or B occurs" is s. Show the only value for $P(B)$ is $(s - r)(1 - r)^{-1}$.
14. At a local sweet shop, 10% of all customers buy ice cream, 2% buy fudge, and 1% buy ice cream and fudge. If a customer selected at random bought fudge, what is the probability the customer bought an ice cream? If a customer selected at random bought ice cream, what is the probability the customer bought fudge?
15. For any two events A and B, show that $P(A|A \cap (A \cap B)) = 1$.
16. A production lot contains 1000 microchips, of which 10% are defective. Two chips are successively drawn at random without replacement. Determine the probability
    (A) Both chips selected are nondefective.
    (B) Both chips are defective.
    (C) The first chip is defective and the second chip is nondefective.
    (D) The first chip is nondefective and the second chip is defective.

17. Suppose the sampling scheme in exercise 16 was with replacement; that is, the first chip is returned to the lot before the second chip is drawn. Show how the probabilities computed in exercise 16 change.
18. Spare power supply units for a communications terminal are provided to the government from three different suppliers A_1, A_2, and A_3. Thirty percent come from A_1, 20% come from A_2, and 50% come from A_3. Suppose these units occasionally fail to perform according to their specifications and the following has been observed: 2% of those supplied by A_1 fail, 5% of those supplied by A_2 fail, and 3% of those supplied by A_3 fail. What is the probability any one of these units provided to the government will perform *without* failure?
19. In a single day, ChipyTech Corporation's manufacturing facility produces 10,000 microchips. Suppose machines *A*, *B*, and *C* individually produce 3000, 2500, and 4500 chips daily. The quality control group has determined the output from machine A has yielded 35 defective chips, the output from machine B has yielded 26 defective chips, and the output from machine C has yielded 47 defective chips.
    (A) If a chip was selected at random from the daily output, what is the probability it is defective?
    (B) What is the probability a randomly selected chip was produced by machine A? By machine B? By machine C?
    (C) Suppose a chip *was* randomly selected from the day's production of 10,000 microchips and it was found to be defective. What is the probability it was produced by machine A? By machine B? By machine C?
20. Given the evidence-to-hypothesis relationship in Figure 3.7, show that Bayes' rule is the basis for the following equations.

    (A) $$P(H|e_1) = \frac{P(H)P(e_1|H)}{P(H)P(e_1|H) + (1 - P(H))P(e_1|H^c)}$$

    (B) $$P(H|e_1e_2) = \frac{P(H|e_1)P(e_2|H)}{P(H|e_1)P(e_2|H) + (1 - P(H|e_1))P(e_2|H^c)}$$

    (C) $$P(H|e_1e_2e_3) = \frac{P(H|e_1e_2)P(e_3|H)}{P(H|e_1e_2)P(e_3|H) + (1 - P((H|e_1e_2)|e_1))P(e_3|H^c)}$$

21. Consider the value function in Figure 3.10. Sketch the value function subject to the following value increments. The smallest value increment occurs between yellow and red; the value increment between red and green is one and a half times the smallest value increment; the value increment between green and blue is two times the smallest value increment; the value increment between blue and black is

three times the smallest value increment. Compare and contrast this value function with the value function in Figure 3.10.

22. Consider Figure 3.18. Determine the exponential constant for this value function if the midvalue for the mechanical device's repair time is 15 hours.
23. Review and give examples of a nominal scale, an ordinal scale, a cardinal interval scale, a cardinal ratio scale.
24. If a utility function $U(x)$ is monotonically decreasing (i.e., less is better) such that $U(x_{\min}) = 1$ and $U(x_{\max}) = 0$ show that the expected utility is equal to the probability p that $x_{\min}$ occurs.
25. Suppose a lottery X has a range of outcomes bounded by x_1 and x_2. Suppose the probability of any outcome between x_1 and x_2 is uniformly distributed. If $U(x) = a - be^{-x}$, where a and b are constants, show that the certainty equivalent x_{CE} is

$$x_{CE} = -\ln\left(\frac{e^{-x_1} - e^{-x_2}}{x_2 - x_1}\right)$$

26. Suppose $U(x)$ is a monotonically increasing exponential utility function of the form given in Equation 3.43. Show that the certainty equivalent is given by Equation 3.50.
27. Suppose $U(x) = x^2$ over the interval $0 \leq x \leq 1$ and that $x = 0$ with probability p and $x = 1$ with probability $1 - p$. Show that $E(U(x)) > U(E(X))$. What do you conclude about the risk attitude of this decision-maker?
28. Prove Theorems 3.8 and 3.9.
29. Show that $1 - V_1(x, \rho) = V_2(x, -\rho)$, where $V_1(x, \rho)$ is the increasing exponential value function with parameter ρ given by Equation 3.21 and $V_2(x, -\rho)$ is the decreasing exponential value function with parameter $-\rho$ given by Equation 3.22. Show this general property holds for the power-additive utility functions defined by Equations 3.58 and 3.59.
30. Show that the power-additive utility function is the same for monotonically increasing or decreasing preferences when $v = V_Y(y) = 1/2$.

4

A Risk Analysis Framework in Engineering Enterprise Systems

4.1 Introduction

Engineering enterprise systems is an emerging discipline. It encompasses and extends traditional systems engineering to create an enterprise of cooperating systems and services that deliver capabilities to globally distributed user communities, through a rich network of information and communications technologies. Enterprise systems operate ubiquitously in environments that offer cross-boundary access to a wide variety of services, applications, and information repositories.

Engineering enterprise systems is a sophisticated and complex undertaking. Enterprise systems are increasingly being engineered by combining many separate systems, services, and applications which, as a whole, provide an overall capability otherwise not possible. Today, we are in the early stage of understanding how systems engineering, engineering management, and the social sciences join to create systems that operate and evolve in enterprise environments. This chapter introduces enterprise systems, challenges associated with their engineering, and ways to model and measure risks affecting enterprise capabilities such that performance objectives are achieved.

4.2 Perspectives on Engineering Enterprise Systems

Today's systems are continually increasing in scale and complexity. More and more defense systems, transportation systems, financial systems, and human services systems network ubiquitously across boundaries and seamlessly interface with users, information repositories, applications, and services. These systems are an enterprise of people, processes, technologies, and organizations.

A distinguishing feature of enterprise systems is not only their technologies but the way users interface with them and one another. How to design and engineer these systems and their interfaces from human, social, political, and managerial dimensions (Allen et al., 2004) are new challenges. To address these challenges, engineering and social sciences are joining in ways not previously attempted when planning and evolving the design, development, and operation of these large-scale and highly networked systems.

This section discusses the enterprise problem space and systems thinking within that space. The following are excerpts from a perspectives paper on enterprise engineering, written by George Rebovich, Jr., of The MITRE Corporation.*

The Enterprise

In a broad context, an enterprise is an entity comprised of interdependent resources that interact with one another and their environment to achieve goals (Rebovich, 2005). A way to view an enterprise is illustrated in Figure 4.1. In Figure 4.1, resources include people, processes, organizations, technologies, and funding. Interactions include coordinating functions or operations, exchanging data or information, and accessing applications or services.

Historically, systems engineering has focused on the technologies that have enabled the development of the piece parts—the systems and subsystems embedded in the enterprise. Modern systems thinkers (Gharajedaghi, 1999) are increasingly taking a holistic view of an enterprise. Here, an enterprise can be characterized by:

- A multiminded, sociocultural entity comprised of a voluntary association of members who can choose their goals and means
- An entity whose members share values embedded in a (largely common) culture
- An entity having the attributes of a purposeful entity
- An entity whose performance improves through alignment of purposes across its multiple levels

Many enterprises have a nested nature. At every level, except at the very top and bottom levels, an enterprise itself is part of a larger enterprise and contains subenterprises, each with its own people, processes, technologies, funding, and other resources. Nesting within an enterprise can be illustrated by a set of US Air Force programs shown in Figure 4.2. Here, the family of Airborne Early Warning and Control (AEW&C) systems is an enterprise, which is nested in the Command and Control (C2) Constellation enterprise, which is nested in the Air Force C2 enterprise.

* Permission has been granted to excerpt materials from the paper "Enterprise Systems Engineering Theory and Practice, Volume 2: Systems Thinking for the Enterprise: New and Emerging Perspectives," authored by Rebovich, George, Jr., MP 050000043, November 2005.

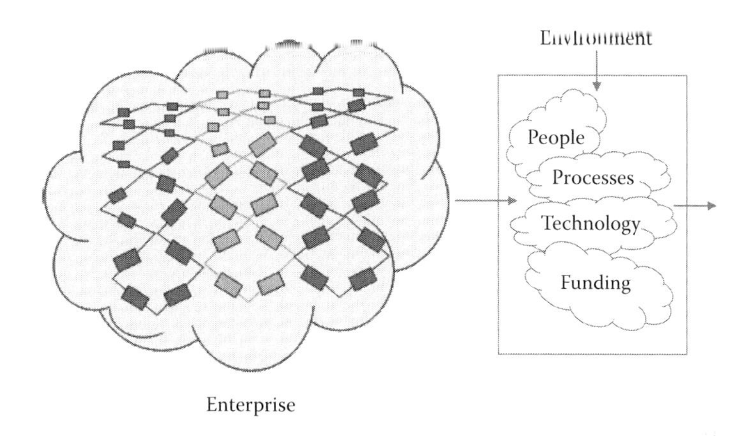

FIGURE 4.1
An enterprise and its environment.

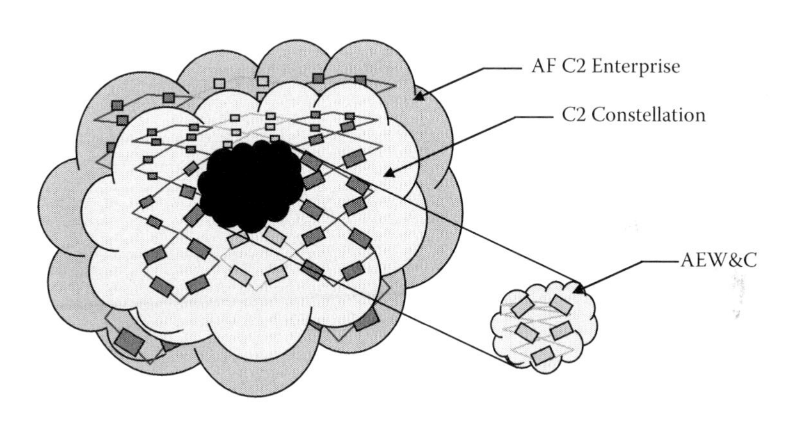

FIGURE 4.2
Nested nature of enterprises.

Alignment of purposes across the levels of the enterprise can improve overall enterprise performance. The subenterprises contribute to the outcomes or goals of the containing enterprise. This view has profound implications for how systems engineers must think about their activities—that they are inexorably linked to the enterprise and its operations as a whole.

For example, at the AEW&C system program level, the view must be that an AEW&C system builds an air picture that serves the higher goal of achieving situation awareness within the C2 Constellation. This requires the AEW&C systems engineer to ask (and answer) how the AEW&C piece parts being developed serve situation awareness in the C2 Constellation in addition to how they serve the AEW&C system specification.

At the next level, the view must be that the C2 Constellation develops integrated capabilities to serve the higher goal of providing net-centric C2 for the Air Force C2 Enterprise. The implication is that the systems engineer must address how the C2 Constellation piece parts serve the Air Force C2 Enterprise, in addition to how they serve the C2 Constellation.

At the highest level in this example, the view must be that the Air Force C2 Enterprise develops Air Force net-centric capabilities to serve the higher goal of providing net-centric C2 for the Joint/Coalition C2 Enterprise. The implication is that the systems engineer must address how the Air Force C2 Enterprise piece parts serve joint and coalition net-centric C2 in addition to how they serve the Air Force C2.

This discussion leads to an operational definition of enterprise from the perspective of an individual (system engineer or other participant) or team in the enterprise. It aims to answer the question, "what is my (our) enterprise?" The enterprise, then, can be viewed as a set of interdependent elements (systems and resources) that a participating actor or actors either control or influence.

This definition of enterprise is a virtual construct that depends on the make-up, authority, and roles of the participating actors in a community of interest. For example, the program team of a system managed by one organization may have virtual control of most engineering decisions being made on the system's day-to-day development activities. If the system is required to be compliant with technical standards developed by an external agency, the program team may have representation on the standards team, but that representation is one voice of many and so the standard is a variable the program team can influence but not necessarily control. The implication is that all actors or teams in an enterprise setting should know their enterprise and be aware of which enterprise elements or variables they control, which they influence, and which they neither control nor influence.

Engineering a system always involves asking good questions and following the implications of their answers. The following offers a series of questions (Rebovich, 2005) that assist in understanding the complexity of an enterprise.

- What is my enterprise? What elements of it do I control? What elements do I influence? What are the elements of my environment that I do not control or influence but which influence me?
- How can a balance be achieved between optimizing at the system-level with enabling the broader enterprise, particularly if it comes at the expense of a smaller system?
- How can different perspectives be combined into one view to enable alignment of purposes across the enterprise?
- Would a change in performance at system or subsystem levels result in changes at the enterprise level? If so, how, and is it important? How would a new enterprise level requirement be met and how would it influence systems below it?

- How can complementary relations in opposing tendencies be viewed to create feasible wholes with seemingly unfeasible parts? How can they be viewed as being separate, mutually interdependent dimensions that can interact and be integrated into an "and" relationship?
- Are dependencies among variables in a system or enterprise such that the ability to make progress in one variable occurs at the expense of others? How can dependencies among variables within an enterprise be identified, monitored, managed accordingly?

Engineering an enterprise system necessitates addressing these and related questions as aids in identifying and managing risks that threaten the successful delivery of enterprise capabilities. In this context, enterprise risk management requires an integration of people, processes, and tools to ensure an enterprise-wide understanding of capability risks,* their potential consequences, interdependencies, and rippling effects within and beyond enterprise boundaries. Ultimately, enterprise risk management aims to establish and maintain a holistic view of risks across the enterprise, so capabilities and performance objectives are achieved via risk-informed resource and investment decisions.

4.3 A Framework for Measuring Enterprise Capability Risk

This section presents an analysis framework for measuring enterprise capability risk. It can be used to address questions that include the following: *What events threaten the delivery of capabilities needed to successfully advance enterprise goals and mission outcomes? If these events occur, how serious are their impacts? How can the progress of risk management plans be monitored? How can risk be considered in resource planning and investment decision-making?* Questions such as these arise when planning, executing, and managing the engineering of large-scale, enterprise-wide systems. Addressing these questions involves not only engineering and technology dimensions but human–social–system interactions as well.

Enterprise risk management differs from traditional practice (Garvey, 2008) in the expanse of the consequence space within which risks affect enterprise goals, mission outcomes, or capabilities. In a traditional case, the consequence space is usually focused on the extent risks negatively affect a system's cost,

* Societal consequences of risks realized from engineering systems (e.g., nuclear, transportation, financial systems) have been studied and published by Murphy and Gardoni (2006). Their work relates notions of capability risks to their potential impacts on the capacity of socio-political structures to operate and on the ability of individuals to function within their respective social-political environments.

schedule, and technical performance (discussed in Chapter 1). Enterprise risk management necessitates broadening the scope of this space. Identifying and evaluating higher-level effects (or consequences) on capabilities and services are critical considerations in decisions on where to allocate resources to manage enterprise risks.

A Capability Portfolio View

One way management plans for engineering an enterprise is to create capability portfolios of technology programs and initiatives that, when synchronized, will deliver time-phased capabilities that advance enterprise goals and mission outcomes. Thus, a *capability portfolio is a time dynamic organizing construct* to deliver capabilities across specified epochs (Garvey, 2008).

Creating capability portfolios is a complex management and engineering analysis activity. In the systems engineering community, there is a large body of literature on *portfolio analysis* for investment decision management applied to the acquisition of advanced systems. This topic, however, is beyond the scope of this book. Instead, the following is focused on applying risk management practices within a generic model of capability portfolios, already defined to deliver capabilities to an enterprise. Figure 4.3 illustrates such a model.

In Figure 4.3, the lowest level is the family of capability portfolios. What does a capability portfolio look like? An example is shown in Figure 4.4. Figure 4.4 presents an inside look at a capability portfolio from a *capability-to-functionality* view. Figure 4.4 derives from a capability portfolio for network operations (OSD, 2005). This is one among many capability portfolios designed to deliver capabilities to the Department of Defense (DOD) Global Information Grid.*

Seen in Figure 4.4, a capability portfolio can be represented in a hierarchical structure. At the top is the capability portfolio itself. Consider this the Tier 1 level. The next tier down the hierarchy presents capability areas, such as network management and information assurance. These Tier 2 elements depict the functional domains that characterize the capability portfolio. Tier 3 is the collection of capabilities the portfolio must deliver by a specified epoch (e.g., 20*xx*). Here, a capability can be defined as *the ability to achieve an effect to a standard under specified conditions using multiple combinations of means and ways to perform a set of tasks* (OSD, 2005). Tier 4 is the functionality that must be integrated to achieve capability outcomes.

For example, consider the capability portfolio shown in Figure 4.4. The Tier 3 capability *ability to create and produce information in an assured environment* refers to the ability to collect data and transform it into information,

* The Department of Defense (DOD) defines the Global Information Grid (GIG) as a globally interconnected, end-to-end set of information capabilities, associated processes, and personnel for collecting, processing, storing, disseminating, and managing information (Government Accountability Office (GAO), 2004).

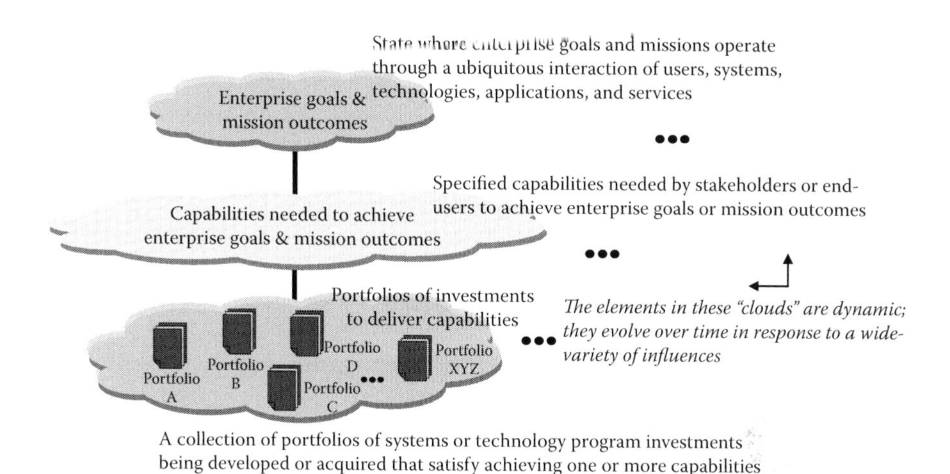

FIGURE 4.3
An enterprise and its capability portfolios.

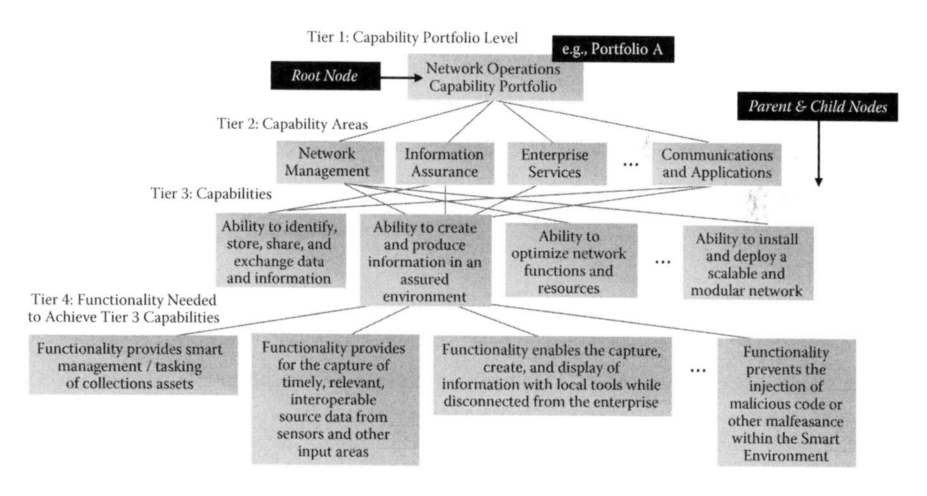

FIGURE 4.4
A capability portfolio for network operations (OSD, 2005).

while providing end-to-end protection to assure the availability of information and validating its integrity (OSD, 2005). Suppose this capability advances toward outcome goals when functionality is delivered that ensure the *capture of timely, relevant, interoperable source data from sensors and other input areas.* Suppose this functionality contributes to this capability's outcome when the *time for information change to be posted and/or subscribers notified is less than 1 minute* (OSD, 2005).

Next, we will use this information and show how a hierarchical representation of a capability portfolio can be used as a modeling framework within which risks can be assessed and capability portfolio risk measures derived. In preparation for this, we first consider a capability portfolio from a supplier–provider context.

Supplier–Provider Perspective

Once a capability portfolio's hierarchy and its elements are defined, it is managed by a team to ensure its collection of technology programs and technology initiatives combine in ways to deliver one or more capabilities to the enterprise. Thus, one can take a *supplier–provider* view of a capability portfolio. This is illustrated in Figure 4.5.

In Figure 4.5, a capability portfolio can be viewed as the provider charged with delivering time-phased capabilities to the enterprise. Technology programs and technology initiatives align to, and synchronized with, the capability portfolio to supply the functionality needed to achieve the provider's capability outcomes.

The supplier–provider view offers a way to examine a capability portfolio from a risk perspective. Look again at Figures 4.3, 4.4, and 4.5. We have enterprise goals and mission outcomes dependent on capability portfolios successfully delivering required capabilities. Next, we have capability portfolios dependent on programs and technologies successfully delivering functionality that enables these capabilities. Thus, major sources of risk originate from the suppliers to these capability portfolios.

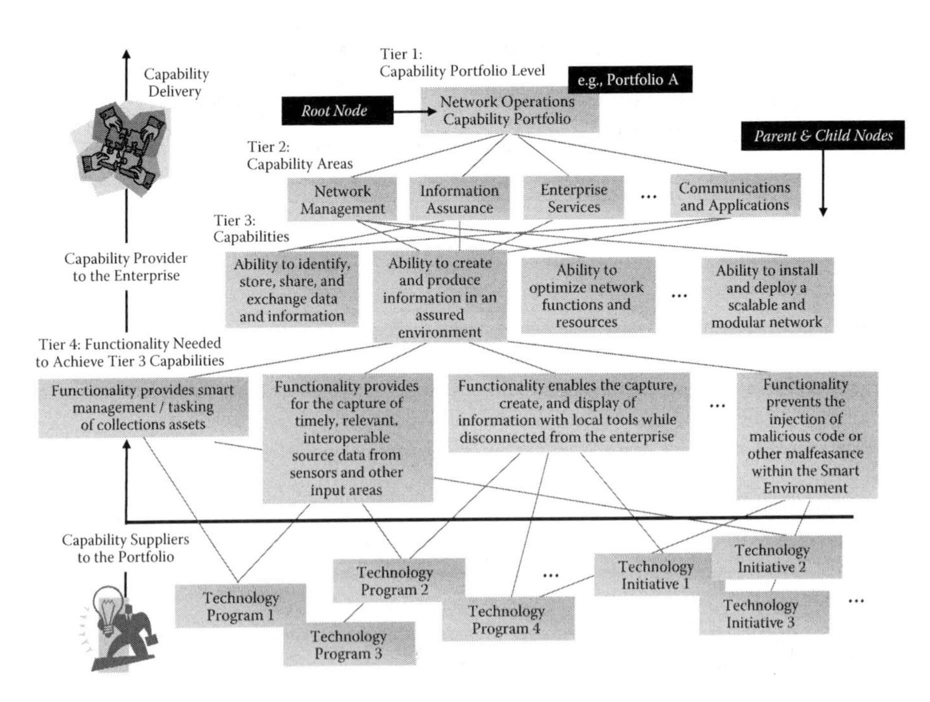

FIGURE 4.5
A supplier–provider view of a capability portfolio.

Supplier risks include unrealistic schedule demands placed on them by portfolio needs or placed by suppliers on their vendors. Supplier risks include premature use of technologies, including the deployment of technologies not adequately tested. Dependencies among suppliers can generate a host of risks, especially when a problem with one supplier generates a series of problems with others. Economic conditions can threaten business stability or the business viability of suppliers and vendors. Unfavorable funding or political influences outside an enterprise can adversely affect its capability portfolios, its suppliers, or the supplier–vendor chains in ways that threaten the realization of enterprise goals and mission outcomes.

These issues are important risk considerations to any engineering system. However, they are a more present and persistent concern in the management of risk in engineering enterprise systems, especially those acquired by supplier–provider models. The following will show how a hierarchical representation of a capability portfolio can serve as a modeling and analytical framework within which risks can be assessed and capability portfolio risk measures derived.

4.4 A Risk Analysis Algebra

When a capability portfolio can be represented in a hierarchical structure it offers a modeling framework within which risks can be assessed and capability risk measures derived. The following illustrates this idea using the hierarchies in Figures 4.4 and 4.5. What is meant by capability risk? In the context of a capability portfolio, we define *capability risk as a measure of the chance and the consequence that a planned capability, defined within a portfolio's envelope, will not meet intended outcomes by its scheduled delivery date* (Garvey, 2008).

First, we will design algebraic rules for computing risk measures within a segment of a capability portfolio's hierarchy. Then, we will show how to extend these computations to operate across a capability portfolio's fully specified hierarchy. This will involve a series of roll-up calculations. Shown will be risk measure computations that originate from leaf nodes, which will then roll-up to measure the risks of parent nodes, which will then roll-up to measure the risk of the capability portfolio itself (i.e., the root node).

When a capability portfolio can be represented in the form of a hierarchy, decision-makers can be provided the trace basis and the drivers behind all risk measures derived for any node at any level in the hierarchy. From this, management has visibility and supporting rationales for identifying where resources are best allocated to reduce (or eliminate) risk events that threaten the success of the portfolio's goals and capability outcome objectives.

In a capability portfolio's hierarchical structure, each element in the hierarchy is referred to as a node. The top-most node is the *root node*. In Figures 4.4

and 4.5, the root node represents the capability portfolio itself, which, in this case, is the Network Operations Capability portfolio. A *parent node* is one with lower-level nodes coming from it. These lower-level nodes are called *child nodes* to that parent node. Nodes that terminate in the hierarchy are called *leaf nodes*. Leaf nodes are terminal nodes in that they have no children coming from them.

In the context of a capability portfolio hierarchy, leaf nodes are terminal nodes that originate from supplier nodes. Here, leaf nodes are *risk events* associated with supplier nodes. Thus, the risk measures of leaf nodes drive the risk measures of supplier nodes. The risk measures of supplier nodes drive the risk measures of their parent nodes. The risk measures of parent nodes drive the risk measures of their parent nodes and so forth. Hence, risk measures computed for all nodes originate from risk measures derived for leaf nodes. This *ripple-in-the-pond* effect is reflective of capability portfolio risk management from a supplier–provider perspective.

Risks that trace to suppliers are a major source of risk to the portfolio's ability to deliver capability to the enterprise. However, it is important to recognize that suppliers are not the only source of risk. Risks external to a capability portfolio's supplier–provider envelope are very real concerns. Risk sources outside this envelope must also be considered when designing and implementing a formal risk analysis and management program for a capability portfolio or family of capability portfolios.

Figure 4.6 shows a Tier 3 capability from the portfolio in Figure 4.5. For convenience, we have numbered the nodes as shown. Figure 4.6 shows three supplier nodes responsible for contributing to functionality node 3.22—one of four functions needed for Capability 3.2 to be delivered as planned. Functionality node 3.22 is a parent node to the supplier nodes EWXT, QSAT, and S-RAD.

Shown in Figure 4.6, two of these supplier nodes are technology programs. One supplier node is a technology initiative. In practice, this distinction can be important. A technology program is often an engineering system acquisition—one characterized by formal contracting, well-defined requirements, and adherence to engineering standards and program management protocols. A technology initiative is often targeted at developing a specific technology for an engineering system or a user community. An example might be the development of advanced encryption technology for the information assurance community.

Whether supplier nodes are technology programs or technology initiatives, they exist in a capability portfolio because of their contributions to parent nodes. From the portfolio perspectives in Figures 4.5 and 4.6, functionality nodes are the parent nodes to these supplier nodes. Here, the contributions of supplier nodes integrate in ways that enable functionality nodes. Functionality nodes integrate in ways that enable their corresponding capability nodes—the collection of capabilities the portfolio is expected to successfully deliver to the enterprise.

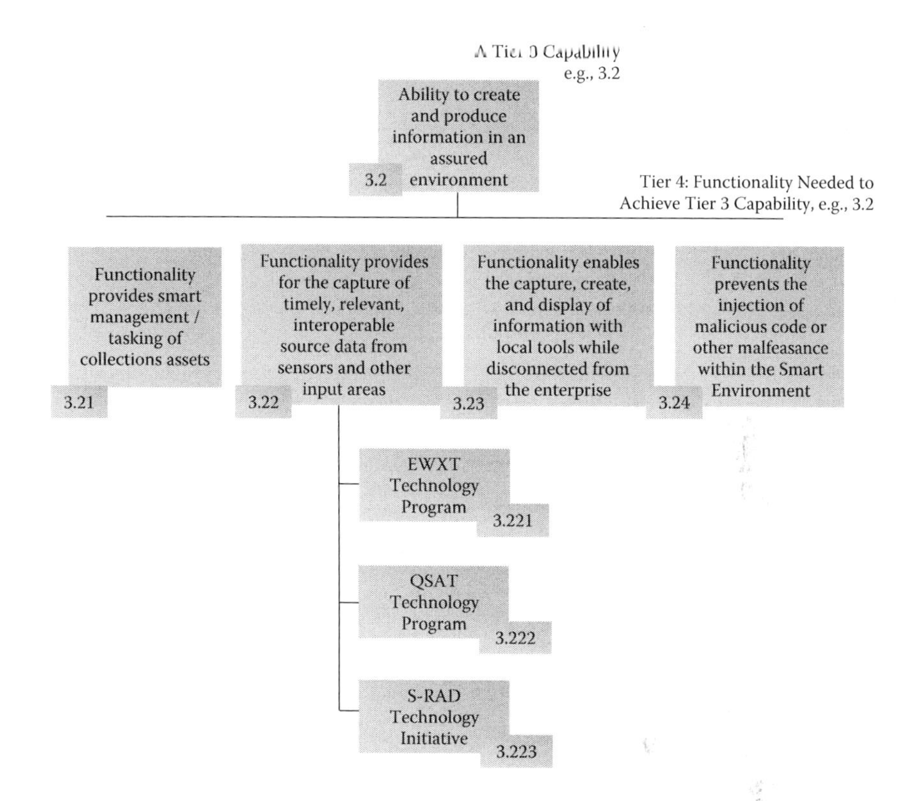

FIGURE 4.6
A Tier 3 Capability from the portfolio in Figure 4.5.

At the supplier level, we define *contribution* by a supplier node as *that which advances the portfolio's ability to provide capability that meets the needs of the portfolio's user communities.* A supplier's contribution to its parent functionality node could be in many forms and include technologies, engineering analyses, and software applications. At the supplier level, risk events can have adverse consequences on the cost, schedule, or technical performance of the supplier's contribution to its parent functionality node.

Risk events can also negatively affect a supplier node's programmatic activities. These activities include the technical or program-related work products that support the supplier's business, engineering, management, or acquisition practices needed to advance the outcome objectives of the supplier's contribution to its parent functionality node. Technical or program-related work products include architecture frameworks, engineering analyses, organizational structures, governance models, and acquisition management plans.

In addition, supplier nodes can be negatively affected by political risks, business risks, economic risks, and the integrity of supply chains. These risks not only threaten suppliers but they can *directly* threaten functionality or capability nodes at those levels in the portfolio's hierarchy. Thus, risk

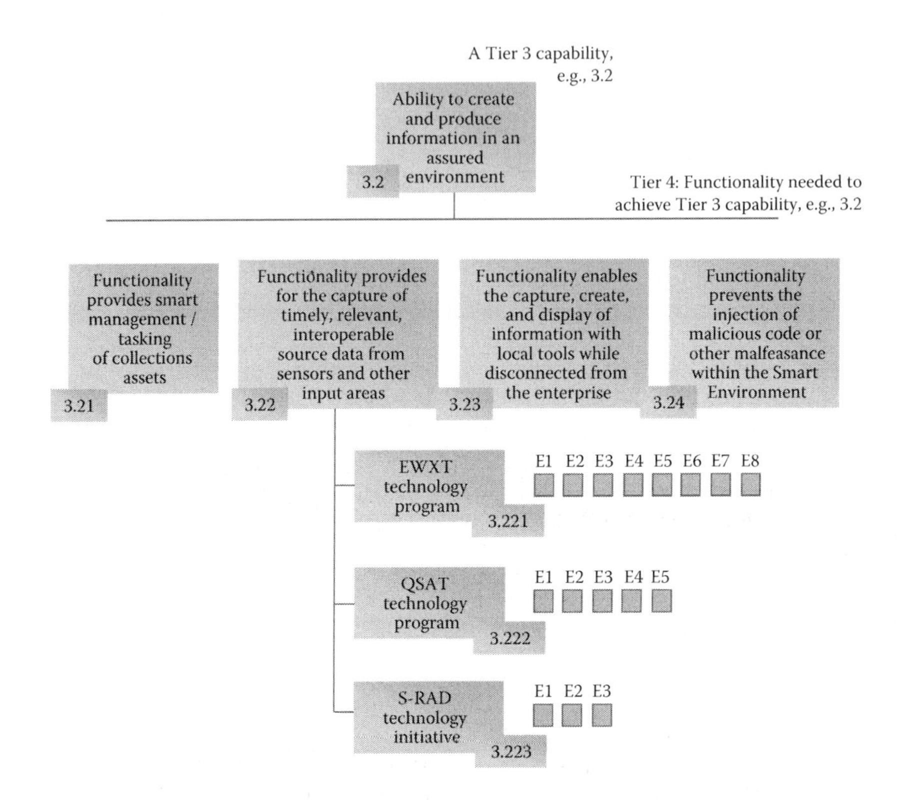

FIGURE 4.7
Capability 3.2 supplier risk set.

events from a capability portfolio perspective are of multiple types with the potential for multiconsequential impacts on parent nodes located at any level in the hierarchy.

Figure 4.7 shows leaf nodes intended to represent supplier node risk events. These leaf nodes are labeled E1, E2, E3, and so on. They denote risk events that, if they occur, would negatively affect the supplier node's contribution to its parent functionality node (Functionality 3.22, in this case). Risks that threaten supplier node's contributions to functionality 3.22 have *ripple-in-the-pond* effects on the portfolio's delivery expectations for Capability 3.2. As we will see, risks that affect Capability 3.2 can have horizontal and vertical effects elsewhere in the portfolio. Next, we look at eight risk events associated with the EWXT technology program node shown in Figure 4.8.

In Figure 4.8, each EWXT risk event is given a color. The color reflects a score or measure of the risk event's potential impact to Functionality 3.22, if it occurs. In Figure 4.8, each risk event happens to be either RED (R) or YELLOW (Y). Suppose the basis for each color derives from a function of each risk event's occurrence probability and its impact or consequence. Definition 4.1 presents this function, called *risk score*.

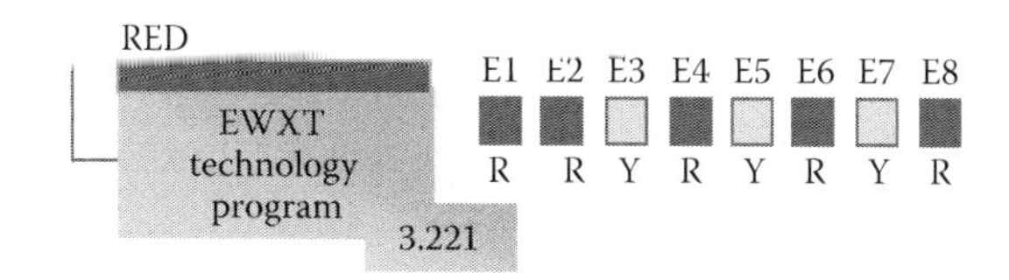

FIGURE 4.8
EWXT technology program risk set (R = Red, Y = Yellow).

Definition 4.1: Risk Event Risk Score*

The risk score of risk event E is given by the additive value function

$$Risk\ Score(E) = RS(E) = u_1 V_{\text{Probability}}(E) + u_2 V_{\text{Impact}}(E) \tag{4.1}$$

subject to the considerations in Theorem 3.6. In Equation 4.1, the first term is a value function for the risk event's occurrence probability. The second term is a value function for the severity of the risk event's impact, if it occurs. The coefficients u_1 and u_2 are nonnegative weights such that $0 \le u_1 \le 1$, $0 \le u_2 \le 1$, and $u_1 + u_2 = 1$. In Equation 4.1, these value functions can be designed to produce measures along either a 0 to 1 or a 0 to 100 cardinal interval scale, as discussed in Chapter 3.

In Equation 4.1, suppose the linear value function in Figure 4.9 is assumed to represent the risk event's occurrence probability. Non-linear relationships are also possible, as shown in Figures 3.19 or 3.36. Suppose Table 4.1 presents a constructed scale (introduced in Chapter 3) for assessing the value of the second term in Equation 4.1.

Returning to Figure 4.8 and applying Equation 4.1, suppose the risk scores of the EWXT technology program's risk events $E1, E2, E3, \ldots, E8$ are computed by the function

$$Risk\ Score(Ei) = RS(Ei) = u_1 V_{\text{Probability}}(Ei) + u_2 V_{\text{Impact}}(Ei)$$

where $i = 1, 2, 3, \ldots, 8$ in this case. Suppose the values of these risk scores are as shown in Figure 4.10, where risk event E1 has a risk score of 85, risk event E2 has a risk score of 90, risk event E3 has a risk score of 60, and so forth. Given the eight risk scores for the EWXT technology program, shown in Figure 4.10, what is an overall measure of the risk that supplier node EWXT poses to functionality node 3.22†? The following is one way to formulate this measure using a rule called the "max" average.

Definition 4.2: Max Average

The max average of $\{x_1, x_2, x_3, \ldots, x_n\}$, where $0 \le x_i \le 100$, $i = 1, 2, 3, \ldots, n$, is

$$\text{Max Ave} = \lambda m + (1 - \lambda)\text{Average}\{x_1, x_2, x_3, \ldots, x_n\} \tag{4.2}$$

* Equation 4.1 is one of many ways to formulate a *Risk Score* measure. The reader is directed to Garvey (2008) for additional approaches to formulating this measure.

† In general, identified risks and their risk scores are temporal. They reflect the risk situation known at a given point in time and thus should be regularly reviewed and updated accordingly.

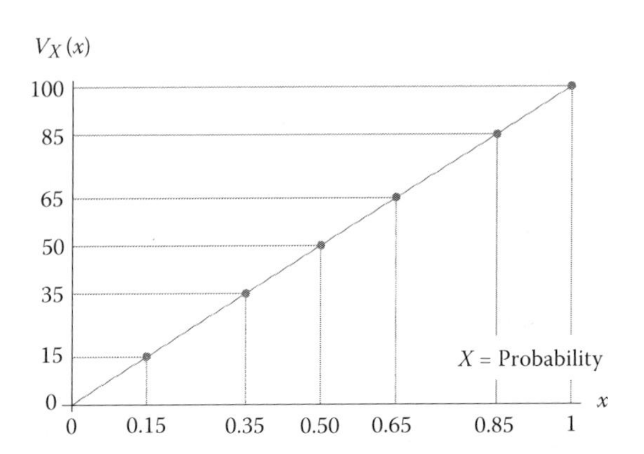

FIGURE 4.9
A value function for occurrence probability.

where $m = \text{Max}\{x_1, x_2, x_3, \ldots, x_n\}$, $0 \le \lambda \le 1$, and λ is a weighting function.*

Suppose the capability portfolio's management decided to use the weighting function in Figure 4.11. In the context of this discussion, the x_i's in Equation 4.2 equate to the risk scores in Figure 4.10. Thus, from Equation 4.2, we have

$$RS_{3.221} = \lambda(90) + (1-\lambda)\text{Average}\{85, 90, 60, 75, 48, 73, 50, 79\}$$

where $m = \text{Max}\{85, 90, 60, 75, 48, 73, 50, 79\} = 90$. From Figure 4.11, it follows that $\lambda = 0.70$; therefore,

$$RS_{3.221} = (0.70)(90) + (0.30)(70) = 84$$

where $RS_{3.221}$ is the risk score of the EWXT technology program, denoted by node 3.221 in Figure 4.10. Thus, the EWXT technology program (a supplier node) has a high risk score. In accordance with the color rating scale in Figure 4.10, EWXT falls in the RED (R) color band. The results of this discussion are shown in Figure 4.12.

In summary, the EWXT technology program contributes a high degree of risk to functionality node 3.22 which, in turn, contributes to the risk of capability node 3.2. In this case, it can be shown that EWXT risk events E1, E2, E4, E6, and E8 are responsible for 93% of the EWXT technology program's risk score. These five risk events signal areas in the EWXT program where increased management focus and risk mitigation planning may be warranted.

* An example weighting function is shown in Figure 4.11. The shape of the weighting function can have a significant influence on scores generated by the max average rule. In practice, its shape should be designed to model the team's (or decision-maker's) preferences for how much the maximum score should influence the overall score.

TABLE 4.1

A Sample Constructed Scale: Supplier Node Impacts

Ordinal Scale (Score)	Definition: Risk Event Impacts on a Supplier Node's Contribution to its Parent Node	Cardinal Interval Scale (Score)
5	A risk event that, if it occurs, impacts the supplier node to the extent that its contribution to its parent node is severely degraded or compromised. The nature of the risk is such that outcome objectives for the supplier node's contribution are either not met or are extremely unacceptable (e.g., fall well-below minimum acceptable levels).	80 to 100
4	A risk event that, if it occurs, impacts the supplier node to the extent that its contribution to its parent node is marginally below minimum acceptable levels. The nature of the risk is such that outcome objectives for the supplier node's contribution are moderately unacceptable.	60 to < 80
3	A risk event that, if it occurs, impacts the supplier node to the extent that its contribution to its parent node falls well-below stated objectives but remains enough above minimum acceptable levels. The nature of the risk is such that outcome objectives for the supplier node's contribution are borderline acceptable.	40 to < 60
2	A risk event that, if it occurs, impacts the supplier node to the extent that its contribution to its parent node falls below stated objectives but falls well-above minimum acceptable levels. The nature of the risk is such that outcome objectives for the supplier node's contribution are reasonably acceptable.	20 to < 40
1	A risk event that, if it occurs, impacts the supplier node to the extent that its contribution to its parent node is negligibly affected. The nature of the risk is such that outcome objectives for the supplier node's contribution are completely acceptable, but regular monitoring for change is recommended.	0 to < 20

Measuring Up: How Supplier Risks Affect Functionality

The preceding discussion presented one way to derive a risk score measure of the EWXT technology program. However, EWXT is just one of three supplier nodes to functionality node 3.22. What about the other supplier nodes? How might their risk measures combine into an overall measure of risk to functionality node 3.22? What ripple effects do supplier risks have on all dependent higher-level nodes in the capability portfolio's hierarchy? The following will address these and related questions.

Suppose risk measures for the other two supplier nodes to functionality node 3.22 are shown in Figure 4.13. These are the QSAT technology program

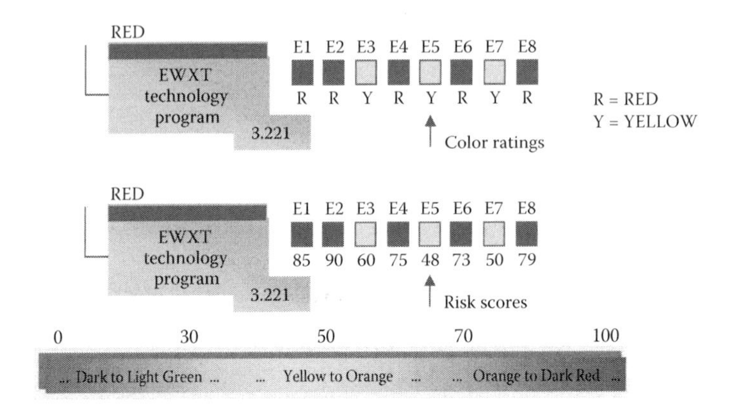

FIGURE 4.10
Example risk scores for EWXT program risks (R = Red, Y = Yellow)*.

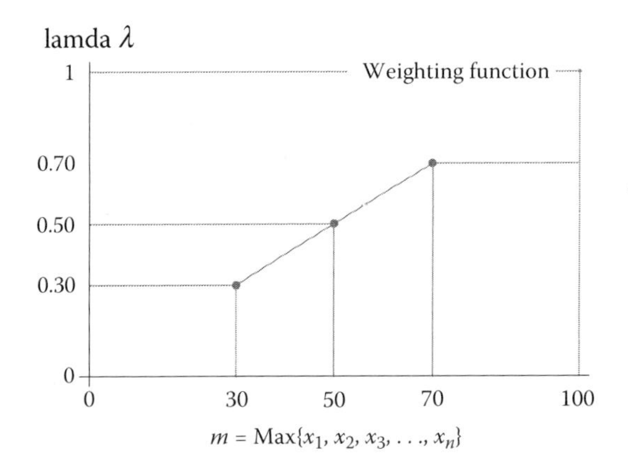

FIGURE 4.11
An example max average weighting function. See Garvey (2008) for other forms of this function.

and the S-RAD technology initiative. Suppose their risk measures were also derived by the max average rule given by Equation 4.2. From this, how can the risk measures from all three supplier nodes, in Figure 4.13, combine into an overall measure of risk to Functionality 3.22? One way is to apply a variation of the max average rule to the set of risk scores derived for the supplier nodes. We will call this variation the *critical average*.

Definition 4.3: Critical Average

Suppose a parent node has n child nodes. Suppose $\{x_1, x_2, x_3, \ldots, x_n\}$ is a set of child node scores, where $0 \le x_i \le 100$ for $i = 1, 2, 3, \ldots, n$. If A is a subset of $\{x_1, x_2, x_3, \ldots, x_n\}$ that contains only the scores of the child nodes deemed

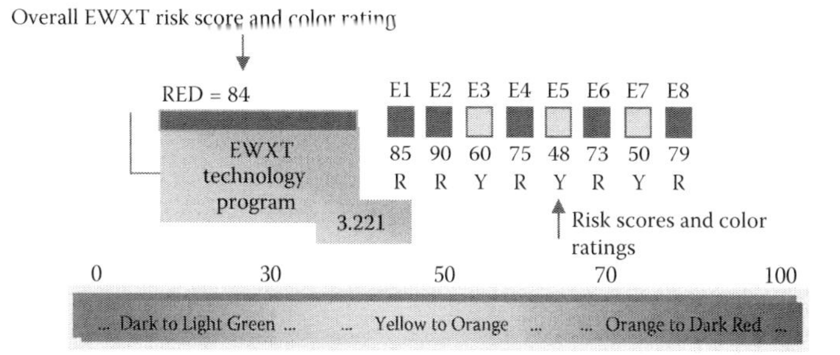

FIGURE 4.12
Overall EWXT program risk score and color rating (max ave, R = Red, Y = Yellow).

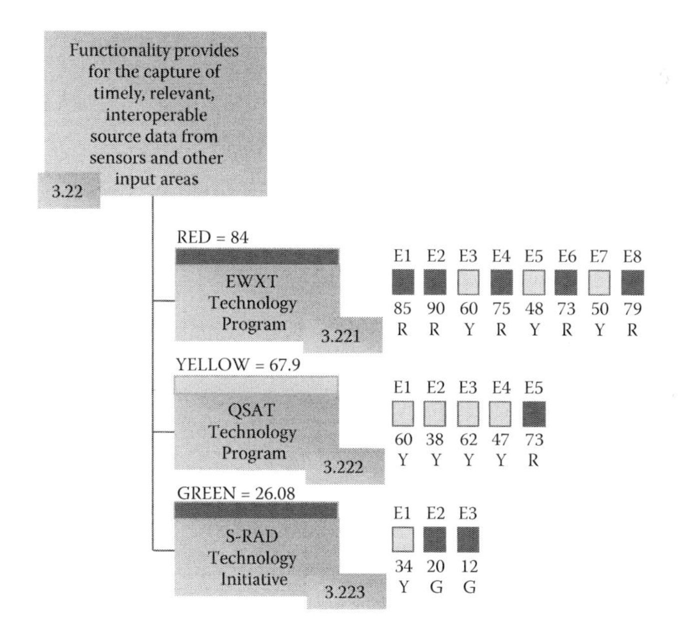

FIGURE 4.13
Supplier node risk measures to Functionality 3.22.

*critical** to the parent node, then the critical average of the set $\{x_1, x_2, x_3, \ldots, x_n\}$ is given by

$$\text{Crit Ave} = \lambda \text{Max}\{A\} + (1-\lambda)\text{Average}\{x_1, x_2, x_3, \ldots, x_n\} \tag{4.3}$$

where λ is a weighting function based on the maximum of A.

To apply the critical average to the supplier nodes in Figure 4.13, suppose the EWXT technology program is the only critical supplier to Tier 4

* A child node is *critical* to its parent node if the parent node's outcome objectives are severely degraded, or not achieved, without its child node's contribution.

functionality node 3.22. From Definition 4.3, the risk score of functionality node 3.22 is

$$RS_{3.22} = \lambda \text{Max}\{RS_{3.221}\} + (1 - \lambda)\text{Average}\{RS_{3.221}, RS_{3.222}, RS_{3.223}\} \quad (4.4)$$

where λ is a weighting function. Suppose we use a weighting function similar to Figure 4.11. Given this, from the risk scores in Figure 4.13 and Equation 4.4 we have

$$RS_{3.22} = (0.70)(84) + (1 - 0.70)\text{Average}\{84, 67.9, 26.08\} = 76.6$$

With $RS_{3.22} = 76.6$, it follows that Tier 4 functionality node 3.22 has a high risk score and falls in the RED (R) color band—in accordance with the color rating scale in Figure 4.10. The magnitude of this score is driven by (1) the criticality of the EWXT technology program's contribution to functionality node 3.22 and (2) the risk scores of EWXT risk events E1, E2, E4, E6, and E8.

Findings from this analysis identifies where management attention is needed with respect to reducing Tier 4 functionality node 3.22 risks. Addressing the threat posed by the EWXT technology program to functionality node 3.22 will lessen the potential of unwanted effects at higher dependency levels across the capability portfolio's hierarchy. The results of this discussion are shown in Figure 4.14.

Measuring Up: How Functionality Risks Affect Capability

The preceding presented ways to derive a measure of Tier 4 functionality node 3.22 risk as a function of its supplier node risks. Shown in Figure 4.7, functionality node 3.22 is one of the four functionality nodes to Tier 3 capability node 3.2. What about the other functionality nodes? How might their risk score measures combine into an overall measure of Tier 3 capability node 3.2 risk? The following addresses these questions.

Suppose risk scores for Tier 4 functionality nodes 3.21, 3.23, and 3.24 are given in Figure 4.15. Suppose they were computed as a function of the risk scores of their supplier nodes (not shown) in the same way the risk score of functionality node 3.22 was derived. Suppose Tier 4 functionality nodes 3.23 and 3.24 are critical for Tier 3 capability node 3.2 to achieve its outcome objectives. From Definition 4.3, the risk score of capability node 3.2 is

$$RS_{3.2} = \lambda \text{Max}\{RS_{3.23}, RS_{3.24}\} + (1 - \lambda)\text{Average}\{RS_{3.21}, RS_{3.22}, RS_{3.23}, RS_{3.24}\} \quad (4.5)$$

where λ is a weighting function. Suppose we use a weighting function similar to Figure 4.11. Given this, from the risk scores in Figure 4.15 and Equation 4.5 we have

$$RS_{3.2} = (0.70)\text{Max}\{42.4, 82.9\} + (1 - 0.70)\text{Average}\{35, 76.6, 42.4, 82.9\} = 75.8$$

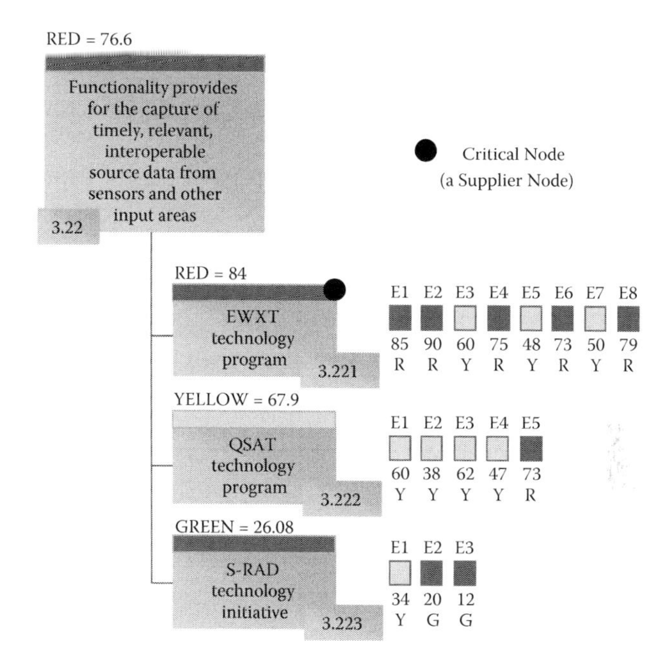

FIGURE 4.14
Risk measure derived for Functionality 3.22.

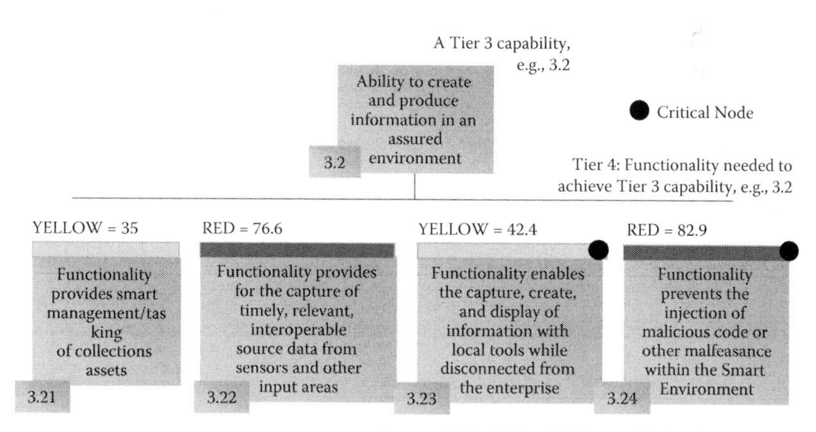

FIGURE 4.15
Risk measures for Capability 3.2 functionality nodes.

With $RS_{3.2} = 75.8$, it follows that Tier 3 capability node 3.2 has a high risk score and falls in the RED (R) color band—in accordance with the color rating scale in Figure 4.10. The magnitude of this score is driven by the criticality of Tier 4 functionality nodes 3.23 and 3.24 to Tier 3 capability node 3.2 and their individual risk scores.

Findings from this analysis identifies where management attention is needed to reduce capability node 3.2 risks. Addressing the threat posed by Tier 4 functionality nodes 3.23 and 3.24 to Tier 3 capability node 3.2 will lessen the potential of unwanted effects at higher dependency levels across the capability portfolio's hierarchy. The results of this discussion are shown in Figure 4.16.

Measuring Up: How Capability Risks Affect the Capability Portfolio

The preceding presented ways to derive a measure of Tier 3 capability node 3.2 risk as a function of the risk score measures of its Tier 4 functionality nodes. Shown in Figure 4.17, capability node 3.2 is one of the four capability nodes to the Tier 2 Information Assurance (IA) capability area. What about the other capability nodes? How might their risk score measures combine into an overall measure of Tier 2 Information Assurance risk? The following addresses these questions.

Suppose risk scores for Tier 3 capability nodes 3.1, 3.3, and 3.4 are given in Figure 4.18. Suppose they were computed as a function of the risk scores of their functionality nodes (not shown) in the same way the risk score of capability node 3.2 was derived. Suppose Tier 3 capability nodes 3.1, 3.3, and 3.4 are critical for the Tier 2 IA capability area to achieve its outcome objectives. From Definition 4.3, the risk score of the IA capability area is

$$RS_{IA} = \lambda \text{Max}\{RS_{3.1}, RS_{3.3}, RS_{3.4}\} + (1-\lambda)\text{Average}\{RS_{3.1}, RS_{3.2}, RS_{3.3}, RS_{3.4}\} \quad (4.6)$$

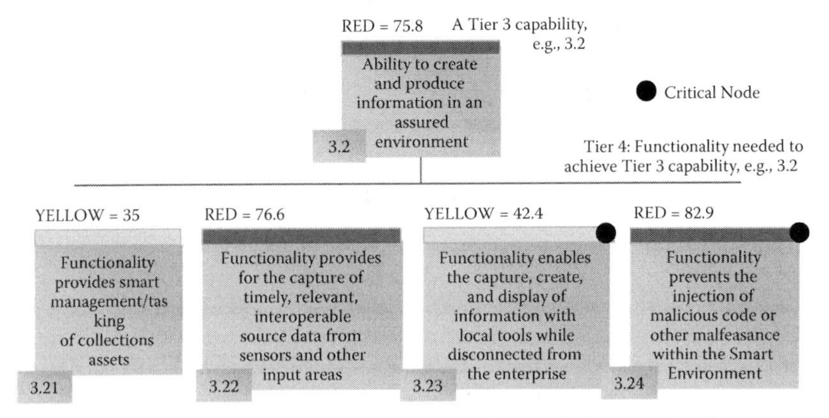

FIGURE 4.16
Risk measure for Capability 3.2: Critical average rule.

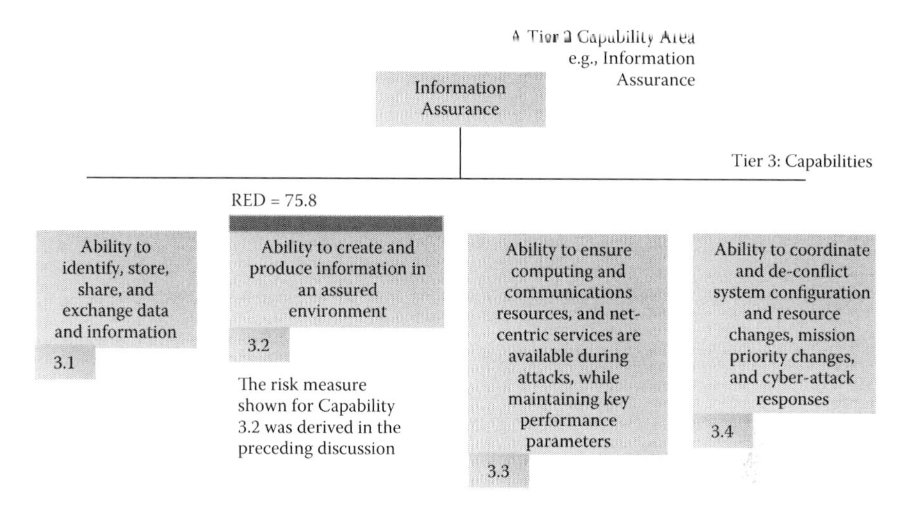

FIGURE 4.17
Information assurance: A Tier 2 capability area.

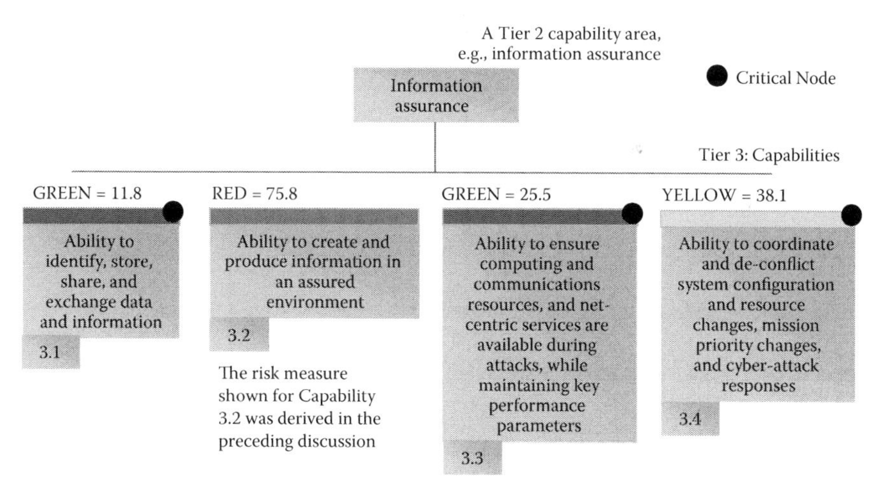

FIGURE 4.18
Information assurance: Capability risk measures.

where λ is a weighting function. Suppose we use a weighting function similar to Figure 4.11. Given this, from the risk scores in Figure 4.18 and Equation 4.6 we have

$$\begin{aligned} RS_{IA} &= (0.381)\text{Max}\{11.8, 25.5, 38.1\} \\ &\quad + (1-0.381)\text{Average}\{11.8, 75.8, 25.5, 38.1\} = 38 \end{aligned}$$

With $RS_{IA} = 38$, it follows that the Tier 2 Information Assurance capability area node has a moderate score and falls in the YELLOW (Y) color band—in accordance with the color rating scale in Figure 4.10. The magnitude of this score suggests management watch the IA capability area for unwanted changes that might increase its current level of threat to portfolio outcomes. The results of this discussion are shown in Figure 4.19.

The preceding presented ways to derive a risk score measure of the IA capability area as a function of the risk score measures of its Tier 3 capability node risks. Shown in Figure 4.20, IA is one of four Tier 2 capability areas to

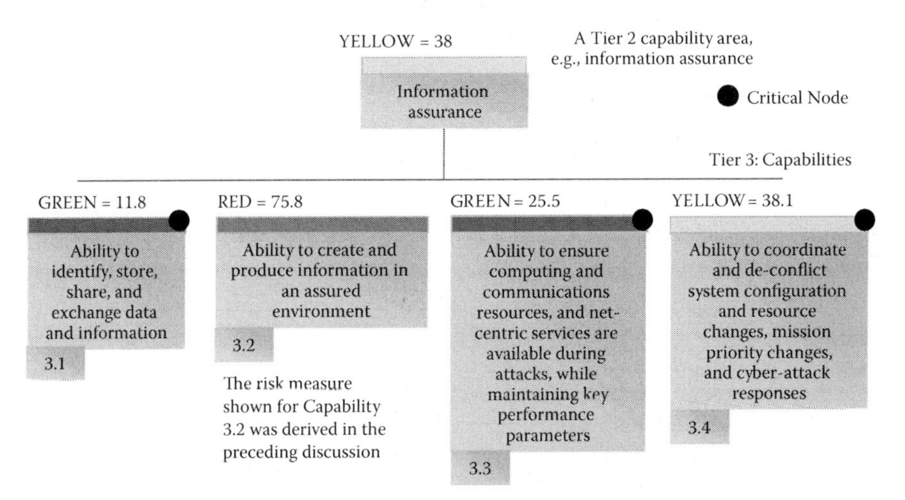

FIGURE 4.19
Information assurance: Capability risk measures.

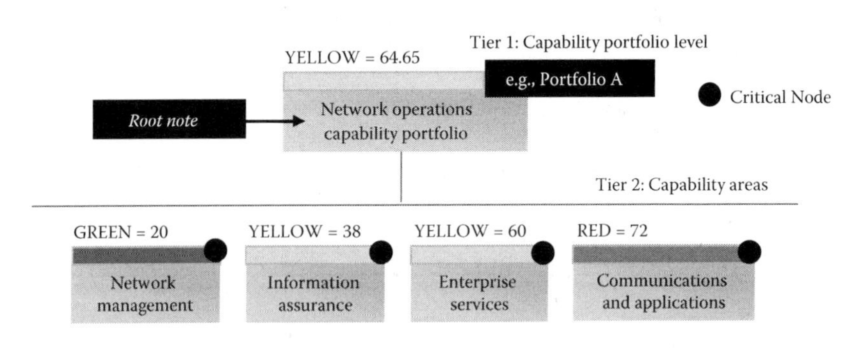

FIGURE 4.20
Network operations capability portfolio-level risk measure.

Tier 1—the Network Operations Capability Portfolio. What about the other Tier 2 capability area nodes? How might their risk score measures combine into an overall measure of risk to the Network Operations Capability Portfolio? The following addresses these questions.

Suppose risk score measures for the other three Tier 2 capability area nodes Network Management, Enterprise Services, and Communications and Applications are given in Figure 4.20. In Figure 4.20, all four Tier 2 capability area nodes are identified as critical to the Network Operations Capability Portfolio (the Tier 1 root node). Given this, the critical average rule will equal the max average rule. Applying the max average rule, with the weighting function in Figure 4.11, to the Tier 2 capability area risk scores in Figure 4.20 results in an overall Tier 1 risk score of 64.65 for the Network Operations Capability Portfolio. Thus, this portfolio has a moderate risk score and falls in the YELLOW (Y) color band—in accordance with the color rating scale in Figure 4.10. The magnitude of this score suggests management keep watch on this portfolio for unwanted changes that might increase its current level of threat to the portfolio's outcome objectives.

The preceding discussion presented an algebra designed to measure risk, at any node in a capability portfolio, when risk events originate from a capability portfolio's supplier levels. Computational rules were defined and illustrated to show how risk score measures were derived, in part, from a series of roll-up calculations. Risk score measures derived from leaf nodes were rolled-up to measure the risks of parent nodes. Risk score measures derived for parent nodes were rolled-up to measure the risk of the capability portfolio.

In the context of this formalism, the *number* of risk events associated with a supplier node does not fully drive the magnitude of its risk score measure. Consider the max average rule. This rule is purposefully designed to weight more heavily risk events, in a set of events, with higher risk score measures than those in the set with lower risk score measures. Although risk scores of all risk events associated with a supplier node are included in the max average, *their effect on the node's overall risk score is controlled by the shape or form of the weighting function* λ. Because of this, each risk event does not necessarily contribute equally to the supplier node's overall risk score measure. A supplier node with a set of five risk events can have a higher risk score than one with a set containing more than five risk events *and vice versa*.

Thus, with the max average rule it is important to design the shape or form of its weighting function to capture the team's (or decision-maker's) preferences for the degree the maximum risk score should influence the overall risk score. One weighting function is shown in Figure 4.11. Many other shapes are possible (Garvey, 2008).

The max average rule applied in the context of Figure 4.12 operates, under certain conditions, as a decision-maker's "alert function." In Figure 4.12, the supplier node's risk score measure is 84 given the eight risks E1 through E8. Suppose management actions were taken such that risks E3 through E8 were eliminated from this supplier node's risk set. With this, the EWXT technology program would now have a risk score measure of 89.25.

Why did this supplier node's risk score increase despite the elimination of all but two of its risks? The answer includes the following: (1) management actions eliminated E3 through E8—but they did not eliminate the two most serious risks, E1 and E2, from the node's risk set; and (2) the max average rule operates only on the risk set presented; so even though E3 through E8 were eliminated, the max average rule only "sees" a supplier node with two serious risks E1 and E2.

The fact that the risk score increased is noteworthy, but not as important as the result that the node remained in the RED risk color band in this example. Thus, the max average rule can be *tuned* to alert management when a supplier node still faces a high degree of risk because of the presence of even just a few very serious risks—despite the elimination of less serious ones from the set.

What about risks to capabilities when risk events originate from sources or conditions outside of supplier nodes? How can these risks be considered in a capability portfolio risk assessment? Risks that threaten capabilities to be delivered by a capability portfolio can originate from sources other than those that affect only the portfolio's suppliers. These events can directly attack one or more capability nodes in a capability portfolio's hierarchy. For example, uncertainties in geo-political landscapes may impact operational demands on capabilities that stress planned performance.

Dependencies between capability portfolios in families of portfolios, such as those that constitute an enterprise, are also potential risk sources. Here, outcome objectives for capabilities delivered by one capability portfolio may depend on the performance of capabilities delivered by another capability portfolio. Identifying risk events from non-supplier sources and capturing their contribution to the risk measure is an important consideration in the risk assessment and analysis process.

The risk analytic framework described in this chapter provides ways to track and report risks faced by capability nodes, as a function of the many sources of risk affecting the nodes and ultimately the capability portfolio. In practice, it is recommended that supplier and nonsupplier measures of capability risk be separately derived, tracked, and reported to the capability portfolio's management team. In addition, each risk should be tagged according to its type (or nature) and tracked in the portfolio's overall risk event population. If this is done, then a variety of management indicators can be developed. These include (1) the frequency with which specific types of risk affect capability nodes, and (2) the degree to which a capability node's risk score measure is driven by supplier versus nonsupplier source conditions, including understanding the nature and drivers of these conditions. We end this discussion with a summary of the information needed to implement capability portfolio risk analyses and management.

4.5 Information Needs for Portfolio Risk Analysis

Risk management in an enterprise capability portfolio context has unique and thought challenging information needs. These needs can be grouped into two categories. The first category addresses capability value and the second one addresses supplier contributions, criticality, and risks as they relate to enabling the portfolio to deliver capability.

Information needs that address *capability value* include the following:

- For each Tier 3 capability, as shown in Figure 4.4, what standard (or outcome objective) must each capability meet by its scheduled delivery date?
- For each Tier 3 capability, what is the source basis for its standard (or outcome objective)? Does it originate from user-driven needs, policy-driven needs, model-derived values, a combination of these, or from other sources?
- For each Tier 3 capability, what extent does the standard (or outcome objective) for one capability depend on others to meet their standards (or outcome objectives)?

Information needs that *address supplier contributions, criticality, and risks* include the following:

- For each Tier 3 capability, which technology programs and technology initiatives are the suppliers contributing to that capability?
- For each Tier 3 capability, what (specifically) are the contributions of its suppliers?
- For each Tier 3 capability, how do supplier contributions enable the capability to achieve its standard (or outcome objective)?
- For each Tier 3 capability, which technology programs and technology initiatives are critical contributors to enable the capability to achieve its standard (or outcome objective)?
- What risks originate from (or are associated with) suppliers that, if they occur, negatively affect their contributions to capability?

A similar set of information needs can be crafted for risk events that originate from nonsupplier-related sources or conditions.

Measuring, tagging, and tracking risk events in the ways described aids management in identifying courses of action. Specifically, whether options exist to attack risks directly at their sources or to engage them by deliberate intervention actions—actions aimed at lessening or eliminating their potential capability consequences.

Process tailoring, socialization, and establishing governance protocols are critical considerations in enterprise engineering risk management. Ensuring these aspects succeed is time well-spent. With this, effective and value-added engineering management practices can be institutionalized—practices that enable capability portfolio outcomes, and ultimately those of the enterprise, to be achieved via risk-informed resource and investment management decisions.

The approach presented for enterprise capability portfolio risk management provides a number of beneficial and actionable insights. These include the following:

- Identification of risk events that threaten the delivery of capabilities needed to advance goals and capability outcome objectives.
- A measure of risk for each capability derived as a function of each risk event's occurrence probability and its consequence.
- An analytical framework and logical model within which to structure capability portfolio risk assessments—one where assessments can be combined to measure and trace their integrative effects on engineering the enterprise.
- Through the framework, ways to model and measure risk as capabilities are time-phased across incremental capability development approaches.
- Decision-makers provided the trace basis and the event drivers behind all risk measures derived for any node at any level of the capability portfolio's hierarchy. With this, capability portfolio management has visibility and supporting rationales for identifying where resources are best allocated to reduce (or eliminate) risk events that threaten achieving enterprise goals and capability outcome objectives.

4.6 The "Cutting Edge"

This chapter presented an analytical framework and computational model for assessing and measuring risk in the engineering of enterprise systems. It illustrated one way to represent, model, and measure risk when engineering an enterprise from a capability portfolio perspective.

Few protocols presently exist for measuring capability risk in the context of capability portfolios. Additional research is needed on such protocols and how to customize them to specific supplier-provider relationships. Here, further concepts from graph theory might be used to visualize and model a capability portfolio's supplier-provider topology. New computational algebras might then be designed to generate measures of capability risk unique to that portfolio's dependency relationships.

Protocols are also needed to capture and measure horizontal and vertical dependencies among capabilities and suppliers within capability portfolios and across families of capability portfolios that make an enterprise. With this, the ripple effects of failure in one capability (or supplier) on other dependent capabilities (or suppliers) or portfolios could be formally measured. Developing ways to capture and measure these effects would enable designs to be engineered that minimize dependency risks. This might lessen or even avoid potentially cascading negative effects that dependencies can have on the timely delivery of enterprise services to consumers.

Additional research areas at the "cutting edge" include the following:

- How time-phasing capability delivery to consumers should be factored into risk assessment, measurement, and management formalisms.
- How to approach risk measurement and management in enterprises that consists of dozens of capability portfolios with hundreds of supplier programs. For this, the idea of representing large-scale enterprises by *domain capability portfolio clusters* might be explored and a new concept of *portfolio cluster risk management* might be developed.
- How to design decision analytic methodologies to measure risk criticality that captures each risk's multiconsequential impacts and dependencies across enterprise-wide capabilities.

The materials in this chapter aimed to bring conceptual understandings of the enterprise engineering problem space into view. Along with this, risk management theory and practice for engineering enterprises can evolve. This topic falls at the interface between risk management methods for engineering traditional systems with those needed for engineering enterprises. Recognizing this interface and then addressing its challenges is an essential step towards discovering new methods and new practices uniquely designed to successfully manage risk in engineering an enterprise.

Questions and Exercises

1. In the following figure, suppose program node N_j has 10 risk events Ei $(i = 1, 2, 3, \ldots, 10)$. Suppose each risk event's occurrence probability and its impacts (if it occurs) are scored as shown in Figure 4.21. Given this information, answer the following questions.

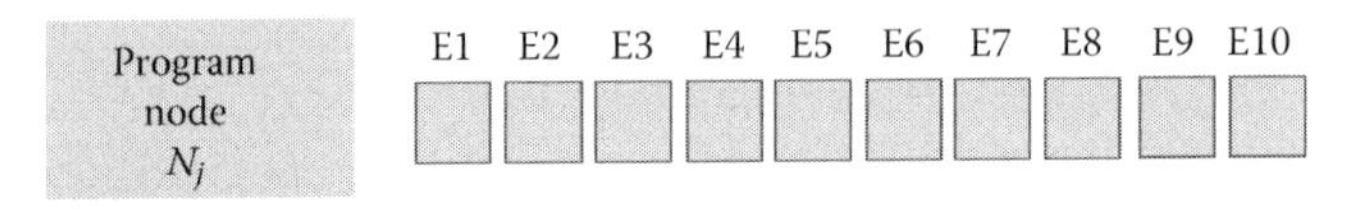

Risk Event	Event Probability (%)	Cost Impact (%)	Schedule Impact Months	Technical Perf Impact Level (1–5)	Programmatic Impact Level (1–5)
1	95	12	4	4	4
2	65	5	2	3	2
3	55	15	12	4	5
4	50	18	14	2	4
5	90	12	10	5	1
6	15	4	3	1	3
7	25	2	1	3	1
8	80	11	16	2	4
9	85	4	9	5	3
10	75	3.5	6	3	2

FIGURE 4.21
Exercise 1.

(A) Use Definition 4.1 to compute the risk score measure of each risk event. Assume equal weights in computing each risk score measure.

(B) Develop a most-to-least critical risk ranking of these risk events using the computed risk score measures.

(C) Create a scatter plot of the results in (A) and (B) similar to the scatter plot shown in Figure 3.42.

2. Using the risk scores computed for the risk events in Exercise 1, determine the risk score of program node N_j using

(A) The mean of the risk scores.

(B) The maximum of the risk scores.

(C) The max average of the risk scores; assume the weight function shown in Figure 4.11.

(D) What drives the differences between the risk scores in (A), (B), and (C)? Discuss when a mean, maximum, or a max average risk score rule might be more appropriate than the others.

3. From Definitions 4.2 and 4.3, determine when the max average and critical average are equal.

4. In Figure 4.22, show and explain why critical average and max average rules both generate a risk measure of 64.65 for the node labeled Network Operations Capability portfolio.

5. Suppose Figure 4.23 presents a portion of a capability portfolio defined as part of engineering an enterprise system. Given the information shown, apply the risk analysis algebra in this chapter to

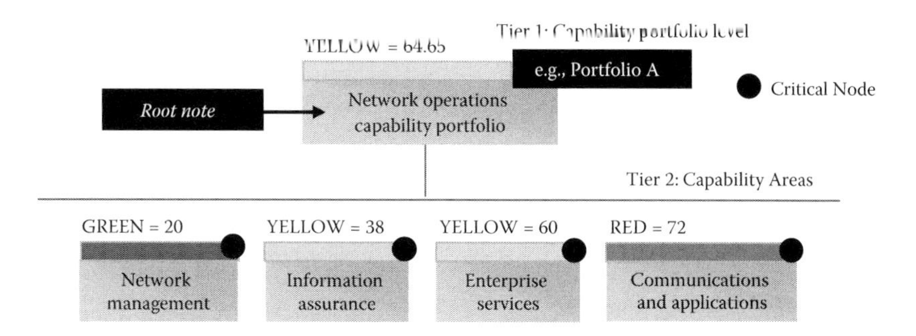

FIGURE 4.22
Exercise 4.

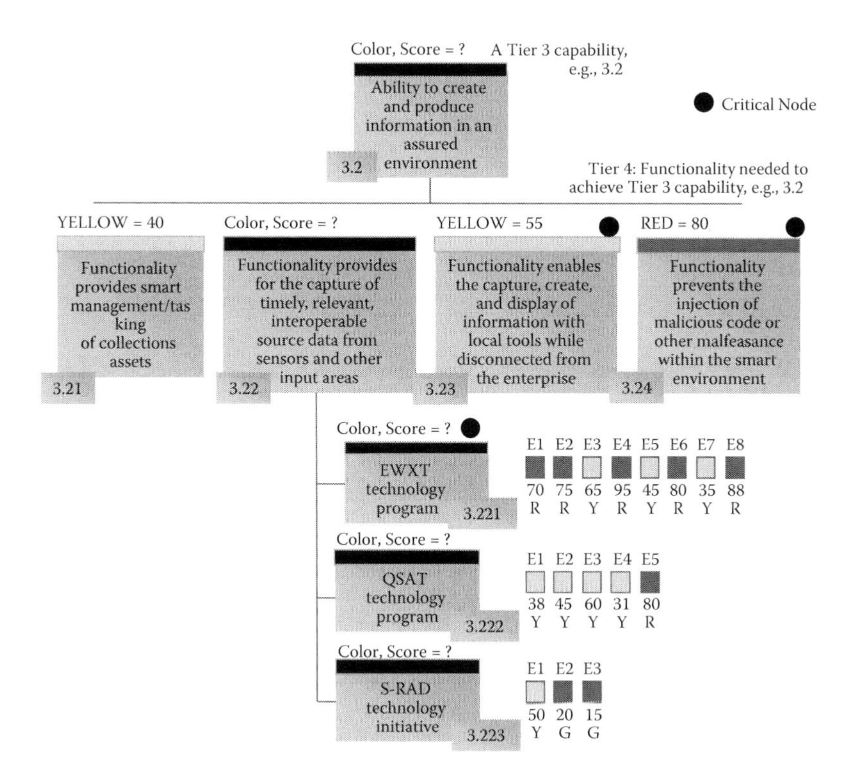

FIGURE 4.23
Exercise 5.

derive a risk measure for Capability 3.2. What risks are driving this measure?

The following are questions for advanced research investigations.

6. **Risk Measurement**

   This chapter presented a framework and algebra for assessing and measuring capability risk in the context of capability portfolios defined for engineering an enterprise. The discussion illustrated one way among possible ways to represent, model, and measure risk when engineering an enterprise from a capability portfolio perspective. Few protocols presently exist for measuring capability risk in the context of capability portfolios.

   Think about such protocols and how they might be designed to address various types of supplier–provider relationships. Here, concepts from network or graph theory might be used to visualize and model a capability portfolio's supplier–provider topology. Computational rules might then be designed to algebraically generate measures of capability risk unique to that capability portfolio's topology.

7. **Capturing Dependencies**

   Related to the above, protocols are needed to capture and measure horizontal and vertical dependencies among capabilities and suppliers within capability portfolios and across families of capability portfolios that make up an enterprise. With this, the ripple effects of failure in one capability (or supplier) on other dependent capabilities (or suppliers) or portfolios could be formally measured.

   Think about ways to capture, model, and measure these effects so designs can be engineered that minimize dependency risks and their potentially cascading negative effects on the timely delivery of enterprise services to consumers.

8. **Time-Phase Considerations**

   Think about how the time phasing of capability delivery to consumers of enterprise services might be planned and then factored into risk assessment, measurement, and management formalisms.

9. **Enterprise Scale Considerations**

   Think about how to approach risk measurement and management in enterprises that consist of dozens of capability portfolios with hundreds of supplier programs. Here, a new idea of representing large-scale enterprises by *domain capability portfolio clusters* might be explored and a new concept of *portfolio cluster risk management* might be developed.

10. **Risk-Adjusted Benefit Measure**

    Consider a capability portfolio being managed through a supplier–provider approach, as discussed in this chapter. Suppose a capability portfolio manager must select investing in suppliers that offer the most benefit to achieving capability, in terms of desired outcomes.

Think about how to measure investment benefit but how to adjust this measure to account for risks each supplier may face in delivering their contribution to the capability portfolio's desired capability outcomes. How might a portfolio manager optimally select the most risk-favorable subset of suppliers from a set of competing choices.

11. **Governance**

Think about ways to structure engineering management and oversight protocols for an enterprise risk management process. Define necessary process participants, the decision chain of authority and its operations, and the roles of stakeholders in the process as consumers of enterprise services.

5

An Index to Measure Risk Correlationships

5.1 Introduction

Chapter 4 described a way to structure the risk management problem space in engineering enterprise systems from a capability portfolio perspective. A representation of this space by a supplier–provider metaphor in the form of a mathematical graph was developed. This graph is a logical topology of nodes that depict supplier–provider relationships unique to a capability portfolio. Capturing dependencies between nodes is a critical aspect in the analysis and management of risk in engineering enterprise systems.

In this book, we posit two types of dependencies that affect risk in engineering capabilities for an enterprise system. One is risk inheritance; that is, how risk-dependent are capabilities so that threats to them can be discovered *before* contributing programs (e.g., suppliers) degrade, fail, or are eliminated. The other is operational dependence; that is, what is the effect on the operability of capability if, *due to the realization of risk*, one or more contributing programs (e.g., suppliers) or supplier–provider chains degrade, fail, or are eliminated? The first type of dependency is addressed in this chapter. The second type of dependency is discussed in Chapter 6.

This chapter introduces a new engineering risk management metric called the *risk correlationship index*. The risk correlationship (RCR) index measures the *strength of the influence* of risk inheritance between supplier programs and across supplier–provide chains that comprise a capability portfolio. Risk inheritance, if present, can increase the threat that risks with one supplier program may adversely affect others and ultimately their collective contributions to the associated capabilities. The purpose of the RCR index is to signal where risk-reducing opportunities exist to minimize dependency risks that, if realized, have cascading negative effects on the ability of an enterprise to deliver capabilities and services to users.

5.2 RCR Postulates, Definitions, and Theory

The development of the RCR index is based on a set of postulates. These postulates define the index in terms of its behavior in the context of engineering enterprise systems by capability portfolios. The RCR postulates are stated in the language of a mathematical graph that represents the supplier–provider metaphor, as discussed in Chapter 4. As such, these postulates assume a parent–child relationship between a capability node (C-node) and the set of supplier program nodes (P-nodes) that contribute to enabling that capability. First, we begin with a definition of *risk inheritance.*

Definition 5.1: Risk Inheritance

A RCR exists between program nodes if and only if one program node inherits one or more risk events from one or more other program nodes.

RCRs only directly exist between P-nodes; that is, only P-nodes can directly inherit risk events. RCRs indirectly exist between C-nodes when P-nodes associated with them have RCRs with other P-nodes in the capability portfolio.

Postulate 5.1: Capability Node Risk Score

A capability node's risk score is a function of the risk scores of its supplier program nodes.

Postulate 5.2: Capability Node RCRs are Indirect

A capability node's risk correlationships are indirect. They result and derive only from RCRs that directly exist between supplier program nodes, from across the capability portfolio, to that capability node.

Postulate 5.3: Inheritance Bounds

The risk score of a program node with noninherited risk events that then inherits one or more risks from one or more other program nodes cannot be lower than its risk score prior to the inheritance.

Postulate 5.4: Probability Invariant with Inheritance

A risk event's occurrence probability is invariant with respect to inheritance.

Postulate 5.5: Impacts Can Vary with Inheritance

An inherited risk event's impact (or consequence) on a receiving program node can be different from its impact (or consequence) on the sending program node.

Postulate 5.6: Impacts Assessed Against Capability

A risk event inherited by a program node shall have its impacts assessed in terms of how the risk, if it occurs, has negative consequences on that program node's ability to deliver its contribution to its associated capability node.

Postulate 5.7: Inherited Risk Events Have Resolution Priority

Inherited risk events are "first" targets for resolution or elimination by management.

Inherited risk events have, by their nature, extended their threat to other programs beyond their source program nodes. This complicates coordination, collaboration, and risk resolution planning between management and stakeholders across all levels of the portfolio. Impacts on multiple stakeholders, users, and outcome goals of affected program nodes must be jointly and carefully considered when planning, executing, and managing resolution strategies for inherited risks.

Notation 5.1

Let E denote a risk event. Let E_{ix} denote the xth risk event in the set of y risk events for program node P_i, where $x = 1, 2, 3, \ldots, y$.

Notation 5.2

Let $[[E]]$ denote a risk event *inherited* by one program node from another program node. Let $[[E_{i,j,k}]]$ denote that program node P_i inherits from program node P_j the risk event k, $k = 1, 2, 3, \ldots, \xi$.

Notation 5.3

Let $\sim I$ denote a set of *noninherited risk events*. For example, the set

$$\sim I = \{E_{i1}, E_{i2}, E_{i3}, \ldots, E_{iy}\}$$

signifies that program node P_i contains a set of $x = 1, 2, 3, \ldots, y$ nonherited events.

Notation 5.4

Let I denote a set of *inherited risk events*. For example, the set

$$I = \{[[E_{1,2,6}]]\}$$

signifies that program node P_1 inherits from program node P_2 the risk event E_6. The set

$$I = \{[[E_{1,2,6}]], [[E_{1,2,4}]], [[E_{1,2,5}]], [[E_{1,3,9}]]\}$$

signifies that program node P_1 inherits from program node P_2 the risk events are E_6, E_4, and E_5 and program node P_1 inherits from program node P_3 the risk event E_9. The first subscript can be dropped when it is evident from a mathematical graph of P-node relationships (shown in Figure 5.5) which P-nodes are receiving (inheriting) risk events from those that are sending them (indicated by the second subscript).

Notation 5.5

Let $RS(E_{ix})$ denote the risk score of the xth risk event in the set of risk events for program node P_i, where $x = 1, 2, 3, \ldots, y$.

Notation 5.6

Let $RS([[E_{i,j,k}]])$ denote the risk score of risk event k that program node P_i inherits from program node P_j, where $k = 1, 2, 3, \ldots, \xi$.

Definition 5.2: Risk Event Risk Score*

Subject to the conditions in Theorem 3.6, the risk score of risk event E is given by the additive value function

$$Risk\ Score(E) = RS(E) = u_1 V_{\text{Probability}}(E) + u_2 V_{\text{Impact}}(E) \tag{5.1}$$

In Equation 5.1, the first term is a value function for the risk event's occurrence probability. The second term is a value function for the severity of the risk event's impact, if it occurs. The coefficients u_1 and u_2 are nonnegative weights such that $0 \le u_1 \le 1, 0 \le u_2 \le 1$, and $u_1 + u_2 = 1$. In Equation 5.1, these value functions can be designed to produce measures along either a 0 to 1 or a 0 to 100 cardinal interval scale, as discussed in Chapter 3.

In Chapter 3, Figure 3.19 illustrated a family of value functions that can be created for a risk event's occurrence probability. From Chapter 1 recall that a risk event E is a special kind of probability event, in that E has both a nonzero and not certain occurrence probability; that is, $0 < P(E) < 1$. However, a value function for the occurrence probability of a *risk event* may indeed be designed to equal 0 in the limit; that is, as $P(E)$ approaches 0 from the right we have

$$\lim V(P(E)) \equiv \lim V_{\text{Probability}}(E) = 0 \text{ as } P(E) \to 0^+$$

Similar considerations apply to value functions designed to capture a risk event's impact.

Definition 5.3: Node Risk Score (Noninheritance)

The risk score of the ith program node's set of *noninherited* risk events is denoted by $RS(P_i \mid \sim I)$ and is computed by the max average of the set $\sim I$,

* Equation 5.1 is one of many ways to formulate a *Risk Score* measure. The reader is directed to Garvey (2008) for additional approaches to formulating this measure.

which contains the noninherited risk event risk scores specific to program node P_i. The max average rule was introduced in Chapter 4 (Definition 4.1).

Definition 5.4: Node Risk Score (Inheritance)

The risk score of the *i*th program node's set of *inherited* risk events is denoted by $RS(P_i|I)$ and is computed by the max average of the set *I*, which contains the inherited risk event risk scores specific to program node P_i.

Definition 5.5: Program Node Risk Score (Mixed Case)

The risk score of the *i*th program node's set of *noninherited* and *inherited* risk events is denoted by $RS(P_i| \sim I \wedge I)$ and is defined as follows:

$$RS(P_i| \sim I \wedge I) = \begin{cases} RS(P_i| \sim I) \text{ if } Z_1 \text{ is true} \\ MaxAve(RS(P_i| \sim I), RS(P_i|I)) \text{ if } Z_2 \text{ is true} \end{cases} \quad (5.2)$$

where Z_1 is when $RS(P_i|I) \leq RS(P_i| \sim I)$ and Z_2 is when $RS(P_i| \sim I) \leq RS(P_i|I)$.

The analytical philosophy behind the rule given by Definition 5.5 is as follows. If a program node's set of inherited risk events has a higher overall risk score than the node's set of noninherited risk events, then the node's overall risk score should be driven by the impacts of inheritance. Alternatively, if a program node's set of noninherited risk events has a higher overall risk score than the node's set of inherited risk events, then the node's overall risk score should be driven by the impacts of noninheritance—as these are the more threatening events to the program node's ability to deliver its contribution to its capability node.

The risk score rule in Definition 5.5 is one of many possible forms, with many design variations possible. For instance, certain conditions (when triggered) may warrant invoking the max rule in Equation 5.2. When evaluating the design of a risk score rule, the magnitudes of the values produced should always reflect the risk tolerance level of the program or decision maker receiving its results.

Definition 5.6: Capability Node Risk Score

The risk score *RS* of a capability node is the max average of its individual program node risk scores that include (if present) the influence of risk event inheritance in accordance with Definition 5.5.

Definition 5.7: Program Node RCR Index

If program node P_i contains noninherited risk events and risk events inherited from one or more other program nodes in the capability portfolio, then the RCR index of P_i is defined as follows:

$$0 \leq RCR(P_i) = \frac{RS(P_i \mid \sim I \wedge I) - RS(P_i \mid \sim I)}{RS(P_i \mid \sim I \wedge I)} \leq 1 \tag{5.3}$$

Definition 5.8: Capability Node RCR Index

If capability node C_z contains one or more program nodes that have RCRs with other program nodes in the capability portfolio, then the RCR index of capability node C_z is defined as follows:

$$0 \leq RCR(C_z) = \frac{RS(C_z \mid \sim I \wedge I) - RS(C_z \mid \sim I)}{RS(C_z \mid \sim I \wedge I)} \leq 1 \tag{5.4}$$

where $RS(C_z \mid \sim I \wedge I)$ is the risk score of capability node C_z computed by the max average of its set of P-node risk scores derived from Equation 5.2. The term $RS(C_z \mid \sim I)$ is the risk score of capability node C_z computed by the max average of its set of P-node risk scores that do not include the influence of risk event inheritance.

Finally, recall from Postulate 5.2 that capability node RCRs are indirect. They result and derive only from RCRs that exist directly between supplier program nodes in the portfolio. A capability node's RCR index is a response measure—one that derives from the influence of risk event inheritance between P-nodes that comprise that specific capability. This is illustrated in Figure 5.1, whose interpretation is discussed next.

Suppose Figure 5.1 illustrates a capability portfolio that consists of three capability nodes (C-nodes) and four supplier program nodes (P-nodes). The left-most assemblage of nodes shows an alignment of P-nodes under each C-node. Let this alignment indicate which supplier programs are responsible to deliver technologies that enable the associated capability. For example, in Figure 5.1 capability node Cap_2 is dependent on two supplier program nodes P_2 and P_3 for Cap_2 to achieve its intended outcomes.

In Figure 5.1, the right-most graph shows arrows between the C-nodes and the P-nodes. These arrows signal that RCRs exist between them. This is called an RCR graph. In an RCR graph, risk correlationships are always indicated by arrows.* Figure 5.2 shows the RCR graph in Figure 5.1 with the

* In this chapter, arrows indicate that (1) a RCR exists between two nodes, and (2) the direction of the risk inheritance in terms of the inheritance sending node and the inheritance receiving node. When arrows on a graph indicate risk inheritance, then it is called an RCR graph. In Chapter 6, arrows will indicate the direction of operational dependencies between nodes in feeder–receiver relationship. Such a graph, discussed in Chapter 6, is called a functional dependency network analysis (FDNA) graph.

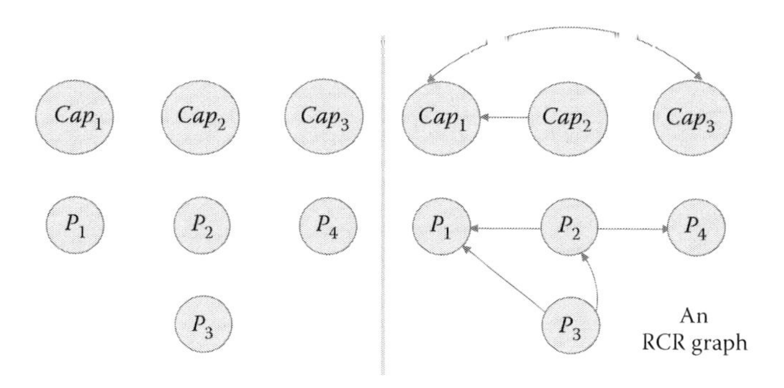

FIGURE 5.1
An RCR graph: Program-to-capability node RCRs.

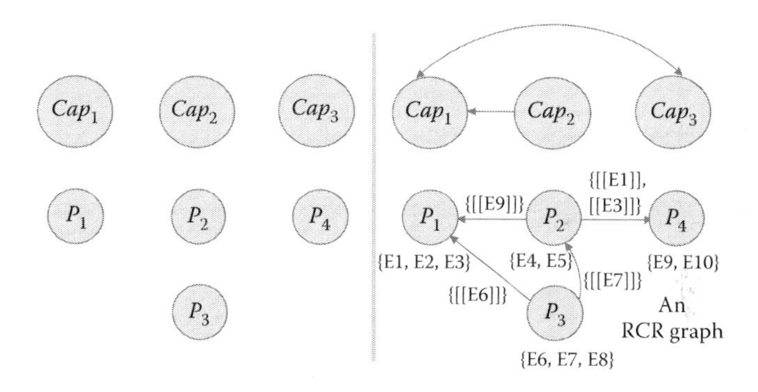

FIGURE 5.2
An RCR graph: A risk event inheritance view.

inheritance flow of risk events from one node to another node. On the right side of Figure 5.2, a collection of risk events is shown under each program node. Recall that a risk event is one that, if it occurs, has unwanted consequences for the respective program node's ability to deliver its contribution to its respective capability node.

From Definition 5.1, a RCR exists between P-nodes if and only if one P-node inherits one or more risk events from one or more other P-nodes. When RCRs are present, they only exist directly between P-nodes. In accordance with Postulate 5.2, RCRs between C-nodes are indirect. They reflect the presence of RCRs between P-nodes that comprise the specific capability. In Figures 5.1 and 5.2, this is indicated by the arched arrow above the C-nodes.

The RCR index for P-nodes behaves in accordance with the following properties. These properties follow from the preceding postulates and definitions.

Property 5.1

If program node P_i has a mix of noninherited and inherited risk events and $RS(P_i|I) \leq RS(P_i|\sim I)$, then $RCR(P_i) = 0$.

Proof

Program node P_i is given to have a mix of noninherited and inherited risk events. It is also given that $RS(P_i|I) \leq RS(P_i|\sim I)$. From Equation 5.2, it follows that $RS(P_i|\sim I \wedge I) = RS(P_i|\sim I)$; thus,

$$RCR(P_i) = \frac{RS(P_i|\sim I \wedge I) - RS(P_i|\sim I)}{RS(P_i|\sim I \wedge I)} = \frac{RS(P_i|\sim I) - RS(P_i|\sim I)}{RS(P_i|\sim I)} = 0$$

Property 5.1 reflects the following: if the RCR between P_i and another program node is zero, then inheritance has no contribution to the risk score of P_i. Property 5.1 reflects Postulate 5.3 which states: *The risk score of a P-node with noninherited risk events that then inherits one or more risks from one or more other P-nodes cannot be lower than its risk score prior to the inheritance.*

Property 5.2: RCR Index Bounds

The RCR index falls within the interval $0 \leq RCR(P_i) \leq 1$.

Property 5.2 reflects the behavior that the closer the risk correlationship index is to 0 the lesser the influence of risk inheritance on program node P_i. Alternatively, the closer the risk correlationship index is to 1 the greater the influence of risk inheritance on program node P_i. Thus, the RCR index is simply a strength of influence measure. Proving Property 5.2 is an exercise for the reader.

Property 5.3

If program node P_i contains a mix of noninherited and inherited risk events and if $RS(P_i|\sim I) < RS(P_i|I)$ then $RS(P_i|\sim I) < RS(P_i|\sim I \wedge I) < RS(P_i|I)$.

5.3 Computing the RCR Index

The following problems illustrate computing the RCR index.

PROBLEM 5.1

From the information in Figure 5.3, compute the RCR index for program node P_1.

Solution

From Definition 5.3 we have

$$RS(P_1 | \sim I) = MaxAve\,(67, 43, 44, 21, 50, 55) = 60.29$$

From Definition 5.4 we have

$$RS(P_1 | I) = MaxAve\,([[95]]) = 95$$

From Definition 5.5, and since $RS(P_1 | \sim I) \leq RS(P_1 | I)$, the overall risk score of P_1 is

$$RS(P_1 | \sim I \wedge I) = MaxAve\,(60.29, 95) = 89.7935$$

In the above max average calculations suppose the weighting function in Figure 4.11 was used. From Definition 5.7 we have

$$RCR(P_1) = \frac{RS(P_1 | \sim I \wedge I) - RS(P_1 | \sim I)}{RS(P_1 | \sim I \wedge I)} = \frac{89.7935 - 60.29}{89.7935} = 0.32857$$

Therefore, approximately 33% of the overall risk score of P_1 is contributed by P_3. This finding would be noteworthy to report to the engineering management team.

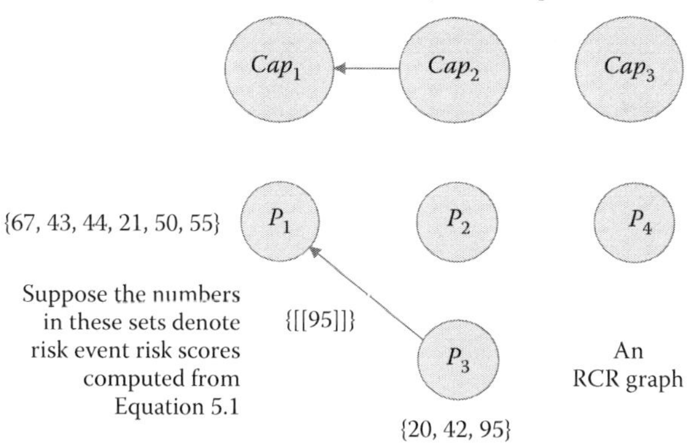

FIGURE 5.3
Problem 5.1 RCR graph.

PROBLEM 5.2

From the information in Figure 5.4, compute the RCR index for program node P_1.

Solution

From Definition 5.3 we have

$$RS(P_1 | \sim I) = MaxAve\,(95) = 95$$

From Definition 5.4 we have

$$RS(P_1 | I) = MaxAve\,([[67]], [[43]], [[44]], [[21]], [[50]], [[51]]) = 60.29$$

From Definition 5.5, and since $RS(P_1 | I) < RS(P_1 | \sim I)$, the overall risk score of P_1 is

$$RS(P_1 | \sim I \wedge I) = RS(P_1 | \sim I) = 95$$

In the above max average calculations suppose the weighting function in Figure 4.11 was used. From Definition 5.7 we have

$$RCR(P_1) = \frac{RS(P_1 | \sim I \wedge I) - RS(P_1 | \sim I)}{RS(P_1 | \sim I \wedge I)} = \frac{95 - 95}{95} = 0$$

Therefore, in accordance with Postulate 5.3, the contribution from P_3 has no effect on the overall risk score of P_1 (in this case). Recall that Postulate 5.3

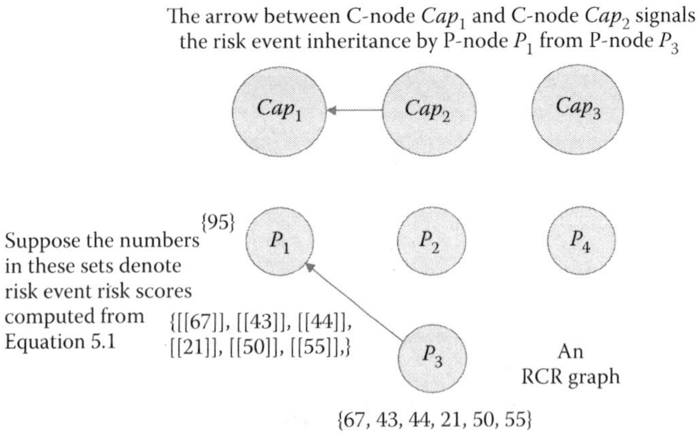

FIGURE 5.4
Problem 5.2 RCR graph.

states: *The risk score of a program node with noninherited risk events that then inherits one or more risks from one or more other program nodes cannot be lower than its risk score prior to the inheritance.*

PROBLEM 5.3

Compute the RCR indexes for all nodes in Figure 5.5, given the data in the left three columns of Table 5.1.

Solution

Figure 5.5 shows 10 supplier program nodes providing contributions to 3 capability nodes. Here, program nodes within a capability node need not be unique to that capability. In practice, the same program node may appear beneath multiple capability nodes if the technology program represented by that node is supplying multiple contributions to those capabilities. For example, program node P_2 might be the same technology program as program node P_6, but it is supplying multiple contributions to capability node Cap_1 and Cap_2.

In Figure 5.5, observe that program node P_1 has three noninherited risk events E_{11}, E_{12}, and E_{13} and three inherited risk events $[[E_{71}]]$, $[[E_{51}]]$, and $[[E_{52}]]$. These inherited risk events come from program nodes P_7 and P_5. Suppose

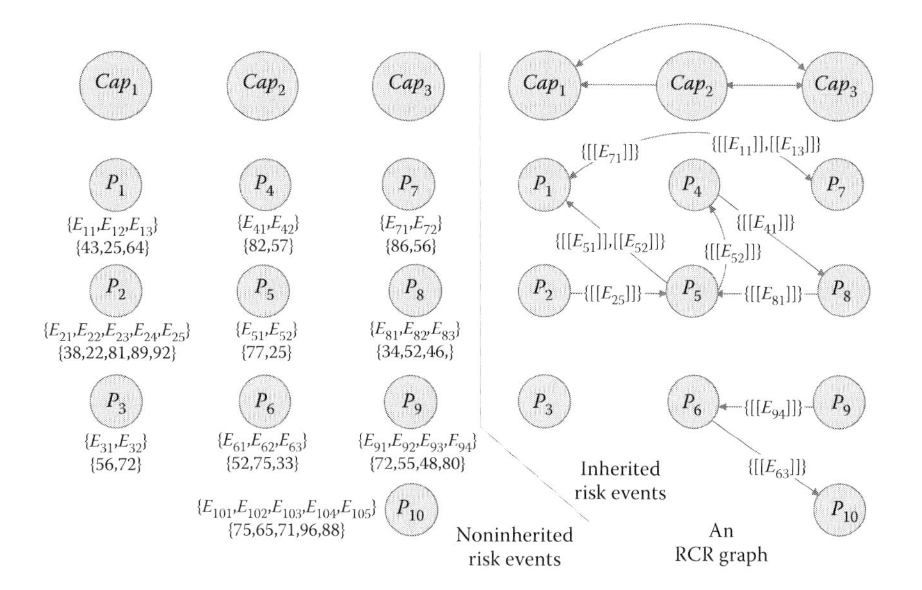

FIGURE 5.5
Problem 5.3. Noninherited and inherited risk events and their risk scores.

TABLE 5.1

Problem 5.3 Data and Computations

Program Node	Noninherited Risk Event Risk Scores	Inherited Risk Event Risk Scores	$RS(P_i \mid \sim I)$	$RS(P_i \mid I)$	$RS(P_i \mid \sim I \wedge I)$	$RCR(P_i)$
P_1	{43, 25, 64}	{86, 77, 25}	56.80	79	75.67	0.249
P_2	{38, 22, 81, 89, 92}	None	83.72	0	83.72	0
P_3	{56, 72}	None	69.60	0	69.60	0
P_4	{82, 57}	{25}	78.25	25	78.25	0
P_5	{77, 25}	{92, 34}	69.20	83.30	81.185	0.148
P_6	{52, 75, 33}	{80}	68.50	80	78.275	0.125
P_7	{86, 56}	{43, 64}	81.50	60.22	81.50	0
P_8	{34, 52, 46}	{82}	48.16	82	76.924	0.374
P_9	{72, 55, 48, 80}	None	75.125	0	75.125	0
P_{10}	{75, 65, 71, 96, 88}	{33}	90.90	33	90.90	0

risk scores for these six risk events are given in Table 5.1 and were computed by Equation 5.1; that is,

$$\{RS(E_{11}), RS(E_{12}), RS(E_{13})\} = \{43, 25, 64\}$$

$$\{RS([[E_{71}]]), RS([[E_{51}]]), RS([[E_{52}]])\} = \{86, 77, 25\}$$

From Definitions 5.3 and 5.4, program node P_1's risk scores are as follows:

$$\begin{aligned} RS(P_1 \mid \sim I) &= MaxAve(\{RS(E_{11}), RS(E_{12}), RS(E_{13})\}) \\ &= MaxAve(\{43, 25, 64\}) = 56.8 \end{aligned}$$

$$\begin{aligned} RS(P_1 \mid I) &= MaxAve(\{RS([[E_{71}]]), RS([[E_{51}]]), RS([[E_{52}]])\}) \\ &= MaxAve(\{86, 77, 25\}) = 79 \end{aligned}$$

From Definitions 5.5, the risk score of program node P_i, when P_i is characterized by a set of noninherited and inherited risk events is

$$RS(P_i \mid \sim I \wedge I) = \begin{cases} RS(P_i \mid \sim I) & \text{if } RS(P_i \mid I) \le RS(P_i \mid \sim I) \\ MaxAve(RS(P_i \mid \sim I), RS(P_i \mid I)) & \text{if } RS(P_i \mid \sim I) \le RS(P_i \mid I) \end{cases}$$

From the above, since $RS(P_1 \mid \sim I) < RS(P_1 \mid I)$ the combined risk score of program node P_1 is

$$RS(P_1 \mid \sim I \wedge I) = MaxAve(56.8, 79) = 75.67$$

Thus, the RCR index of P_1 is

$$RCR(P_1) = \frac{RS(P_1 | \sim I \wedge I) - RS(P_1 | \sim I)}{RS(P_1 | \sim I \wedge I)} = \frac{75.67 - 56.8}{75.67} = 0.249$$

The results for the other nine program nodes were computed in a similar manner and are summarized in Table 5.1. To the capability portfolio manager and to the portfolio's individual program managers, inherited risks are first targets for resolution or elimination. Seen in Figure 5.5, inherited risk events extend their threat to other programs beyond their originating program nodes.

Inheritance complicates program-to-portfolio cross-coordination, collaboration, and resolution planning. Multiple stakeholders, users, and outcome goals of affected program and capability nodes must be jointly considered when planning and executing resolution strategies for inherited risks. Thus, from a criticality perspective, inherited risk events are signaled as prime targets for early management attention and intervention.

Table 5.2 shows the computational results of each capability node's risk score as a function of the relationships in Figure 5.5 and the derived program node risk scores shown in Table 5.1. For example, the risk score and the risk correlationship index for capability node C_2 are formulated from Definition 5.8, Figure 5.5, and Table 5.1 as follows:

$$\begin{aligned} \textit{Risk Score}\ (C_2) &= RS(C_2) \\ &= \textit{MaxAve}(\{RS(P_4 | \sim I \wedge I), RS(P_5 | \sim I \wedge I), RS(P_6 | \sim I \wedge I)\}) \\ &= RS(C_2 | \sim I \wedge I) = \textit{MaxAve}(\{78.25, 81.185, 78.275\}) = 80.601 \end{aligned}$$

Here, $RS(C_2 | \sim I \wedge I)$ is the risk score of capability node C_2 computed over its set of P-node risk scores that include the influence of inheritance (Table 5.1). The RCR index of capability node C_2 is then computed, from Equation 5.4, as follows:

$$RCR(C_z) = \frac{RS(C_z | \sim I \wedge I) - RS(C_z | \sim I)}{RS(C_z | \sim I \wedge I)}$$

$$RCR(C_2) = \frac{RS(C_2 | \sim I \wedge I) - RS(C_2 | \sim I)}{RS(C_2 | \sim I \wedge I)} = \frac{80.601 - \textit{MaxAve}\ (\{78.25, 69.2, 68.5\})}{80.601}$$

$$RCR(C_2) = \frac{RS(C_2 | \sim I \wedge I) - RS(C_2 | \sim I)}{RS(C_2 | \sim I \wedge I)} = \frac{80.601 - 76.37}{80.601} = 0.0525$$

TABLE 5.2

The Influence of Risk Inheritance on C-Node Risk Scores

Capability Node	Program Node Risk Score Set	Capability Node Risk Score $RS(C_z)$	$RCR(C_z)$
C_1	{75.67, 83.72, 69.6}	81.503	0.02315
C_2	{78.25, 81.185, 78.275}	80.601	0.0525
C_3	{81.5, 76.924, 75.125, 90.90}	87.964	0.0245

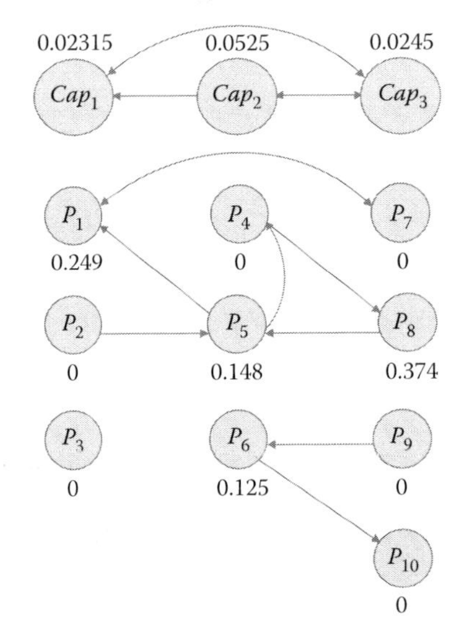

FIGURE 5.6
Problem 5.3 nodes: An RCR index view.

Observe that a capability node's risk correlationships are *indirect*. They derive only from risk correlationships that exist *directly* between supplier program nodes in the portfolio. So, a capability node's RCR index is really a *response measure*—one that derives from the effects of risk event inheritance between P-nodes that comprise the supplier dimensions of the capability portfolio.

Figure 5.6 presents the nodal topology of Problem 5.3 visualized by RCR indices. With this topology management can view time-history changes to these indices and see where high RCR indexes exist or remain between program and capability nodes. A rank-order from highest-to-lowest RCR index by program and capability nodes affected by inheritance can be generated and monitored over time.

5.4 Applying the RCR Index: A Resource Allocation Example

This section illustrates an application of the RCR index to resource allocation decisions. Here, we integrate an operations research optimization algorithm into the theory of risk inheritance and the RCR index. The aim is to demonstrate a formal and analytically traceable investment decision protocol. This protocol works to identify which combination of program nodes offer the maximum reduction in capability risk, when risk resolution assets are allocated from a constrained risk resolution budget. The identified nodes will be the best among other program node candidates for management intervention and the investment of limited risk resolution funds. The optimization algorithm discussed below falls into a class known as constrained optimization algorithms. It is informally known as the *knapsack model*.

The Knapsack Optimization Algorithm

The knapsack problem is a classic problem in operations research. One form of this problem can be described as follows. Suppose you have a finite collection of items you want to pack into your knapsack. Suppose the knapsack has limited capacity so it is not possible to include all items. Suppose each item has a certain value (or utility) to you. Given this, which items can be included in the knapsack such that the value of its collection of items is maximized but does not exceed the knapsack's capacity?

A classic knapsack problem can be mathematically formulated as follows:

$$\text{Maximize } v_1x_1 + v_2x_2 + v_3x_3 + \cdots + v_nx_n$$
$$\text{subject to } w_1x_1 + w_2x_2 + w_3x_3 + \cdots + w_nx_n \leq K$$

where x_i for $i = 1, 2, 3, \ldots, n$ takes the value 0 if item x_i is not included in the knapsack and takes the value 1 if item x_i is included in the knapsack. The parameter w_i is the weight (e.g., in pounds) of item x_i and K is the overall weight capacity of the knapsack.

The first equation is called the objective function. The second equation is called the constraint equation. Solving the knapsack problem involves integer programming, a specialized optimization technique. The Microsoft® Excel Solver program can be used to find solutions to the knapsack problem.

Instead of a knapsack, let us think of the problem of choosing which P-nodes to include in a risk resolution portfolio. However, suppose this portfolio is defined by a fixed budget for funding P-node risk resolution plans. The decision problem is to select P-nodes that have the highest risk to their associated capability while not exceeding the overall risk resolution budget. As mentioned above, we can think of this as a knapsack problem. The mathematical set up is as follows.

Let $j = \{1, 2, 3, \ldots, n\}$ be a set indexing the candidate P-nodes (note: subscripts here are local to this knapsack formulation). Let

$$x_j = \begin{cases} 0 & \text{if the } j\text{th P-node is not in the risk resolution portfolio} \\ 1 & \text{if the } j\text{th P-node is in the risk resolution portfolio} \end{cases}$$

Here, we want to

$$\begin{aligned} &\text{Maximize } v_1x_1 + v_2x_2 + v_3x_3 + \cdots + v_nx_n \\ &\text{subject to } c_1x_1 + c_2x_2 + c_3x_3 + \cdots + c_nx_n \leq C \end{aligned}$$

where v_j is the RCR index of jth P-node, c_j is the cost to resolve the risk events that comprise the jth P-node, and C is the total risk resolution budget.

Table 5.3 illustrates this application in the context of ten P-nodes described in Problem 5.3. Table 5.1 presents the risk scores of these P-nodes. Suppose the columns of the Table 5.3 are the ten P-nodes all competing for funding from the fixed risk resolution budget. Suppose the total risk resolution budget is $20 million. However, the cost to resolve the risks in all ten P-nodes is just over $36 million.

Furthermore, suppose the management team decides that the risk resolution budget should only be allocated to P-nodes with risk scores greater than or equal to 70. The reasoning being that P-nodes with risk scores equal to or higher than 70 fall into a "RED" color zone and are those that most threaten capability. Given this, which P-nodes should be funded in the "risk resolution portfolio" such that they collectively offer the maximum reduction in potential risk to capability while not exceeding the $20 million budget.

TABLE 5.3

P-Node Input Matrix

Optimization Input Matrix	**Program Node ID 1**	**Program Node ID 2**	**Program Node ID 3**	**Program Node ID 4**	**Program Node ID 5**
Objective Function: P-Node RCR Index	0.249	0	0	0	0.148
Subject to Constraint: Risk Resolution Cost ($K)	5900	2300	4321	7656	2132

Optimization Input Matrix	**Program Node ID 6**	**Program Node ID 7**	**Program Node ID 8**	**Program Node ID 9**	**Program Node ID 10**
Objective Function: P-Node RCR Index	0.125	0	0.374	0	0
Subject to Constraint: Risk Resolution Cost ($K)	6241	1325	3127	2112	1111

TABLE 5.4

P-Node Solution Matrix

Program Node ID 1	Program Node ID 2	Program Node ID 3	Program Node ID 4	Program Node ID 5
1	0	0	0	1
Program Node ID 6	**Program Node ID 7**	**Program Node ID 8**	**Program Node ID 9**	**Program Node ID 10**
1	1	1	0	1

To find the optimal collection of P-nodes to include in the risk resolution portfolio, given the above conditions, we can model this situation as a "knapsack" problem. Here, the coefficients of the objective function are the RCR indexes of the P-nodes. The coefficients of the constraint equation are the resolution costs of these P-nodes. For example, in Table 5.3, the first P-node has an RCR index of 0.249 and a risk resolution cost of $5.9 million. Next we use the Microsoft® Excel Solver program to solve this optimization problem. The results from Excel Solver are shown in Table 5.4.

In Table 5.4, the P-nodes indicated by a "1" in the gray box of the solution matrix are those to be funded. The P-nodes indicated by a "0" in the solution matrix are those not to be funded. Here, P-nodes 1, 5, 6, 7, 8, and 10 are the optimal collection of P-nodes to fund in the risk resolution portfolio. Program nodes 2, 3, 4, and 9 are not funded and hence are excluded from the risk resolution portfolio.

The mix of funded P-nodes, shown in Table 5.4, is the optimal combination of P-nodes to allocate funds that (1) offer the largest risk reduction to the capability portfolio and (2) comes as close as possible, while not exceeding, the total risk resolution budget of $20 million. The risk resolution costs for all P-nodes selected for investment sum to $19.836 million.

The analysis approach discussed illustrates a formal way to allocate limited risk resolution resources to P-nodes considered most threatening to capability. Analytical approaches such as these are valuable "first-filters" that support management decision-making. They are not replacements for human judgment or creative intervention management. The organization's leadership should always look at results such as these and consider additional trade-offs, options, or creative ways to address critically threatening risks, given risk resolution budget constraints. One way to use this analysis is to let it form the basis for deliberating why, where, and when increased resources are needed. This approach reveals not only those P-nodes whose risks can be included in a risk resolution funding portfolio but those that, without relaxing constraints or finding workarounds, should be excluded.

5.5 Summary

Developing ways to represent and measure capability dependencies in engineering enterprise systems is a critically important aspect of enterprise risk management. The importance of this problem is many-fold. The primary concern is enabling the management team to study the ripple effects of failure in one capability on other dependent capabilities across an enterprise. Offering ways to study these effects enables engineers to design for minimizing dependency risks that, if realized, have cascading negative effects on the ability of an enterprise to deliver services to consumers.

The problem discussed in this chapter focused on one form of dependency—risk inheritance between supplier programs in a capability portfolio. An index was developed that measures the influence of risk inheritance among supplier programs and capabilities. The RCR index identifies and captures the effects of inheritance on the risk that program nodes will fail to deliver their contributions to the capabilities that depend on them.

Questions and Exercises

1. Compute the RCR indexes for all nodes in Figure 5.7.
2. Compute the RCR indexes for all nodes in Figure 5.8.
3. Compute the RCR indexes for all nodes in Figure 5.9.
4. Prove Property 5.2.
5. If $RS(P_i \mid\sim I) \leq RS(P_i \mid I)$ show that $RS(P_i \mid\sim I \wedge I) = \frac{1}{2}(1-\lambda)RS(P_i \mid\sim I)$ $+\frac{1}{2}(1+\lambda)RS(P_i \mid I)$ where $0 \leq \lambda \leq 1$.

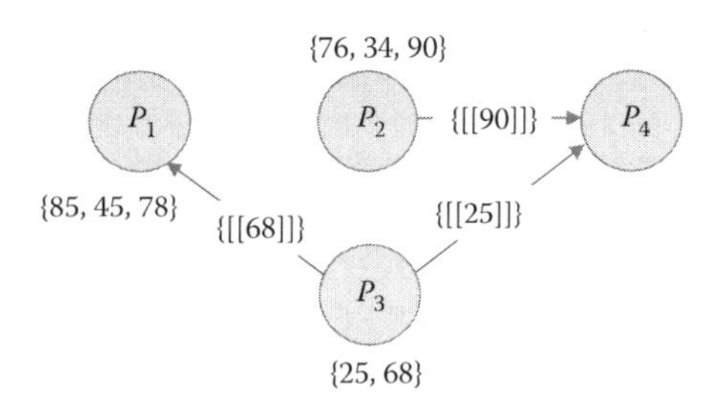

FIGURE 5.7
RCR graph for Exercise 1.

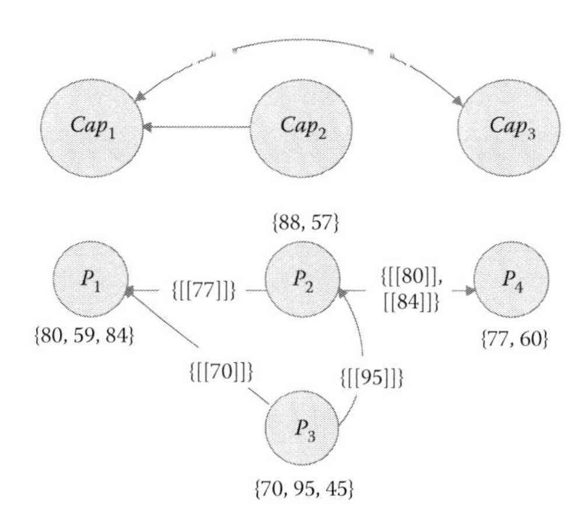

FIGURE 5.8
RCR graph for Exercise 2.

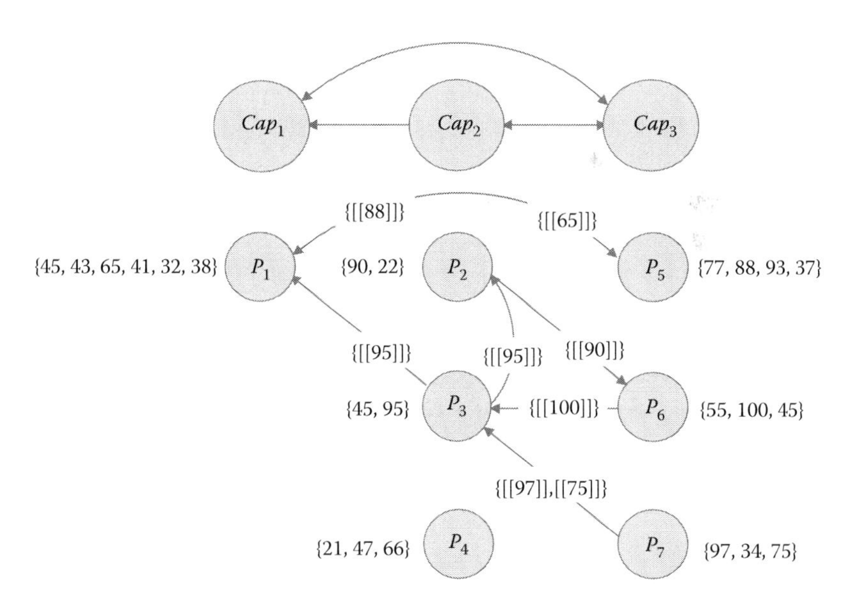

FIGURE 5.9
RCR graph for Exercise 3.

6

Functional Dependency Network Analysis

6.1 Introduction

Critical considerations in engineering systems, systems of systems, and enterprise systems are identifying, representing, and measuring the effects of dependencies between suppliers of technologies and providers of services the consumers. The importance of this topic is many-fold. Primary, is to study the ripple effects of failure in one capability on other dependent capabilities across an enterprise as a highly connected system. Providing mechanisms to anticipate these effects early in design enables engineers to minimize dependency risks that, if realized, can have cascading unwanted effects on the ability of a system to deliver its services.

One way to approach dependency analysis is by using mathematical graph theory. Graph theory offers a visual representation of complex dependencies between entities and enables the design of formalisms to measure and trace the effectiveness of these relationships as they affect many parts and paths in a graph. Likewise, an enterprise can be represented as a network of dependencies between contributing systems, infrastructures, or organizational entities. They can be expressed as nodes on a graph depicting direction, strength, and criticality of feeder–receiver dependency chains or relationships. With these formalisms, algorithms can be designed to address questions such as:

> What is the effect on the operability of enterprise capability if, due to the realization of risks, one or more contributing programs or supplier–provider chains degrade, fail, or are eliminated? How tolerant is the operability of an enterprise if, due to the realization of risks, one or more contributing programs or supplier–provider chains degrade or fail?

Functional dependency network analysis (FDNA) is a new approach to address these and related questions (Garvey and Pinto, 2009). FDNA is a methodology and a calculus created to measure the ripple effects of degraded operability in one or more entities or feeder–receiver chains on enterprise capabilities, due to the realization of risks. FDNA measures the magnitude of system inoperability if such entities or chains are damaged and whether the loss is unacceptable. From this, algorithms can then be designed to derive optimal portfolios of investments that strengthen entities most critical to maintaining a system's effectiveness in the presence of uncertain or adverse conditions.

6.2 FDNA Fundamentals

FDNA is a way to measure inflows and outflows of value across a topology of feeder–receiver node dependency relationships. Figure 6.1 illustrates such a topology. Figure 6.1 shows a mathematical graph of a capability portfolio* with dependency relationships between feeder nodes and receiver nodes. A mathematical graph is a set of points and a set of lines, with each line connecting two points. The points of a graph are known as vertices or nodes. The lines connecting the vertices are known as edges or arcs. The lines of graphs can have directedness. Arrows on one or both endpoints indicate directedness. Such a graph is said to be directed. "A graph or directed graph together with a function which assigns a positive real number to each line is known as a network" (Weisstein, MathWorld). As seen in Figure 6.1, an FDNA graph is a directed graph and later we show it is a network.

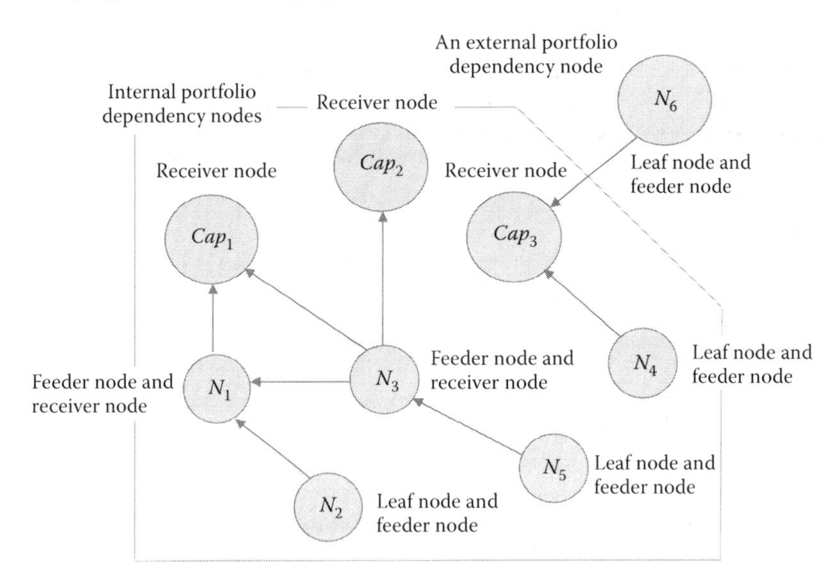

FIGURE 6.1
An FDNA graph of a capability portfolio. An FDNA graph is a logical topology in that it represents how value (utils) transmits through the graph between nodes and not necessarily how nodes are physically connected.

* Chapter 4 discusses engineering enterprise systems from a capability portfolio perspective.

A graph may also be viewed in terms of parent–child relationships. A parent node is one with lower level nodes coming from it. These are called child nodes to that parent node. Nodes that terminate in a graph are called leaf nodes. Leaf nodes are terminal nodes; they have no children coming from them.

Similar to parent–child dependency relationships, an FDNA graph shows dependencies between receiver nodes and feeder nodes. A receiver node is one that depends, to some degree, on the performance of one or more feeder nodes. A feeder node is one that "feeds" contributions to one or more receiver nodes. A node can be a feeder and a receiver node.

A mathematical graph is an important way to visualize dependency relationships between nodes. In FDNA, it is used to study the transmission of value between nodes, so their operability can be mathematically expressed and measured.

Mathematically Expressing Dependence and Operability

In FDNA, a dependency exists between nodes when the operability achieved by one node relies, to some degree, on the operability achieved by other nodes. Operability is a state in which a node is functioning at some level of performance. The level of performance achieved by a node can be expressed by a measure of value, worth, or "the utility it yields" (Bernoulli, 1738).

In FDNA, a node's measure of value is called its operability level. Value is analogous to a von Neumann-Morgenstern (vNM) utility, a dimensionless number expressed in "utils." In FDNA, we define what a node produces as its measure of performance (MOP) and the value (util) of what is produced as its operability level or its measure of effectiveness (MOE).

In a dependency relationship between nodes, contributions to the dependent node from other nodes are context specific to the nature of the supplying nodes. Contributions result from the achievement of outputs by nodes that reflect their performance.

For example, suppose node N_i produces and supplies coolant fluid to various engine manufacturers. A measure of performance for this node might be the rate at which it produces coolant fluid. Suppose a production rate of 9000 gallons per hour means this node is performing at half its full operability level. If 100 utils means a node is fully operable (wholly effective) and 0 utils means a node is fully inoperable (wholly ineffective), then a performance level of 9000 gallons per hour implies node N_i has an operational effectiveness of 50 utils of value. We can continue to assess the values (utils) of other performance levels achieved by N_i to form what is known as a value function.* Figure 6.2 illustrates one such a function.

* Chapter 4 presents an extensive discussion on value function theory.

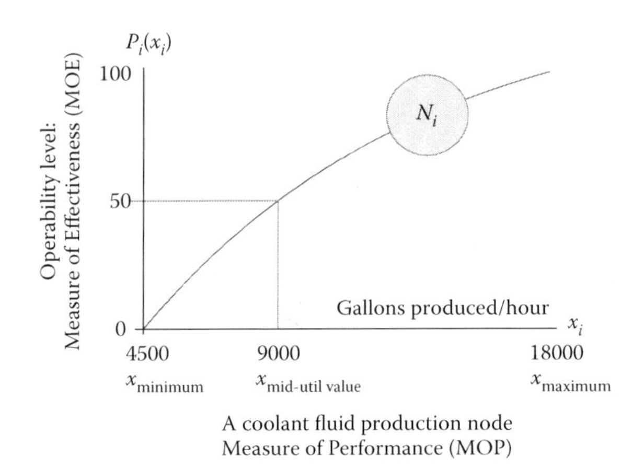

FIGURE 6.2
A node's value function.

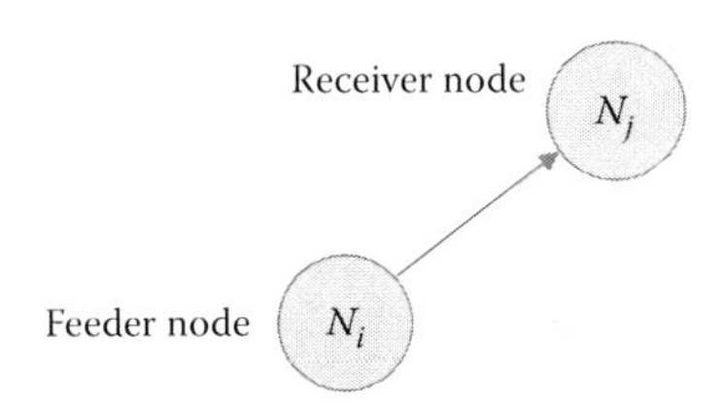

FIGURE 6.3
An FDNA graph of a single feeder–receiver node pair.

In general, a value function is a real-valued mathematical function that models a node's measure of value, worth, or utility associated with the levels of performance it achieves. The following presents how value functions can be used as a basis to model and measure the transmission of value between nodes, across a topology of multinodal dependencies. We start with Figure 6.3.

Figure 6.3 is an FDNA graph of two nodes N_i and N_j. The arrow indicates a dependency relationship exists between them. The arrow's direction means the operability of receiver node N_j relies, to some degree, on the operability of feeder node N_i. The notation *ij* is adopted in FDNA to indicate *i* is the index of a feeder node to a receiver node of index *j*. How can an operability dependency between these nodes be mathematically expressed by value functions?

A linear value function is one way to express the operability dependency of receiver node N_j on feeder node N_i. Figure 6.4 presents such a family. The left side of Figure 6.4 shows a family of operability functions where the operability level of N_j is a *linearly increasing* function of the operability level of N_i. The family shown is given by the equation,

$$P_j = \alpha_{ij} P_i + 100(1 - \alpha_{ij}), \quad 0 \le \alpha_{ij} \le 1, 0 \le P_i, P_j \le 100 \tag{6.1}$$

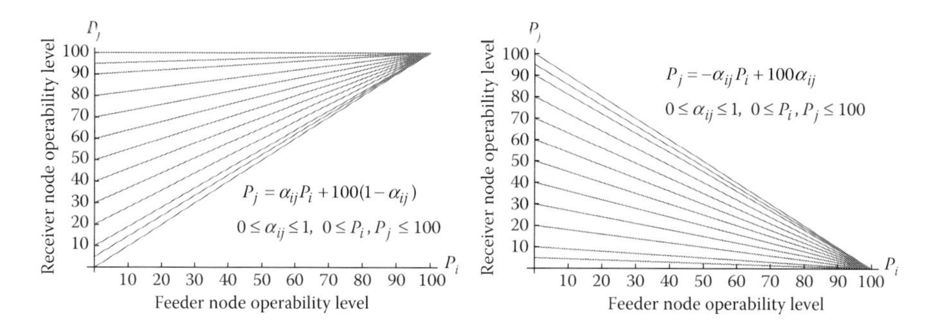

FIGURE 6.4
Operability as linear value functions.

where P_j is the operability level of N_j, P_i is the operability level of N_i, and α_{ij} is the increase in the operability level of N_j with each unit increase in the operability level of N_i.

The right side of Figure 6.4 shows a family of functions where the operability level of N_j is a *linearly decreasing* function of the operability level of N_i. The family shown is given by the equation,

$$P_j = -\alpha_{ij} P_i + 100\alpha_{ij}, \quad 0 \le \alpha_{ij} \le 1, 0 \le P_i, P_j \le 100 \tag{6.2}$$

where P_j is the operability level of N_j, P_i is the operability level of N_i, and $-\alpha_{ij}$ is the decrease in the operability level of N_j with each unit increase in the operability level of N_i.

A nonlinear value function is another way to express the operational dependency of receiver node N_j on feeder node N_i. Figure 6.5 presents such a family. The left side of Figure 6.5 shows a family of operability functions where the operability level of N_j can be a *nonlinear monotonically increasing* function of the operability level of N_i. The family shown is given by the equation,

$$P_j = \frac{100(1 - e^{-P_i/\rho})}{1 - e^{-100/\rho}} \tag{6.3}$$

where P_j is the operability level of N_j, P_i is the operability level of N_i, ρ is a scaling constant, and $0 \le P_i, P_j \le 100$.

The right side of Figure 6.5 shows a family of operability functions where the operability level of N_j can be a *nonlinear monotonically decreasing* function of the operability level of N_i. The family shown is given by the equation,

$$P_j = \frac{100(1 - e^{-(100 - P_i)/\rho})}{1 - e^{-100/\rho}} \tag{6.4}$$

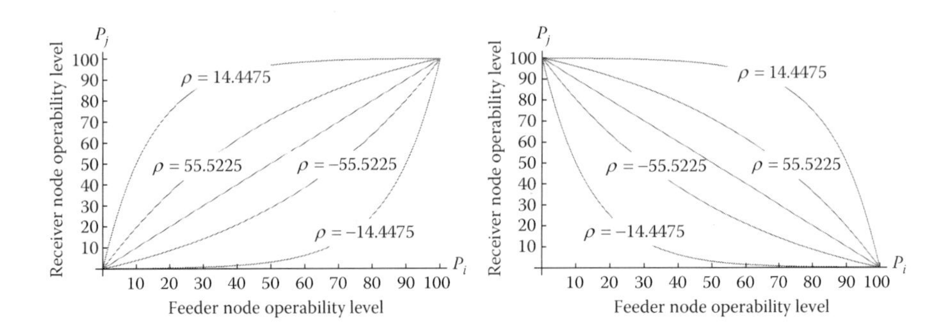

FIGURE 6.5
Operability as nonlinear value functions. *Note*: In decision theory, Equation 6.3 and Equation 6.4 are known as exponential value functions (Kirkwood, 1997).

where P_j is the operability level of N_j, P_i is the operability level of N_i, ρ is a scaling constant, and $0 \le P_i, P_j \le 100$.

For the single feeder–receiver node pair in Figure 6.3, the operability level of receiver node N_j can be expressed as a function f of the operability level of its feeder node N_i. In general, we can write P_j as

$$P_j = f(P_i), \quad 0 \le P_i, P_j \le 100 \tag{6.5}$$

where P_j is the operability level of N_j and P_i the operability level of N_i. Figures 6.4 and 6.5 illustrate how the operability function f might be expressed by linear or nonlinear value functions. How does this approach extend to multiple feeder nodes and in cases where nodes have feeder and receiver roles? We consider these situations next.

Suppose receiver node N_j has dependency relationships on two feeder nodes N_i and N_h, as shown in Figure 6.6. In the figure, the direction of the arrows indicates that the operability of N_j relies, to some degree, on the operability of N_i and N_h. For the feeder–receiver node relationships in Figure 6.6, the operability level of receiver node N_j can be expressed as a function f of the operability levels of its feeder nodes N_i and N_h. In general, we can write P_j as

$$P_j = f(P_i, P_h), \quad 0 \le P_i, P_h, P_j \le 100 \tag{6.6}$$

where P_j is the operability level of N_j, P_i is the operability level of N_i, and P_h is the operability level of N_h.

Suppose receiver node N_j has dependency relationships on h feeder nodes, as shown in Figure 6.7. In the figure, the direction of the arrows indicate that the operability of N_j relies, to some degree, on the operability levels of $N_1, N_2, N_3, \ldots, N_h$. For the feeder–receiver node relationships in Figure 6.7, the operability level of receiver node N_j can be expressed as a function f of

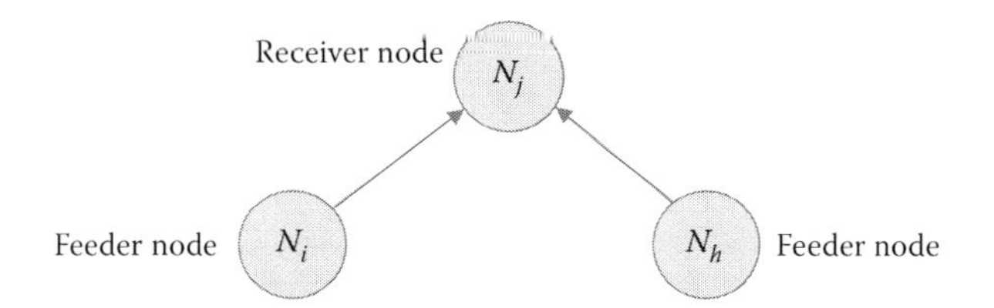

FIGURE 6.6
An FDNA graph with two feeder nodes.

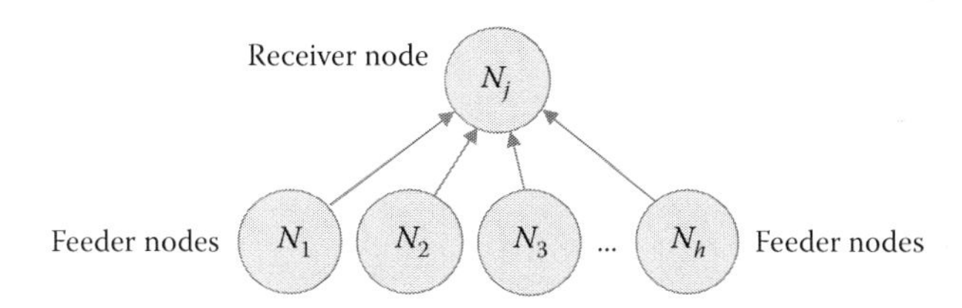

FIGURE 6.7
An FDNA graph with *h* feeder nodes.

the operability levels of its feeder nodes $N_1, N_2, N_3, \ldots, N_h$. In general, we can write P_j as

$$P_j = f(P_1, P_2, P_3, \ldots, P_h), \quad 0 \leq P_1, P_2, P_3, \ldots, P_h \leq 100 \tag{6.7}$$

where P_j is the operability level of N_j and $P_1, P_2, P_3, \ldots, P_h$ are the operability levels of feeder nodes $N_1, N_2, N_3, \ldots, N_h$, respectively. Next, consider the situation where a node is a feeder and a receiver as shown in Figure 6.8.

In Figure 6.8, receiver node N_p has a dependency relationship on node N_j and N_j has dependency relationships on nodes $N_1, N_2, N_3, \ldots, N_h$. In this case, N_j has the dual role of being a feeder node and a receiver node. For the feeder–receiver node relationships in Figure 6.8, the operability level of receiver node N_p can be expressed as a function of the operability levels of its feeder nodes $N_j, N_1, N_2, N_3, \ldots, N_h$. Since N_p has a dependency relationship on N_j we can write the expression $P_p = f(P_j)$. Since N_j has dependency relationships on $N_1, N_2, N_3, \ldots, N_h$ we can write the expression $P_j = g(P_1, P_2, P_3, \ldots, P_h)$. Combining expressions, we have

$$P_p = f(g(P_1, P_2, P_3, \ldots, P_h)), \quad 0 \leq P_1, P_2, P_3, \ldots, P_h \leq 100 \tag{6.8}$$

The equation for P_p was fashioned by a *composition of operability functions*. Function composition is when one function is expressed as a composition

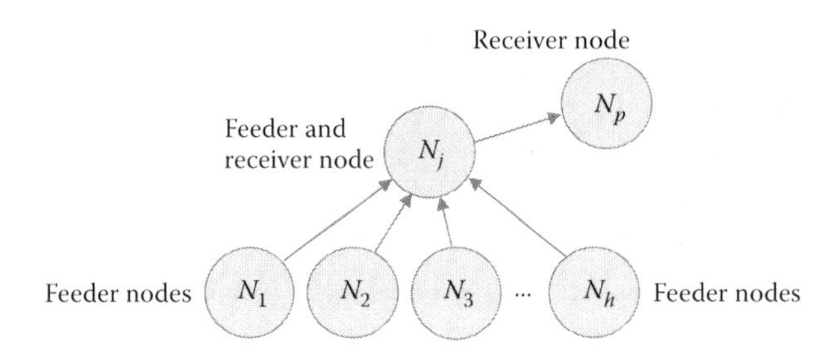

FIGURE 6.8
An FDNA graph with a feeder and receiver node.

(or nesting) of two or more other functions.* Function composition applies in FDNA when the operability function of one node is a composition of the operability functions of two or more other nodes. Next, we discuss forms the operability function f might assume. Selecting a specific form is central to FDNA and a major flexibility feature of the approach.

The first principle in formulating operability functions in FDNA is to understand the context of the graph, the nature of its nodes, and what their relationships and interactions mean. For instance, if the operability levels of all feeder nodes in Figure 6.7 are equally important to the operability level of receiver node N_j, then the operability function might be defined as

$$P_j = f(P_1, P_2, P_3, \ldots, P_h) = \frac{P_1 + P_2 + P_3 + \cdots + P_h}{h} \tag{6.9}$$

If the operability levels of some feeder nodes in Figure 6.7 are more important than others to the operability level of receiver node N_j, then the operability function might be defined as

$$P_j = f(P_1, P_2, P_3, \ldots, P_h) = w_1 P_1 + w_2 P_2 + w_3 P_3 + \cdots + w_h P_h \tag{6.10}$$

where $w_1, w_2, w_3, \ldots, w_h$ are nonnegative weights whose values range from 0 to 1 and where $w_1 + w_2 + w_3 + \cdots + w_h = 1$. If all weights are equal, then Equation 6.10 reduces to Equation 6.9.

In value function theory, the operability function f in Equation 6.10 is known as the linear additive model. This model is representative of a class of rules known as compensatory models. From an FDNA perspective, a compensatory model for f will produce operability measures with properties similar to a mathematical mean.

* In mathematics, if $h = g(y)$ and $y = f(x)$ then h is a function of x and can be written as a composition of the functions g and f; that is, $h = g(f(x)) = r(x)$ for all x in domain X. In general, functions formed through composition are not commutative; thus, the order with which they appear is important in their evaluation.

Another formulation for f might be the minimum function. The minimum function behaves as a weakest link rule. In general, the weakest link rule asserts no chain is stronger than its weakest link (Rescher, 2006). From an FDNA perspective, the weakest link rule means the operability level of a receiver node is equal to the operability level of its weakest performing feeder node. A weakest link formulation of f for the FDNA graph in Figure 6.7 can be written as follows:

$$P_j = f(P_1, P_2, P_3, \ldots, P_h) = \mathrm{Min}(P_1, P_2, P_3, \ldots, P_h) \tag{6.11}$$

Weakest link formulations offer many desirable properties in modeling and measuring inflows and outflows of value (utils) across a topology of feeder–receiver node dependency relationships. The remainder of this chapter will present FDNA by weakest link formulations and show, from this perspective, the many properties that derive from this rule.

There are many ways to express dependence and operability as it relates to nodes and their interactions in a mathematical graph. This topic is of considerable depth and breadth. It is a rich area for further study and research. We close this section with a set of FDNA postulates. These propositions are the foundations of FDNA and the basis for its calculus.

Postulate 6.1: Feeder Node

A feeder node is one that "feeds" contributions to one or more receiver nodes.

Postulate 6.2: Receiver Node

A receiver node is one whose performance depends, to some degree, on the performance of one or more feeder nodes.

Postulate 6.3: Feeder and Receiver Node

A node can be a feeder node and a receiver node.

Postulate 6.4: Leaf Node

Leaf nodes are terminal nodes; they have no subordinate nodes (children nodes) coming from them. A node that is a feeder node and a receiver node is not a leaf node.

Postulate 6.5: Measure of Performance

Each node can be expressed by a measure of its performance (MOP).

Postulate 6.6: Operability

Operability is a state where a node is functioning at some level of performance.

Postulate 6.7: Operability Level, Measure of Effectiveness

Each node can be expressed by a value that reflects its operability level associated with the measure of performance it achieves. A node's operability level is also known as its measure of effectiveness (MOE). A node's operability level is equivalent to a von Neumann-Morgenstern utility measure (utils). The operability level of any FDNA node falls in the interval [0, 100] utils.

Postulate 6.8: Inoperable/Operable Node

A node is fully inoperable (wholly ineffective) if its operability level is 0 utils. A node is fully operable (wholly effective) if its operability level is 100 utils.

Postulate 6.9: Dependency Relationships, Computing Operability Levels

A dependency relationship exists between nodes when the performance level of one node relies, to some degree, on the performance levels of other nodes. In a dependency relationship, a receiver node's operability level is computed by a function of the operability levels of its feeder nodes. This function is called the operability function.

Postulate 6.10: Paths Between Single Component Nodes are Acyclic*

In an FDNA graph, a path with the same first and last node that connects one or more other nodes that each produce one and only one unique product is disallowed. Nodes that produce one and only one product are called *single component nodes.*

6.3 Weakest Link Formulations

Functional dependency network analysis was developed using weakest link rules to express the operability functions of nodes in an FDNA graph. Definition 6.1 presents the general weakest link rule defined for FDNA. Definition 6.1 allows for the inclusion of criticality constraints, where the operability level of a receiver node could be limited by the operability level of a feeder node.

Definition 6.1: FDNA General Weakest Link Rule (GWLR)

If receiver node N_j has dependency relationships on feeder nodes $N_1, N_2, N_3, \ldots, N_h$, then the operability of N_j is given by the general expression

$$P_j = \mathrm{Min}(F(P_1, P_2, P_3, \ldots, P_h), G(P_1, P_2, P_3, \ldots, P_h)) \tag{6.12}$$

where $F(P_1, P_2, P_3, \ldots, P_h)$ is the operability function of N_j based on the strength with which N_j depends on the performance of feeder nodes $N_1, N_2, N_3, \ldots, N_h$.

* Postulate 6.10 is further discussed in Sections 6.6.2 and 6.6.3.

The function $G(P_1, P_2, P_3, ..., P_h)$ is the operability function of N_j based on the criticality with which N_j depends on the performance of feeder nodes $N_1, N_2, N_3, ..., N_h$. This section discusses Definition 6.1, and its variations and forms the functions F and G might assume. First, we introduce a measure called the *limited average*.*

Definition 6.2: Limited Average

The limited average of a finite set of nonnegative real numbers $x_1, x_2, x_3, ..., x_h$ is given by

$$\text{Min}\left(\frac{w_1 x_1 + w_2 x_2 + w_3 x_3 + \cdots + w_h x_h}{w_1 + w_2 + w_3 + \cdots + w_h}, x_1 + \beta_1, x_2 + \beta_2, x_3 + \beta_3, \ldots, x_h + \beta_h\right) \quad (6.13)$$

where $w_1, w_2, w_3, ..., w_h$ are nonnegative weights and $\beta_k \geq 0$ for all $k = 1, 2, 3, ..., h$. The limited average is a constrained weighted average, constrained by each β_k imposed on each x_k. The limited average is equal to Min $(x_1 + \beta_1, x_2 + \beta_2, x_3 + \beta_3, ..., x_h + \beta_h)$ if this is less than or equal to the weighted average of $x_1, x_2, x_3, ..., x_h$. If $\beta_k = 0$ for all $k = 1, 2, 3, ..., h$, then the limited average is equal to Min$(x_1, x_2, ..., x_h)$. If $x_1, x_2, x_3, ..., x_h$ are treated as h feeder nodes to a receiver node N_j, then a limited average form of FDNA GWLR (Definition 6.1) can be defined.

Definition 6.3: FDNA Limited Average Weakest Link Rule (LAWLR)

If receiver node N_j has dependency relationships on h feeder nodes $N_1, N_2, N_3, ..., N_h$, then the operability level of N_j is given by

$$P_j = \text{Min}(F(P_1, P_2, P_3, \ldots, P_h), G(P_1, P_2, P_3, \ldots, P_h)) \quad (6.14)$$

$$F(P_1, P_2, P_3, \ldots, P_h) = \frac{w_1 P_1 + w_2 P_2 + w_3 P_3 + \cdots + w_h P_h}{w_1 + w_2 + w_3 + \cdots + w_h} \quad (6.15)$$

$$G(P_1, P_2, P_3, \ldots, P_h) = \text{Min}(P_1 + \beta_{1j}, P_2 + \beta_{2j}, P_3 + \beta_{3j}, \ldots, P_h + \beta_{hj}) \quad (6.16)$$

where $0 \leq \beta_{kj} \leq 100$ and $k = 1, 2, 3, \ldots, h$. In Definition 6.3, the operability level of receiver node N_j is constrained by each β_{kj} imposed on each N_k. If these constraints are such that $G(P_1, P_2, P_3, ..., P_h) < F(P_1, P_2, P_3, ..., P_h)$ then

$$P_j = \text{Min}(P_1 + \beta_{1j}, P_2 + \beta_{2j}, P_3 + \beta_{3j}, \ldots, P_h + \beta_{hj}) \quad (6.17)$$

* The limited average was created by Dr. Brian K. Schmidt, The MITRE Corporation, 2005. It is one of many rules developed for problems in portfolio optimization and investment analyses (Moynihan, 2005; Moynihan et al., 2008).

otherwise, $$P_j = \frac{w_1P_1 + w_2P_2 + w_3P_3 + \cdots + w_hP_h}{w_1 + w_2 + w_3 + \cdots + w_h} \quad (6.18)$$

If $\beta_{kj} = 0$ utils for all $k = 1,2,3,\ldots,h$, then $P_j = \text{Min}(P_1, P_2, P_3, \ldots, P_h)$. Observe this is the weakest link formulation presented in Equation 6.11. Recall that Equation 6.11 reflects an operability function *f* where the operability level of a receiver node is equal to the operability level of its weakest performing feeder node. If $\beta_{kj} \geq 100$ utils for all $k = 1,2,3,\ldots,h$, then $F(P_1, P_2, P_3, \ldots, P_h) < G(P_1, P_2, P_3, \ldots, P_h)$ and

$$P_j = \frac{w_1P_1 + w_2P_2 + w_3P_3 + \cdots + w_hP_h}{w_1 + w_2 + w_3 + \cdots + w_h}$$

Thus, the parameter β_{kj} is called the *critically constraint* and it need only fall in the interval $0 \leq \beta_{kj} \leq 100$. Next, we present a property on the influence a criticality constraint has on a receiver node's operability level.

Property 6.1

In FDNA LAWLR, if receiver node N_j has dependency relationships on feeder nodes $N_1, N_2, N_3, \ldots, N_h$, then the operability of N_j can never be more than $P_k + \beta_{kj}$ for all $k = 1,2,3,\ldots,h$ and $0 \leq \beta_{kj} \leq 100$.

Proof

In FDNA LAWLR, the operability of receiver node N_j is given by

$$P_j = \text{Min}(F(P_1, P_2, P_3, \ldots, P_h), G(P_1, P_2, P_3, \ldots, P_h))$$

where $$F(P_1, P_2, P_3, \ldots, P_h) = \frac{w_1P_1 + w_2P_2 + w_3P_3 + \cdots + w_hP_h}{w_1 + w_2 + w_3 + \cdots + w_h}$$

and $$G(P_1, P_2, P_3, \ldots, P_h) = \text{Min}(P_1 + \beta_{1j}, P_2 + \beta_{2j}, P_3 + \beta_{3j}, \ldots, P_h + \beta_{hj})$$

In the above, P_j has one of two outcomes. If $P_j = F(P_1, P_2, P_3, \ldots, P_h)$ then

$$F(P_1, P_2, P_3, \ldots, P_h) < \text{Min}(P_1 + \beta_{1j}, P_2 + \beta_{2j}, P_3 + \beta_{3j}, \ldots, P_h + \beta_{hj})$$

If $P_j = G(P_1, P_2, P_3, \ldots, P_h)$ then

$$\text{Min}(P_1 + \beta_{1j}, P_2 + \beta_{2j}, P_3 + \beta_{3j}, \ldots, P_h + \beta_{hj}) < F(P_1, P_2, P_3, \ldots, P_h)$$

Thus, in either outcome, it always follows that

$$P_j \leq \text{Min}(P_1 + \beta_{1j}, P_2 + \beta_{2j}, P_3 + \beta_{3j}, \ldots, P_h + \beta_{hj})$$

for all k where $k = 1,2,3,...,h$ and $0 \le \beta_{kj} \le 100$; hence, the operability level of N_j, denoted by P_j, can never be more than $P_k + \beta_{kj}$ for all $k = 1,2,3,...,h$ and $0 \le \beta_{kj} \le 100$. This property leads to a constraint condition in FDNA LAWLR called maximum criticality.

Definition 6.4: Maximum Criticality

Receiver node N_j has a maximum criticality of dependency on feeder nodes $N_1, N_2, N_3, ..., N_h$ when the operability level of N_j is equal to the operability level of its weakest performing feeder node. This occurs when $\beta_{kj} = 0$ for all $k = 1,2,3,\ldots,h$ and $0 \le \beta_{kj} \le 100$.

PROBLEM 6.1

Answer the following given the FDNA graph in Figure 6.9.

(A) Apply the operability function given by Equation 6.10 to compute the operability level of receiver node N_j if (1) the operability levels of all feeder nodes are equally important to N_j and (2) the operability levels of $N_1, N_2,$ and N_3 are each twice as important to N_j as is the operability level of N_4 and the operability level of N_4 is equally important to N_j as is the operability level of N_5.

(B) Apply the weakest link operability function given by Equation 6.11 to compute the operability level of receiver node N_j.

(C) Compare and contrast the results in (A) with the result in (B).

Solution

(A) (1) Since all feeder nodes are given to be equally important to receiver node N_j, Equation 6.10 can be written as

$$P_j = \frac{1}{5}P_1 + \frac{1}{5}P_2 + \frac{1}{5}P_3 + \frac{1}{5}P_4 + \frac{1}{5}P_5$$

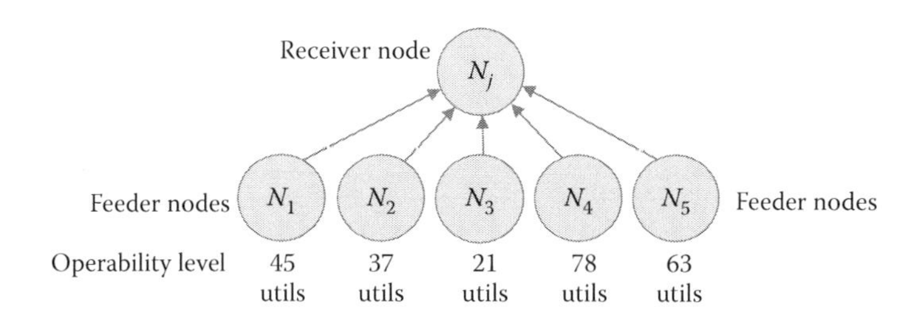

FIGURE 6.9
FDNA graph for Problem 6.1.

where $w_i = 1/5$ for each $i = 1,2,3,4,5$ and $w_1 + w_2 + w_3 + w_4 + w_5 = 1$ as required. Applying this to the FDNA graph in Figure 6.9, the operability level of receiver node N_j is

$$P_j = \frac{45+37+21+78+63}{5} = 48.8 \text{ utils}$$

(A) (2) Here, we are given the operability levels of $N_1, N_2, \text{and}\, N_3$ are each twice as important to N_j as is the operability level of N_4 and the operability level of N_4 is equally important to N_j as is the operability level of N_5. Thus, $w_1 = 2w_4, w_2 = 2w_4, w_3 = 2w_4, \text{and}\, w_4 = w_5$. Since $w_1 + w_2 + w_3 + w_4 + w_5 = 1$ we have $w_1 = w_2 = w_3 = 1/4$ and $w_4 = w_5 = 1/8$. Applying this to the FDNA graph in Figure 6.9, the operability level of receiver node N_j is

$$P_j = \frac{1}{4}P_1 + \frac{1}{4}P_2 + \frac{1}{4}P_3 + \frac{1}{8}P_4 + \frac{1}{8}P_5$$

$$P_j = \frac{1}{4}(45) + \frac{1}{4}(37) + \frac{1}{4}(21) + \frac{1}{8}(78) + \frac{1}{8}(63) = 43.38 \text{ utils}$$

(B) Applying the weakest link operability function we have

$$P_j = \text{Min}(P_1, P_2, P_3, P_4, P_5) = \text{Min}(45, 37, 21, 78, 63) = 21 \text{ utils}$$

(C) The result in (B) is significantly lower than both results in (A) because the operability function is the minimum of the operability levels of the five feeder nodes in Figure 6.9. This means receiver node N_j's operability level is equal to the operability level of its weakest feeder node, which is N_3. Unlike the result in (B), the results in (A) allow the operability level of N_j to be weighted by the importance of the operability levels of all its feeder nodes.

PROBLEM 6.2

Using FDNA LAWLR, consider the graph given in Figure 6.9. Suppose the operability levels of all feeder nodes are equally important to the operability level of receiver node N_j. If the operability level of N_j was determined to equal 25 utils, then (A) is a criticality constraint present? (B) If so, what is its magnitude and which feeder node is constraining the operability level of N_j?

Solution

(A) Under FDNA LAWLR, we can write

$$P_j = 25 = \text{Min}(F(P_1, P_2, P_3, P_4, P_5), G(P_1, P_2, P_3, P_4, P_5))$$

From Definition 6.3, and the operability levels of the feeder nodes in Figure 6.9, we have

$$F(P_1, P_2, P_3, P_4, P_5) = \frac{1}{5}P_1 + \frac{1}{5}P_2 + \frac{1}{5}P_3 + \frac{1}{5}P_4 + \frac{1}{5}P_5 = 48.8$$

From this, it follows that

$$P_j = 25 = \text{Min}(48.8, G(45, 37, 21, 78, 63)) \Rightarrow 25 = G(45, 37, 21, 78, 63)$$

where, in this case, $G(P_1, P_2, P_3, P_4, P_5) < F(P_1, P_2, P_3, P_4, P_5)$. Therefore, in accordance with Definition 6.3, a criticality constraint is present.

(B) To determine the magnitude of this constraint, from Definition 6.3, we have

$$G(45, 37, 21, 78, 63) = \text{Min}(45 + \beta_{1j}, 37 + \beta_{2j}, 21 + \beta_{3j}, 78 + \beta_{4j}, 63 + \beta_{5j})$$

Since $0 \le \beta_{kj} \le 100$ $(k = 1, 2, 3, 4, 5)$, it follows that $25 = G(45, 37, 21, 78, 63)$ only when $\beta_{3j} = 4$. Thus, the magnitude of the criticality constraint is 4 utils, and N_3 is the constraining feeder node to N_j.

This concludes an introduction to weakest link rules for the FDNA operability function. As mentioned earlier, weakest link rules offer many desirable properties in modeling and measuring inflows and outflows of value (utils) across a topology of feeder–receiver node dependency relationships. The next section presents a specific weakest link formulation called the α, β weakest link rule. It was the basis for FDNA's original development (Garvey and Pinto, 2009). The remainder of this chapter presents this rule, its applications, and the rule's many useful properties.

6.4 FDNA (α, β) Weakest Link Rule

This rule is a specific form of the FDNA limited average weakest link rule. The FDNA (α, β) weakest link rule (WLR) is characterized by three parameters. These are the baseline operability level (BOL), the strength of dependency (SOD), and the criticality of dependency (COD).

In FDNA (α, β) WLR, SOD and COD are two types of dependencies that affect a receiver node's operability level. The first type is the strength with which a receiver node's operability level relies, to some degree, on the operability levels of feeder nodes. The second is the criticality of each feeder node's contribution to a receiver node. The SOD and COD parameters capture different but important aspects of feeder–receiver node relationships.

Definition 6.5

The baseline operability level (BOL) of a receiver node is its operability level (utils) when the operability levels of its feeder nodes all equal 0. The BOL is measured with respect to the strength of dependency between a receiver node and its feeder nodes.

The baseline operability level of a receiver node provides a reference point that indicates its operability level prior to (or without) receiving its feeder node contribution, in terms of its dependency relationships on the other nodes and the criticality that may exist with those relationships.

Definition 6.6

Strength of dependency (SOD) is the operability level a receiver node relies on receiving from a feeder node for the receiver node to continually increase its baseline operability level and ensure the receiver node is wholly operable when its feeder node is fully operable.

The strength of dependency with which receiver node N_j relies on feeder node N_i is governed by the strength of dependency fraction α_{ij}, where $0 \le \alpha_{ij} \le 1$. This is shown in Figure 6.10. Notationally, *i* is the index of a feeder node to a receiver node of index *j*.

What would happen if a receiver node's operability level were constrained by its feeder nodes operability levels? What would happen if a receiver node degrades from its baseline operability level without receiving its feeder nodes contributions. In FDNA (α, β) WLR, a parameter called *criticality of dependency* (COD) can address these and related questions.

Definition 6.7

The criticality of dependency (COD) is the operability level β_{kj} (utils) such that the operability level of receiver node N_j with feeder nodes $N_1, N_2, N_3, \ldots, N_h$

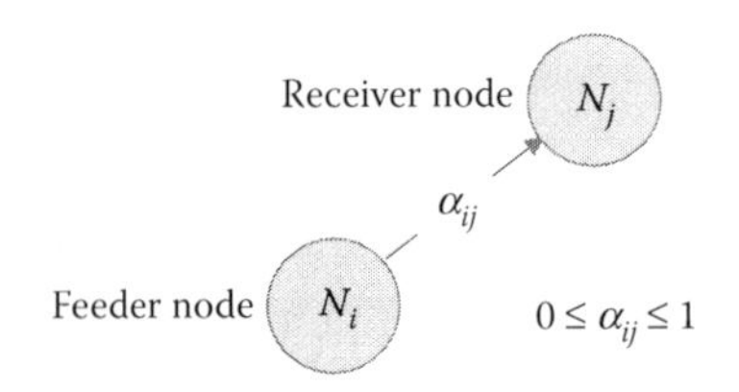

FIGURE 6.10
A single feeder–receiver node pair: strength of dependency fraction α_{ij}.

can never be more than $P_k + \beta_{kj}$ for all $k = 1,2,3,...,h$, where $0 \le \beta_{kj} \le 100$ and P_k is the operability level of feeder node N_k. The parameter β_{kj} is the criticality of dependency constraint between receiver node N_j and its feeder node N_k.

Definition 6.7 is the mathematical characterization of a criticality constraint on a receiver node's operability level, when the receiver node is dependent on one or more feeder nodes.

Definition 6.8

For a single feeder–receiver node pair, shown in Figure 6.11, the criticality of dependency constraint is the operability level β_{ij} (utils) that receiver node N_j degrades to from a reference point operability level $RPOLP_j$, when its feeder node is fully inoperable (wholly ineffective).

Definition 6.8 characterizes a criticality constraint in terms of degradation in a receiver node's operability level. Degradation is measured from an operability level that has meaning with respect to the receiver node's performance goals or requirements. This level is the receiver node's reference point operability level (RPOL). For example, a receiver node's RPOL might be set at its baseline operability level (BOL). Other meaningful operability levels for a receiver node's RPOL are possible. They are context specific to the situation.

Definitions 6.7 and 6.8 offer different ways to view criticality constraints in FDNA (α,β) WLR. Note that a dependency relationship between a receiver node and its feeder node need not always involve a criticality of dependency constraint.

In summary, COD enables the operability level of a receiver node to be constrained by the operability levels of its feeder nodes. This allows a receiver node's operability level to be limited by the performance of one feeder node, even when other feeder nodes to the receiver are fully operable. In FDNA (α,β) WLR, SOD and COD capture different but important effects of feeder–receiver node relationships on their operability levels. Where SOD captures the effects of relationships that improve baseline operability levels, COD captures whether such relationships could involve losses or constraints on these levels. Consider the following narrative.

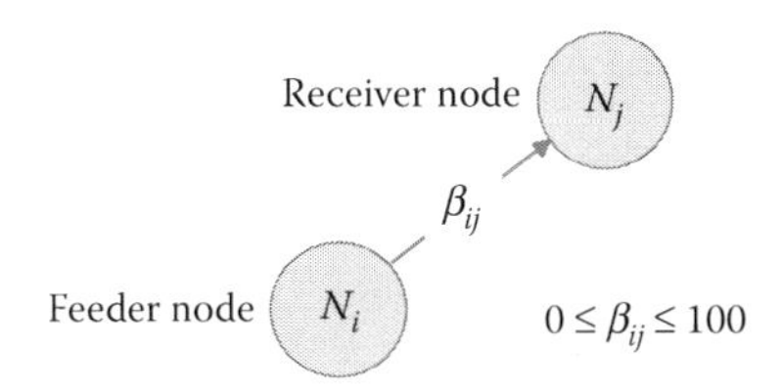

FIGURE 6.11
A single feeder–receiver node pair: criticality of dependency constraint β_{ij}.

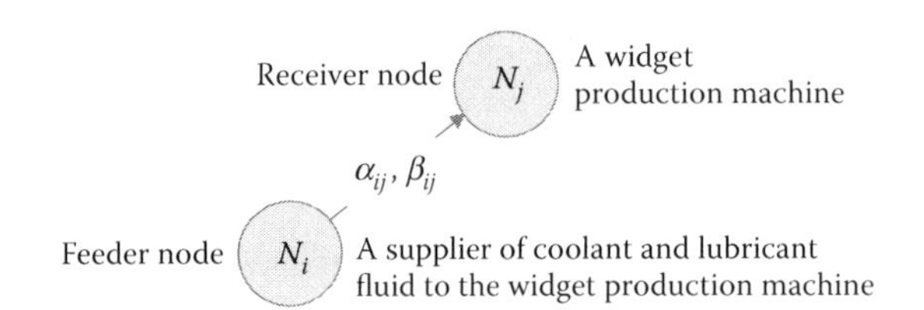

FIGURE 6.12
A widget production machine's dependency on a fluid supplier.

> In Figure 6.12, receiver node N_j is a widget production machine and feeder node N_i supplies coolant and lubricant fluids for the machine's engine. Suppose the machine is fully operable when it is producing 120 widgets per hour; that is, $P_j(x_j = 120) = 100$ utils. Without any supply of fluids from N_i suppose N_j can only produce 80 widgets per hour. Suppose a production rate of 80 widgets per hour is worth 55 utils; that is, $P_j(x_j = 80) = 55$ utils when $P_i = 0$. This implies the baseline operability level of N_j, denoted by $BOLP_j$, is 55 utils.
>
> Suppose the fluids from supplier node N_i are ideal for lowering the operating temperature of the engine and increasing the output of the widget production machine. Without the fluids, the engine's temperature will rise, its parts will wear, and the machine will decline from its baseline operability level of 55 utils and eventually become fully inoperable (wholly ineffective, a value of 0 utils).

This narrative illustrates Definition 6.8 in describing the effect of a criticality constraint on the widget production machine. In this case, the receiver node's RPOL was its BOL. The use of SOD and COD in FDNA (α, β) WLR allows a mix of gain and loss effects that complex feeder–receiver node interactions can have across a topology of multinodal relationships. Next, we present a way to formulate these effects in FDNA (α, β) WLR.

From Definition 6.3, recall that FDNA LAWLR is given by

$$P_j = \text{Min}(F(P_1, P_2, P_3, \ldots, P_h), G(P_1, P_2, P_3, \ldots, P_h))$$

where $F(P_1, P_2, P_3, \ldots, P_h)$ is the operability function of N_j based on the strength with which N_j depends on the performance of feeder nodes $N_1, N_2, N_3, \ldots, N_h$. The function $G(P_1, P_2, P_3, \ldots, P_h)$ is the operability function of N_j based on the criticality with which N_j depends on the performance of feeder nodes $N_1, N_2, N_3, \ldots, N_h$. Recall that F and G were given by Equations 6.15 and 6.16, respectively.

The FDNA (α, β) WLR is a form of the FDNA LAWLR with special expressions for F and G. The following presents these functions for a single feeder–receiver node pair. Then, this is extended to a receiver node with multiple feeder nodes. The formulations presented for F and G are one of many ways to express these functions. They can be tailored to characterize the nature of specific feeder–receiver dependency relationships. As mentioned earlier, this is a rich area for further study and research.

FDNA (α, β) WLR: Single Feeder–Receiver Node Pair

Suppose we have a single feeder–receiver node pair as shown in Figure 6.13. The general weakest link rule, given by Definition 6.1, becomes

$$P_j = \text{Min}(F(P_i), G(P_i)) \tag{6.19}$$

In Equation 6.19, $F(P_i)$ is the operability function for the strength of dependency of N_j on N_i and $G(P_i)$ is the operability function for the criticality of dependency of N_j on N_i. In FDNA(α, β)WLR, $F(P_i)$ is defined by a linear value function of the form

$$F(P_i) = SODP_j = \alpha_{ij} P_i + 100(1 - \alpha_{ij}) \tag{6.20}$$

Figure 6.14 shows the behavior of $F(P_i)$ for various α_{ij} across $0 \le P_i, P_j \le 100$. If $P_i = 0$ utils then $F(0) = 100(1 - \alpha_{ij})$ utils, as shown by the circled values along the vertical axis in Figure 6.14. From Definition 6.5, the baseline operability level of a receiver node is its operability level (utils) when the operability levels of its feeder nodes all equal 0. Thus, in Equation 6.20 the baseline operability level of receiver node N_j is the term $100(1 - \alpha_{ij})$; that is,

$$BOLP_j = 100(1 - \alpha_{ij}) \tag{6.21}$$

where $0 \le \alpha_{ij} \le 1$ and $0 \le P_i, P_j \le 100$.

From Definition 6.6, the strength of dependency is the operability level a receiver node relies on receiving from a feeder node for the receiver node to continually increase its baseline operability level, and ensure the receiver node is wholly operable when its feeder node is fully operable. Equation 6.20 meets these characteristics. The parameter α_{ij} is the strength of dependency fraction. The term $\alpha_{ij}P_i$ is the operability level receiver node N_j relies on receiving from feeder node N_i, for N_j to increase its baseline operability level of $100(1 - \alpha_{ij})$. In Equation 6.20, receiver node N_j is fully operable ($P_j = 100$) when its feeder node is fully operable ($P_i = 100$).

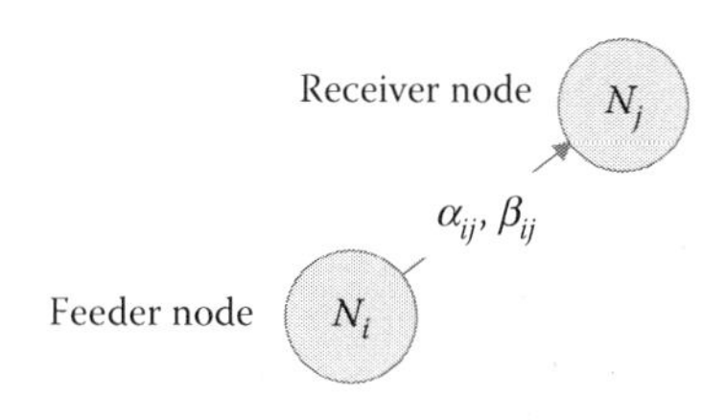

FIGURE 6.13
An FDNA graph of a single feeder–receiver node pair.

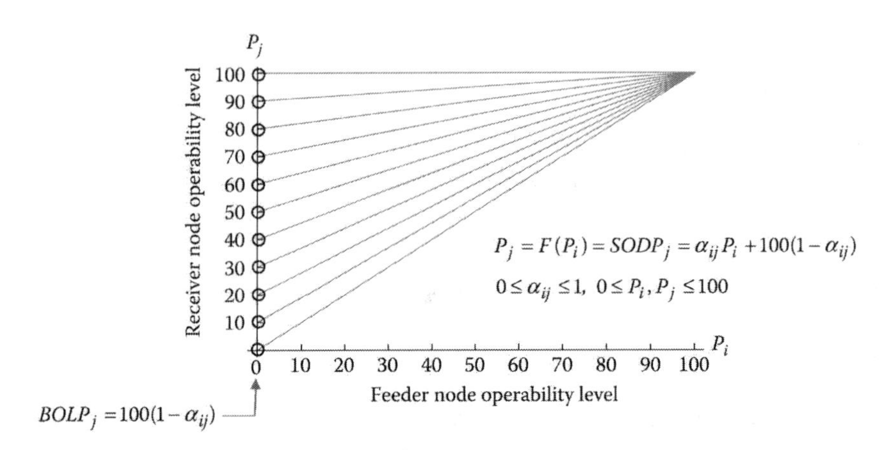

FIGURE 6.14
FDNA (α,β) WLR strength of dependency function $F(P_i)$.

Finally, in Equation 6.20, if the baseline operability level of receiver node N_j is 0 then $\alpha_{ij} = 1$. This is called *maximum strength of dependency*. If the baseline operability level of receiver node N_j is 100, then $\alpha_{ij} = 0$. The greater the value of α_{ij}, the greater the strength of dependency that receiver node N_j has on feeder node N_i and the less N_j's operability level is independent of N_i's level. The smaller the value of α_{ij}, the lesser the strength of dependency that receiver node N_j has on feeder node N_i and the more N_j's operability level is independent of N_i's level.

FDNA (α,β) WLR defines the operability function $G(P_i)$ in accordance with the limited average weakest link rule (Definition 6.3); hence, for the single feeder–receiver node pair in Figure 6.13 we have

$$G(P_i) = CODP_j = P_i + \beta_{ij} \tag{6.22}$$

where $0 \le \beta_{ij} \le 100, 0 \le P_i, P_j \le 100$, and $G(P_i)$ is the operability constraint function of receiver node N_j such that $P_j \le P_i + \beta_{ij}$. This allows the operability level of N_j to be limited, if appropriate, by the performance of N_i.

In summary, for a single feeder–receiver node pair, as shown in Figure 6.13, the FDNA (α,β) WLR is defined as follows:

$$P_j = \mathrm{Min}(F(P_i), G(P_i)) \tag{6.23}$$

$$F(P_i) = SODP_j = \alpha_{ij}P_i + 100(1-\alpha_{ij}) \tag{6.24}$$

$$BOLP_j = 100(1-\alpha_{ij}) \tag{6.25}$$

$$G(P_i) = CODP_j = P_i + \beta_{ij} \tag{6.26}$$

where $0 \le \alpha_{ij} \le 1, 0 \le \beta_{ij} \le 100$ and $0 \le P_i, P_j \le 100$.

PROBLEM 6.3

Given a single feeder–receiver node pair (Figure 6.13) if the baseline operability level of node N_j is 60 utils what operability level must N_i achieve for N_j to reach 80 utils? Assume N_j will not degrade from its baseline operability level if the operability level of N_i is 0. Use the FDNA (α, β) WLR.

Solution

The baseline operability level of N_j is given to be 60 utils. It follows that

$$BOLP_j = 100(1-\alpha_{ij}) = 60 \text{ utils}$$

This implies $\alpha_{ij} = 0.40$. Since N_j will not degrade from its baseline operability level if the operability level of N_i is 0, it follows that $BOLP_j \le \beta_{ij} \le 100$. From FDNA (α, β) WLR

$$P_j = \text{Min}(F(P_i), G(P_i))$$

$$P_j = \text{Min}(0.40P_i + 60, P_i + \beta_{ij})$$

where $BOLP_j \le \beta_{kj} \le 100$ and $0 \le P_i, P_j \le 100$. Receiver node N_j will reach an operability level of 80 utils when $P_j = 0.40P_i + 60 = 80$. This occurs when $P_i = 50$ utils.

Property 6.2

Given the single feeder–receiver node pair in Figure 6.13, if $BOLP_j \le \beta_{ij} \le 100$ then $P_j = SODP_j$.

Proof

Given the single feeder–receiver node pair in Figure 6.13, from Equation 6.23 receiver node N_j has operability function $P_j = \text{Min}(F(P_i), G(P_i))$, where $F(P_i)$ is the operability function for the strength of dependency of N_j on N_i and $G(P_i)$ is the operability function for the criticality of dependency of N_j on N_i. From Equations 6.24–6.26 we have $F(P_i) = SODP_j = \alpha_{ij}P_i + 100(1-\alpha_{ij})$ and $G(P_i) = CODP_j = P_i + \beta_{ij}$. If $BOLP_j \le \beta_{ij} \le 100$, then from Equation 6.25 we have $100(1-\alpha_{ij}) \le \beta_{ij} \le 100$. If $100(1-\alpha_{ij}) \le \beta_{ij} \le 100$ then $\alpha_{ij}P_i + 100(1-\alpha_{ij}) \le P_i + \beta_{ij}$ since $0 \le \alpha_{ij} \le 1$; thus, it follows that $P_j = \text{Min}(\alpha_{ij}P_i + 100(1-\alpha_{ij}), P_i + \beta_{ij}) = \alpha_{ij}P_i + 100(1-\alpha_{ij}) = SODP_j$.

FDNA (α, β) WLR: Receiver Node with Multiple Feeder Nodes

Suppose we have a receiver node with multiple feeder nodes as shown in Figure 6.15. In Figure 6.15, receiver node N_j has dependency relationships on

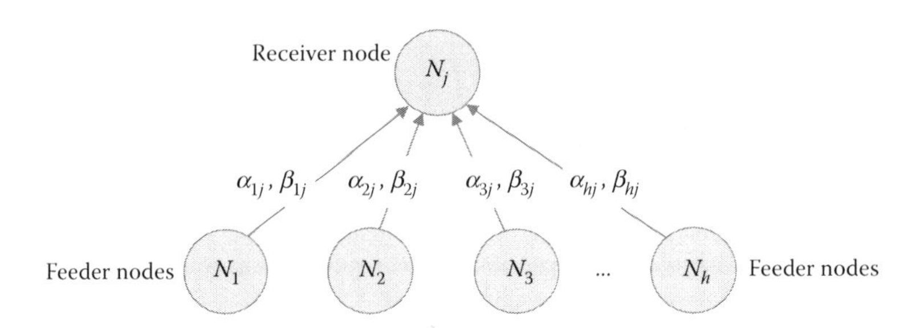

FIGURE 6.15
A receiver node with h feeder nodes.

h feeder nodes $N_1, N_2, N_3, ..., N_h$. From the general weakest link rule (given by Definition 6.1) we have

$$P_j = \text{Min}(F(P_1, P_2, P_3, \ldots, P_h), G(P_1, P_2, P_3, \ldots, P_h)) \tag{6.27}$$

In Equation 6.27, $F(P_1, P_2, P_3, ..., P_h)$ is the operability function of N_j based on the strength with which N_j depends on the performance of feeder nodes $N_1, N_2, N_3, ..., N_h$. The function $G(P_1, P_2, P_3, ..., P_h)$ is the operability function of N_j based on the criticality with which N_j depends on the performance of feeder nodes $N_1, N_2, N_3, ..., N_h$.

In FDNA (α, β) WLR, the function $F(P_1, P_2, P_3, ..., P_h)$ is defined by

$$\begin{aligned} F(P_1, P_2, P_3, \ldots, P_h) &= SODP_j \\ &= \text{Average}(SODP_{1j}, SODP_{2j}, SODP_{3j}, \ldots, SODP_{hj}) \end{aligned} \tag{6.28}$$

where $SODP_{kj} = \alpha_{kj} P_k + 100(1 - \alpha_{kj})$ for $k = 1, 2, 3, \ldots, h$ and $0 \le \alpha_{kj} \le 1$.

In FDNA (α, β) WLR, the baseline operability level of a receiver node with h feeder nodes is defined by

$$BOLP_j = \text{Average}(BOLP_{1j}, BOLP_{2j}, BOLP_{3j}, \ldots, BOLP_{hj}) \tag{6.29}$$

where $BOLP_{kj} = 100(1 - \alpha_{kj})$ for $k = 1, 2, 3, ..., h$ and $0 \le \alpha_{kj} \le 1$.

In FDNA (α, β) WLR, the function $G(P_1, P_2, P_3, ..., P_h)$ is defined in accordance with the limited average weakest link rule; hence, for a receiver node with h feeder nodes

$$\begin{aligned} G(P_1, P_2, P_3, \ldots, P_h) &= CODP_j \\ &= \text{Min}(CODP_{1j}, CODP_{2j}, CODP_{3j}, \ldots, CODP_{hj}) \end{aligned} \tag{6.30}$$

where $CODP_{kj} = P_k + \beta_{kj}$ for $k = 1, 2, 3, \ldots, h$ and $0 \le \beta_{kj} \le 100$.

In summary, for a receiver node with multiple feeder nodes (as shown in Figure 6.15) the FDNA (α, β) WLR is defined as follows:

$$P_j = \text{Min}(F(P_1, P_2, P_3, \ldots, P_h), G(P_1, P_2, P_3, \ldots, P_h)) \tag{6.31}$$

$$F(P_1, P_2, P_3, \ldots, P_h) = SODP_j \tag{6.32}$$

$$SODP_j = \text{Average}(SODP_{1j}, SODP_{2j}, SODP_{3j}, \ldots, SODP_{hj}) \tag{6.33}$$

$$SODP_{kj} = \alpha_{kj} P_k + 100(1 - \alpha_{kj}) \tag{6.34}$$

$$BOLP_j = \text{Average}(BOLP_{1j}, BOLP_{2j}, BOLP_{3j}, \ldots, BOLP_{hj}) \tag{6.35}$$

$$BOLP_{kj} = 100(1 - \alpha_{kj}) \tag{6.36}$$

$$G(P_1, P_2, P_3, \ldots, P_h) = CODP_j \tag{6.37}$$

$$CODP_j = \text{Min}(CODP_{1j}, CODP_{2j}, CODP_{3j}, \ldots, CODP_{hj}) \tag{6.38}$$

$$CODP_{kj} = P_k + \beta_{kj} \tag{6.39}$$

where $0 \leq \alpha_{kj} \leq 1$, $0 \leq \beta_{kj} \leq 100$, and $0 \leq P_k, P_i \leq 100$ for $k = 1, 2, 3, \ldots, h$.

PROBLEM 6.4

Given the FDNA graph in Figure 6.16, use FDNA(α, β) WLR to answer the following.

(A) If $BOLP_{kj} = 0$ and $\beta_{kj} = 0$ utils for $k = 1, 2, 3, 4, 5$ feeder nodes, then show that $P_j = 21$ utils.

(B) If $\alpha_{kj} = 1$ and $\beta_{kj} = 100$ utils for $k = 1, 2, 3, 4, 5$ feeder nodes, then show that $P_j = 48.8$ utils.

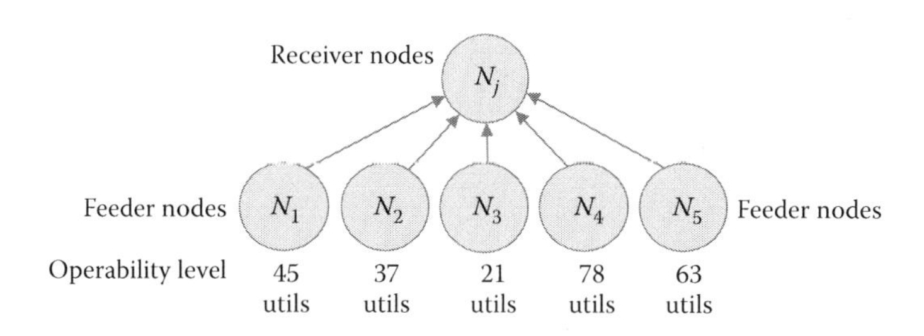

FIGURE 6.16
FDNA graph for Problem 6.4.

(C) Compute the operability level of receiver node N_j given the following FDNA parameters.

$$\alpha_{1j} = 0.35, \beta_{1j} = 30; \quad \alpha_{2j} = 0.15, \beta_{2j} = 45; \quad \alpha_{3j} = 0.90, \beta_{3j} = 50;$$
$$\alpha_{4j} = 0.45, \beta_{4j} = 100; \quad \alpha_{5j} = 0.65, \beta_{5j} = 100$$

(D) Compute the rate by which receiver node N_j changes in operability level with every unit change in the operability level of each feeder node, given the following FDNA parameters.

$$\alpha_{1j} = 0.35, \beta_{1j} = 30; \quad \alpha_{2j} = 0.15, \beta_{2j} = 100; \quad \alpha_{3j} = 0.90, \beta_{3j} = 100;$$
$$\alpha_{4j} = 0.45, \beta_{4j} = 100; \quad \alpha_{5j} = 0.65, \beta_{5j} = 100$$

Solution

(A) Figure 6.16 is an FDNA graph of feeder nodes N_1, N_2, N_3, N_4, N_5, and one receiver node N_j. From Equation 6.12, the operability level of receiver node N_j can be written as

$$P_j = \text{Min}(F(P_1, P_2, P_3, P_4, P_5), G(P_1, P_2, P_3, P_4, P_5))$$

In FDNA (α, β) WLR, we have from Equations 6.31 to 6.39

$$F(P_1, P_2, P_3, P_4, P_5) = SODP_j$$
$$= \text{Average}(SODP_{1j}, SODP_{2j}, SODP_{3j}, SODP_{4j}, SODP_{5j})$$

where $SODP_{kj} = \alpha_{kj} P_k + 100(1 - \alpha_{kj})$ with $BOLP_{kj} = 100(1 - \alpha_{kj})$ and

$$G(P_1, P_2, P_3, P_4, P_5) = CODP_j$$
$$= \text{Min}(CODP_{1j}, CODP_{2j}, CODP_{3j}, CODP_{4j}, CODP_{5j})$$

where $CODP_{kj} = P_k + \beta_{kj}$. Given $BOLP_{kj} = 0$ for $k = 1, 2, 3, 4, 5$ it follows that $\alpha_{kj} = 1$ for all k. Thus, $SODP_j = \text{Average}(P_1, P_2, P_3, P_4, P_5)$. Given $\beta_{kj} = 0$ utils for k = 1, 2, 3, 4, 5, it follows that $CODP_j = \text{Min}(P_1, P_2, P_3, P_4, P_5)$. Combining these results, we have

$$P_j = \text{Min}(\text{Average}(P_1, P_2, P_3, P_4, P_5), \text{Min}(P_1, P_2, P_3, P_4, P_5))$$

From Figure 6.16 it follows that $\text{Min}(P_1, P_2, P_3, P_4, P_5) < \text{Average}(P_1, P_2, P_3, P_4, P_5)$; thus,

$$P_j = \text{Min}(P_1, P_2, P_3, P_4, P_5) = \text{Min}(45, 37, 21, 78, 63) = 21 \text{ utils}$$

Therefore, the operability level of receiver node N_j is equal to the operability level of its weakest performing feeder node, which is N_3 in this case.

This result is consistent with the characterization of maximum criticality given in Definition 6.4. In general, it can be shown that when receiver node N_j is dependent on $k = 1,2,3,...,h$ feeder nodes and $\beta_{kj} = 0$ utils for all k, then $P_j = \text{Min}(P_1, P_2, P_3, \ldots, P_h)$. This is true under the FDNA (α, β) WLR or the FDNA LAWLR.

(B) Given $\alpha_{kj} = 1$ and $\beta_{kj} = 100$ utils for $k = 1,2,3,4,5$ feeder nodes, it follows from part (A) that

$$P_j = \text{Min}(\text{Average}(P_1, P_2, P_3, P_4, P_5),$$
$$\text{Min}(P_1 + 100, P_2 + 100, P_3 + 100, P_4 + 100, P_5 + 100))$$

From Figure 6.16, it follows that

$$\text{Average}(P_1, P_2, P_3, P_4, P_5)$$
$$< \text{Min}(P_1 + 100, P_2 + 100, P_3 + 100, P_4 + 100, P_5 + 100)$$

thus, $P_j = \text{Average}(P_1, P_2, P_3, P_4, P_5)$
$= \text{Average}(45, 37, 21, 78, 63) = 48.8$ utils

(C) Given the FDNA parameters

$$\alpha_{1j} = 0.35, \beta_{1j} = 30; \quad \alpha_{2j} = 0.15, \beta_{2j} = 45; \quad \alpha_{3j} = 0.90, \beta_{3j} = 50;$$
$$\alpha_{4j} = 0.45, \beta_{4j} = 100; \quad \alpha_{5j} = 0.65, \beta_{5j} = 100$$

From Equations 6.33 and 6.34, we can write

$$SODP_j = \frac{\alpha_{1j}}{5}P_1 + \frac{\alpha_{2j}}{5}P_2 + \frac{\alpha_{3j}}{5}P_3 + \frac{\alpha_{4j}}{5}P_4 + \frac{\alpha_{5j}}{5}P_5$$
$$+ 100\left(1 - \frac{\alpha_{1j} + \alpha_{2j} + \alpha_{3j} + \alpha_{4j} + \alpha_{5j}}{5}\right)$$

From Equations 6.38 and 6.39 we can write

$$CODP_j = \text{Min}(P_1 + \beta_{1j}, P_2 + \beta_{2j}, P_3 + \beta_{3j}, P_4 + \beta_{4j}, P_5 + \beta_{5j})$$

In Figure 6.16, $P_1 = 45$ utils, $P_2 = 37$ utils, $P_3 = 21$ utils, $P_4 = 78$ utils, and $P_5 = 63$ utils. Substituting these values, and the given FDNA parameters, in $SODP_j$ and $CODP_j$ we have $SODP_j = 73.25$ utils and $CODP_j = 71$ utils. Thus, the operability level of receiver node N_j is

$$P_j = \text{Min}(SODP_j, CODP_j) = \text{Min}(73.25, 71) = 71 \text{ utils}$$

(D) Given the FDNA parameters

$$\alpha_{1j} = 0.35, \beta_{1j} = 30;\ \alpha_{2j} = 0.15, \beta_{2j} = 100;\ \alpha_{3j} = 0.90, \beta_{3j} = 100;$$
$$\alpha_{4j} = 0.45, \beta_{4j} = 100;\ \alpha_{5j} = 0.65, \beta_{5j} = 100$$

it follows that $P_j = SODP_j$ (why?). From the solution in part (C) we have

$$P_j = SODP_j = \frac{\alpha_{1j}}{5} P_1 + \frac{\alpha_{2j}}{5} P_2 + \frac{\alpha_{3j}}{5} P_3 + \frac{\alpha_{4j}}{5} P_4 + \frac{\alpha_{5j}}{5} P_5 + 100\left(1 - \frac{\alpha_{1j} + \alpha_{2j} + \alpha_{3j} + \alpha_{4j} + \alpha_{5j}}{5}\right)$$

The rate by which receiver node N_j changes in operability level, with every unit change in the operability level of each feeder node, can be determined by computing the partial derivative

$$\frac{\partial P_j}{\partial P_k} \quad \text{for } k = 1, 2, 3, 4, 5$$

From this, it follows that

$$\frac{\partial P_j}{\partial P_1} = \frac{\alpha_{1j}}{5} = 0.07, \frac{\partial P_j}{\partial P_2} = \frac{\alpha_{2j}}{5} = 0.03,$$
$$\frac{\partial P_j}{\partial P_3} = \frac{\alpha_{3j}}{5} = 0.18, \frac{\partial P_j}{\partial P_4} = \frac{\alpha_{4j}}{5} = 0.09, \frac{\partial P_j}{\partial P_5} = \frac{\alpha_{5j}}{5} = 0.13$$

Thus, the rate by which the operability level of N_j changes is most influenced by feeder nodes N_3 and N_5.

Property 6.4

In FDNA(α,β) WLR, if receiver node N_j has dependency relationships on feeder nodes $N_1, N_2, N_3, \ldots, N_h$ then the operability of N_j can never be more than $CODP_j$.

Proof

This proof follows directly from Property 6.1.

Property 6.5

In FDNA (α,β) WLR, if receiver node N_j has dependency relationships on feeder nodes $N_1, N_2, N_3, \ldots, N_h$, then $P_j = SODP_j$, if $\Sigma_{k=1}^{h}(\alpha_{kj}/h)P_k$ is less than or equal to $CODP_j - BOLP_j$.

Proof

From Equations 6.31 to 6.39 we have $P_j = \text{Min}(SODP_j, CODP_j)$, where

$$SODP_j = \text{Average}(SODP_{1j}, SODP_{2j}, SODP_{3j}, \ldots, SODP_{hj})$$

$$CODP_j = \text{Min}(CODP_{1j}, CODP_{2j}, CODP_{3j}, \ldots, CODP_{hj})$$

$$SODP_{kj} = \alpha_{kj}P_k + 100(1-\alpha_{kj})$$

$$CODP_{kj} = P_k + \beta_{kj}$$

and $0 \le \alpha_{kj} \le 1, 0 \le \beta_{kj} \le 100$, and $0 \le P_k, P_j \le 100$ for $k = 1,2,3,\ldots,h$. Thus, we can write the following:

$$SODP_j = \sum_{k=1}^{h} \frac{\alpha_{kj}}{h} P_k + BOLP_j$$

where $BOLP_j = \text{Average}\,(BOLP_{1j}, BOLP_{2j}, BOLP_{3j}, \ldots, BOLP_{hj})$ and $BOLP_{kj} = 100(1-\alpha_{kj})$. From this, it follows that

$$P_j = \text{Min}(SODP_j, CODP_j) = \text{Min}\left(\sum_{k=1}^{h} \frac{\alpha_{kj}}{h} P_k + BOLP_j, CODP_j\right)$$

Now, $P_j = SODP_j$ if

$$\sum_{k=1}^{h} \frac{\alpha_{kj}}{h} P_k + BOLP_j < CODP_j$$

$$\Rightarrow \sum_{k=1}^{h} \frac{\alpha_{kj}}{h} P_k < CODP_j - BOLP_j$$

There is a relationship between Property 6.5 and Property 6.2. In Property 6.5, if $h = 1$ we have a single feeder–receiver node pair (as in Figure 6.13). Then, if we restrict β_{1j} to the interval $BOLP_j \le \beta_{1j} \le 100$ it follows that $P_j = SODP_j$. Hence, Property 6.2 is a special case of Property 6.5.

FDNA (α, β) WLR: Forming the FDNA Dependency Function (FDF)

This discussion illustrates how to form an FDNA dependency function (FDF) for various types of graphs. An approach known as function composition is used. Function composition is when one function is expressed as a composition (or nesting) of two or more other functions. Function composition applies in FDNA when the operability function of one node is a composition of the operability functions of two or more other nodes.

As mentioned earlier, a dependency relationship exists between nodes when the performance level achieved by one node relies, to some degree, on the performance levels achieved by other nodes. Operability is a state where a node is functioning at some level of performance. In FDNA, each node has a measure of its performance (MOP) and a measure of value or worth associated with the performance level achieved. The measure of value or worth is a node's operability level or its measure of effectiveness (MOE). FDNA captures the transmission of performance and value relationships between nodes by way of value functions.* Figure 6.17 illustrates this concept for a single feeder–receiver node pair.

Figure 6.17 shows how performance and value transmits between nodes for a single feeder–receiver node pair. A nesting of value functions captures the transmission of value between nodes for more complex dependency relationships. The following examples illustrate forming the FDF in these cases.

PROBLEM 6.5

Formulate the FDFs for the graph in Figure 6.18. Use FDNA (α, β) WLR.

Solution

Using FDNA (α, β) WLR, from Equation 6.31, we can write

$$P_p = \text{Min}(SODP_p, CODP_p) = \text{Min}(\alpha_{jp} P_j + 100(1 - \alpha_{jp}), P_j + \beta_{jp})$$

where

α_{jp} is the strength of dependency fraction between N_j and N_p

β_{jp} is the criticality of dependency constraint between N_j and N_p

From Equation 6.31, $P_j = \text{Min}(SODP_j, CODP_j)$, where

* A value function is a real-valued mathematical function that models a node's measure of value, worth, or utility over the levels of performance it achieves. Chapter 3 presents a discussion on value functions.

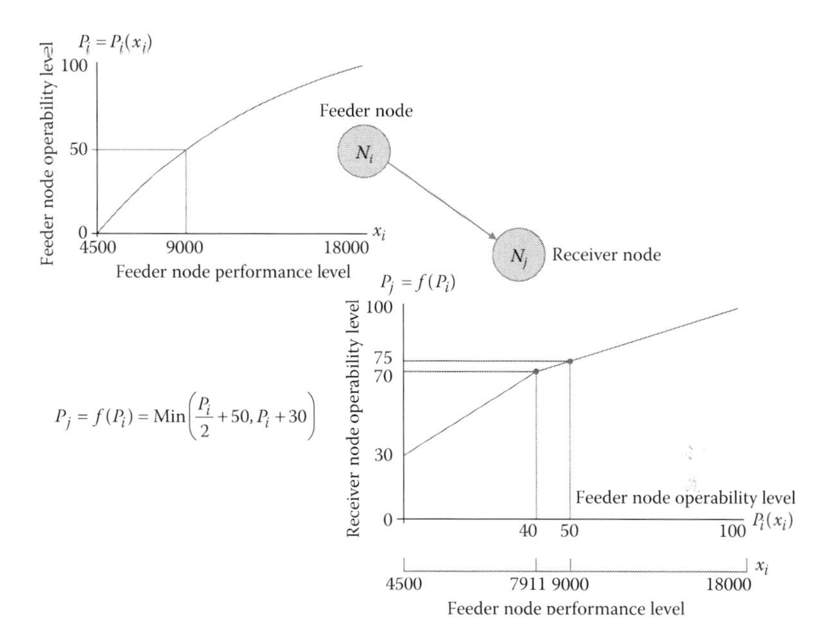

FIGURE 6.17
Transmission of performance and value between nodes.

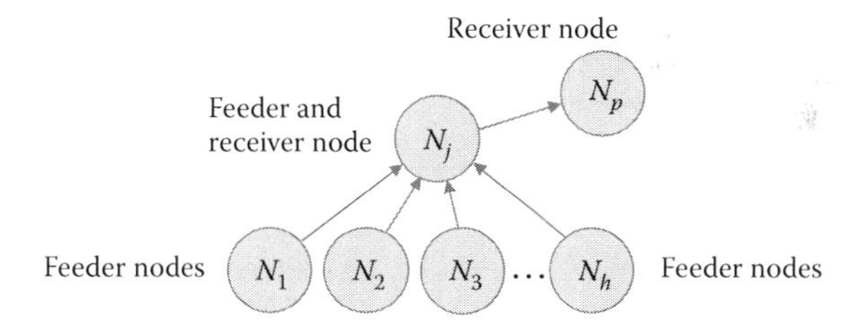

FIGURE 6.18
FDNA graph for Problem 6.5.

$$SODP_j = \frac{\alpha_{1j}}{h}P_1 + \frac{\alpha_{2j}}{h}P_2 + \frac{\alpha_{3j}}{h}P_3 + \cdots + \frac{\alpha_{hj}}{h}P_h + 100\left(1 - \frac{\alpha_{1j} + \alpha_{2j} + \alpha_{3j} + \cdots + \alpha_{hj}}{h}\right)$$

$$CODP_j = \text{Min}(P_1 + \beta_{1j}, P_2 + \beta_{2j}, P_3 + \beta_{3j}, \ldots, P_h + \beta_{hj})$$

where

α_{kj} is the strength of dependency fraction between N_k and N_j

β_{kj} is the criticality of dependency constraint between N_k and N_j

for $k = 1, 2, 3, \ldots, h$. Thus, the FDF for N_p fashioned by a composition of operability functions is

$$P_p = \text{Min}(SODP_p, CODP_p) = \text{Min}(\alpha_{jp}P_j + 100(1-\alpha_{jp}), P_j + \beta_{jp})$$

$$\Rightarrow P_p = \text{Min}(\alpha_{jp}\text{Min}(SODP_j, CODP_j) + 100(1-\alpha_{jp}), \text{Min}(SODP_j, CODP_j) + \beta_{jp})$$

PROBLEM 6.6

Compute the operability level of N_p given the information in Figure 6.19. Use FDNA (α, β) WLR.

Solution

Figure 6.19 is a specific case of the FDNA graph in Figure 6.18 with $h = 5$ feeder nodes. In Figure 6.19, N_j is a receiver node with the same five feeder nodes in Figure 6.16. With this, from Problem 6.4 (C) the operability level of N_j was computed to be

$$P_j = \text{Min}(SODP_j, CODP_j) = \text{Min}(73.25, 71) = 71 \text{ utils}$$

In Figure 6.19, N_j is also a feeder to node N_p. Thus, the operability level of N_p is a function of the operability level of N_j. Using the FDNA (α, β) WLR, the FDF for computing the operability level for N_p is

$$P_p = \text{Min}(SODP_p, CODP_p) = \text{Min}(\alpha_{jp}P_j + 100(1-\alpha_{jp}), P_j + \beta_{jp})$$

where
α_{jp} is the strength of dependency fraction between N_j and N_p
β_{jp} is the criticality of dependency constraint between N_j and N_p
Given $\alpha_{jp} = 0.78$ and $\beta_{jp} = 55$ utils we then have

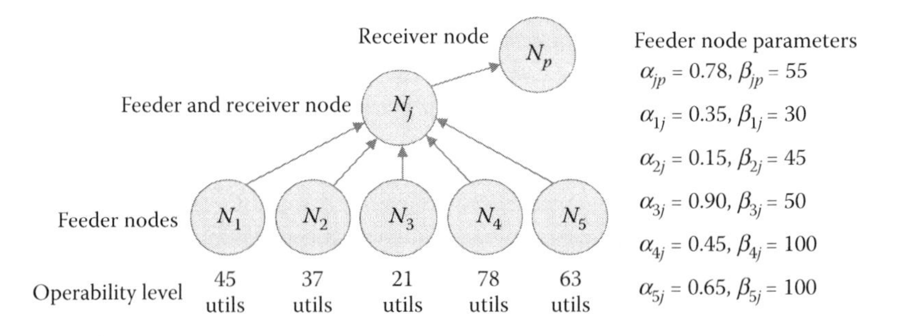

FIGURE 6.19
FDNA graph for Problem 6.6.

$$P_p = \text{Min}(SODP_p, CODP_p) = \text{Min}(\alpha_{jp}P_j + 100(1-\alpha_{jp}), P_j + \beta_{jp})$$

$$= \text{Min}(0.78\,\text{Min}(SODP_j, CODP_j) + 22, \text{Min}(SODP_j, CODP_j) + 55)$$

$$= \text{Min}(0.78\,\text{Min}(73.25, 71) + 22, \text{Min}(73.25, 71) + 55) = 77.38 \text{ utils}$$

PROBLEM 6.7

Formulate the FDFs for the graph in Figure 6.20. Use FDNA (α, β) WLR.

Solution

Using FDNA (α, β) WLR, from Equation 6.31, we can write

$$P_j = \text{Min}\left(\frac{\alpha_{1j}}{2}P_1 + \frac{\alpha_{2j}}{2}P_2 + 100\left(1 - \frac{\alpha_{1j} + \alpha_{2j}}{2}\right), P_1 + \beta_{1j}, P_2 + \beta_{2j}\right)$$

$$P_1 = \text{Min}(\alpha_{21}P_2 + 100(1-\alpha_{21}), P_2 + \beta_{21})$$

where

α_{1j} is the strength of dependency fraction between N_1 and N_j
α_{2j} is the strength of dependency fraction between N_2 and N_j
α_{21} is the strength of dependency fraction between N_2 and N_1
β_{1j} is the criticality of dependency constraint between N_1 and N_j
β_{2j} is the criticality of dependency constraint between N_2 and N_j
β_{21} is the criticality of dependency constraint between N_2 and N_1

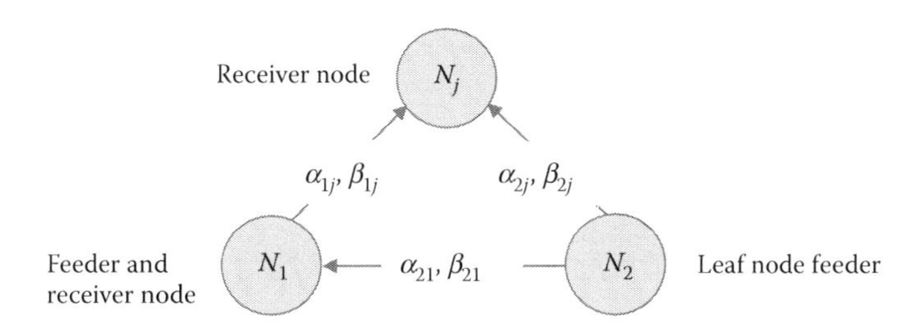

FIGURE 6.20
FDNA graph for Problem 6.7.

PROBLEM 6.8

Formulate the FDFs for the graph in Figure 6.21. Use FDNA (α, β) WLR.

Solution

Using FDNA (α, β) WLR, from Equation 6.31, we can write

$$P_j = \text{Min}\left(\frac{\alpha_{1j}}{2}P_1 + \frac{\alpha_{2j}}{2}P_2 + 100\left(1 - \frac{\alpha_{1j} + \alpha_{2j}}{2}\right), P_1 + \beta_{1j}, P_2 + \beta_{2j}\right)$$

$$P_1 = \text{Min}(\alpha_{31}P_3 + 100(1 - \alpha_{31}), P_3 + \beta_{31})$$

$$P_2 = \text{Min}\left(\frac{\alpha_{12}}{3}P_1 + \frac{\alpha_{32}}{3}P_3 + \frac{\alpha_{42}}{3}P_4 + 100\left(1 - \frac{\alpha_{12} + \alpha_{32} + \alpha_{42}}{3}\right), B\right)$$

$$B = \text{Min}(P_1 + \beta_{12}, P_3 + \beta_{32}, P_4 + \beta_{42})$$

PROBLEM 6.9

Formulate the FDFs for the graph in Figure 6.22. Use FDNA (α, β) WLR. For convenience, let η indicate equivalence to the following FDNA parameters:

$$\eta_1 \equiv \alpha_{1Cap_1}, \beta_{1Cap_1} \quad \eta_2 \equiv \alpha_{3Cap_1}, \beta_{3Cap_1} \quad \eta_3 \equiv \alpha_{3Cap_2}, \beta_{3Cap_2} \quad \eta_4 \equiv \alpha_{6Cap_3}, \beta_{6Cap_3}$$
$$\eta_5 \equiv \alpha_{4Cap_3}, \beta_{4Cap_3} \quad \eta_6 \equiv \alpha_{53}, \beta_{53} \quad \eta_7 \equiv \alpha_{31}, \beta_{31} \quad \eta_8 \equiv \alpha_{21}, \beta_{21}$$

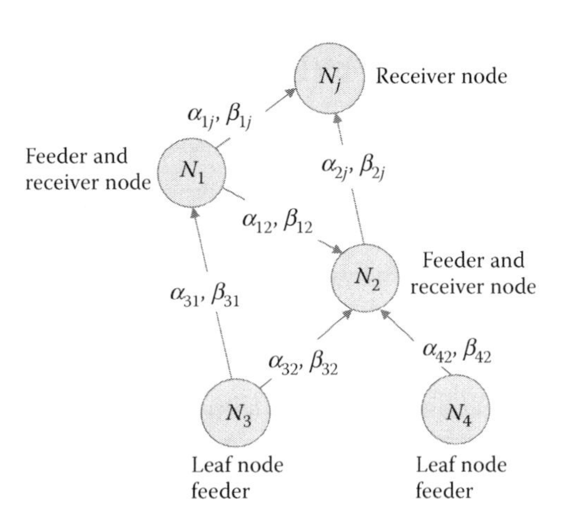

FIGURE 6.21
FDNA graph for Problem 6.8.

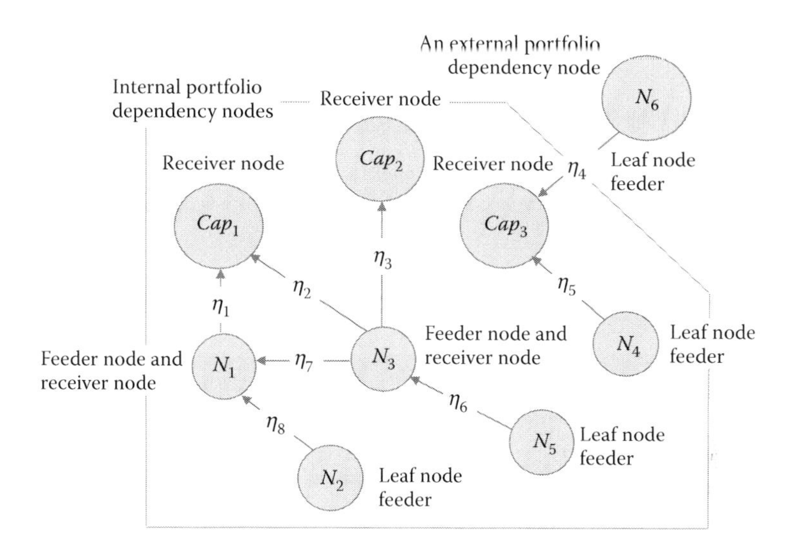

FIGURE 6.22
FDNA graph for Problem 6.9.

Solution

Using FDNA (α, β) WLR, from Equation 6.31 we can write

$$P_1 = \text{Min}\left(\frac{\alpha_{21}P_2}{2} + \frac{\alpha_{31}P_3}{2} + 100\left(1 - \left(\frac{\alpha_{21} + \alpha_{31}}{2}\right)\right), P_2 + \beta_{21}, P_3 + \beta_{31}\right)$$

$$P_3 = \text{Min}(\alpha_{53}P_5 + 100(1 - \alpha_{53}), P_5 + \beta_{53})$$

$$P_{Cap_1} = \text{Min}\left(\frac{\alpha_{1Cap_1}P_1}{2} + \frac{\alpha_{3Cap_1}P_3}{2} + 100\left(1 - \left(\frac{\alpha_{1Cap_1} + \alpha_{3Cap_1}}{2}\right)\right), P_1 + \beta_{1Cap_1}, P_3 + \beta_{3Cap_1}\right)$$

$$P_{Cap_2} = \text{Min}(\alpha_{3Cap_2}P_3 + 100(1 - \alpha_{3Cap_2}), P_3 + \beta_{3Cap_2})$$

$$P_{Cap_3} = \text{Min}\left(\frac{\alpha_{4Cap_3}P_4}{2} + \frac{\alpha_{6Cap_3}P_6}{2} + 100\left(1 - \left(\frac{\alpha_{4Cap_3} + \alpha_{6Cap_3}}{2}\right)\right), P_4 + \beta_{4Cap_3}, P_6 + \beta_{6Cap_3}\right)$$

Problem 6.9 illustrates a simple capability portfolio. In practice, portfolios have a highly complex and intricate nodal topology. The power of the FDNA approach, as shown in these examples, is its ability to operate across complex topologies with only two defining parameters: strength and criticality of dependency between dependent nodes.

Property 6.6

In FDNA (α,β) WLR, if receiver node N_j has dependency relationships on h feeder nodes with $\beta_{kj} = 0$, for all $k = 1,2,3,\ldots,h$, then $P_j = \text{Min}(P_1, P_2, P_3, \ldots, P_h)$.

Proof

From Equations 6.31–6.39 we have

$$\begin{aligned} P_j &= \text{Min}(SODP_j, CODP_j) \\ &= \text{Min}(SODP_j, \text{Min}(P_1 + \beta_{1j}, P_2 + \beta_{2j}, P_3 + \beta_{3j}, \ldots, P_h + \beta_{hj}) \end{aligned}$$

If $\beta_{kj} = 0$, for all $k = 1,2,3,\ldots,h$, then

$$\begin{aligned} P_j &= \text{Min}(SODP_j, CODP_j) \\ &= \text{Min}(SODP_j, \text{Min}(P_1 + 0, P_2 + 0, P_3 + 0, \ldots, P_h + 0)) \\ &= \text{Min}(SODP_j, \text{Min}(P_1, P_2, P_3, \ldots, P_h)) \\ &= \text{Min}(P_1, P_2, P_3, \ldots, P_h) \end{aligned}$$

since $0 \le \alpha_{kj} \le 1$ and $0 \le P_k \le 100$ for all $k = 1,2,3,\ldots,h$. Thus, in FDNA (α,β) WLR if $\beta_{kj} = 0$, for all $k = 1,2,3,\ldots,h$, then the operability level of receiver node N_j is equal to the operability level of its weakest performing feeder node. This property is consistent with the meaning of maximum criticality of dependency given in Definition 6.4. Also, if $BOLP_{kj} = 0$ and $\beta_{kj} = 0$, for all $k = 1,2,3,\ldots,h$, then $\alpha_{kj} = 1$ for all $k = 1,2,3,\ldots,h$ and

$$\begin{aligned} P_j &= \text{Min}(\text{Average}(P_1, P_2, P_3, \ldots, P_h), \text{Min}(P_1, P_2, P_3, \ldots, P_h)) \\ &= \text{Min}(P_1, P_2, P_3, \ldots, P_h) \end{aligned}$$

FDNA (α,β) WLR: Ways to Determine α, β

There are various ways to assess the strength and criticality of dependency parameters between nodes in the FDNA (α,β) WLR. This discussion illustrates two protocols. One is for assessing the strength of dependency fraction. The other is for assessing the criticality constraint.

From Definition 6.6, strength of dependency is the operability level a receiver node relies on receiving from a feeder node for the receiver node to continually increase its baseline operability level and ensure the receiver node is wholly operable when its feeder node is fully operable. The strength of dependency with which receiver node N_j relies on feeder node N_i is governed by the strength of dependency fraction α_{ij} where $0 \le \alpha_{ij} \le 1$ as shown in Figure 6.23.

SOD Question Protocol

A receiver node's baseline operability level can be used to determine α_{ij}. One way this can be done is to ask the following: *What is receiver node N_j's baseline* *operability level (utils) prior to (or without) receiving its feeder node N_i's contribution?* If the answer is 0 utils, then $\alpha_{ij} = 1$; if the answer is 50 utils, then $\alpha_{ij} = 0.50$; if the answer is 70 utils, then $\alpha_{ij} = 0.30$. Thus, for a single feeder-receiver node pair as in Figure 6.23, α_{ij} can be solved from the expression

$$BOLP_j = 100(1 - \alpha_{ij}) = x \tag{6.40}$$

where x is the receiver node's baseline operability level prior to (or without) receiving its feeder node's contribution. The greater the value of α_{ij}, the greater the strength of dependency that receiver node N_j has on feeder node N_i and the less N_j's operability level is independent of N_i's level. The smaller the value of α_{ij}, the lesser the strength of dependency that receiver node N_j has on feeder node N_i and the more N_j's operability level is independent of N_i's level. The approach using Equation 6.40 determines α_{ij} from pairwise assessments of feeder–receiver node dependencies.

Suppose we have (1) a receiver node with multiple feeder nodes, as shown in Figure 6.24, and (2) an assessment the receiver node's baseline operability level. Given this, how might the strength of dependency fraction be determined for each feeder–receiver node pair? The following illustrates an approach.

From Equations 6.31 through 6.39, the FDNA dependency function for the graph in Figure 6.24 is

$$P_j = \text{Min}(\text{Average}(SODP_{1j}, SODP_{2j}, SODP_{3j}), CODP_{1j}, CODP_{2j}, CODP_{3j})$$

$$P_j = \text{Min}\left(\begin{array}{l} \dfrac{\alpha_{1j}}{3}P_1 + \dfrac{\alpha_{2j}}{3}P_2 + \dfrac{\alpha_{3j}}{3}P_3 \\ \quad + 100\left(1 - \dfrac{\alpha_{1j} + \alpha_{2j} + \alpha_{3j}}{3}\right), P_1 + \beta_{1j}, P_2 + \beta_{2j}, P_3 + \beta_{3j} \end{array} \right)$$

where

α_{1j} is the strength of dependency fraction between N_1 and N_j

α_{2j} is the strength of dependency fraction between N_2 and N_j

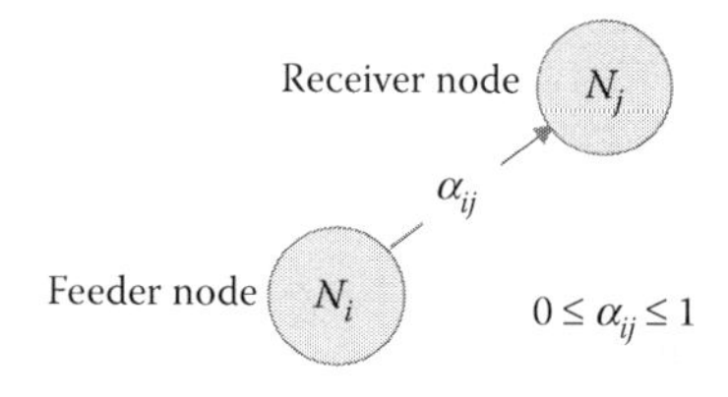

FIGURE 6.23
A single feeder–receiver node pair: strength of dependency fraction α_{ij}.

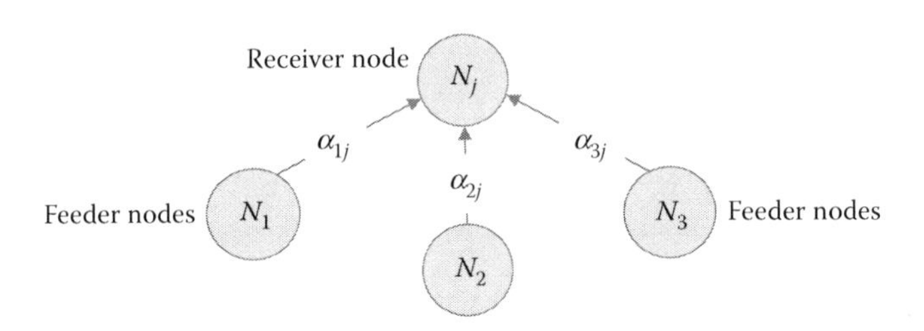

FIGURE 6.24
An FDNA graph with multiple feeder nodes.

α_{3j} is the strength of dependency fraction between N_3 and N_j

β_{1j} is the criticality of dependency constraint between N_1 and N_j

β_{2j} is the criticality of dependency constraint between N_2 and N_j

β_{3j} is the criticality of dependency constraint between N_3 and N_j

In this case, the baseline operability level of receiver node N_j is

$$BOLP_j = 100\left(1 - \frac{\alpha_{1j} + \alpha_{2j} + \alpha_{3j}}{3}\right)$$

Suppose the baseline operability level of receiver node N_j was assessed at 60 utils; then

$$BOLP_j = 100\left(1 - \frac{\alpha_{1j} + \alpha_{2j} + \alpha_{3j}}{3}\right) = 60 \Rightarrow \alpha_{1j} + \alpha_{2j} + \alpha_{3j} = \frac{6}{5}$$

A unique set of values for α_{1j}, α_{2j}, and α_{3j} does not exist. However, suppose the dependency relationships between N_j and N_1 and N_j and N_2 is such that $\alpha_{1j} = 2\alpha_{3j}$ and $\alpha_{2j} = 2\alpha_{3j}$. From this,

$$\alpha_{1j} + \alpha_{2j} + \alpha_{3j} = \frac{6}{5} \Rightarrow 2\alpha_{3j} + 2\alpha_{3j} + \alpha_{3j} = \frac{6}{5}$$

for which $\alpha_{3j} = 6/25$, $\alpha_{2j} = 12/25$, and $\alpha_{1j} = 12/25$; hence,

$$P_j = \text{Min}\left(\frac{12}{75}P_1 + \frac{12}{75}P_2 + \frac{6}{75}P_3 + 60, P_1 + \beta_{1j}, P_2 + \beta_{2j}, P_3 + \beta_{3j}\right)$$

Another approach to determine the strength of dependency fraction is creating a constructed scale. From Chapter 3, constructed scales are often defined when natural scales are not possible or are not practical to use. They are also used when natural scales exist but additional context is desired and hence are used to supplement natural scales with additional information. Table 6.1 illustrates a constructed scale for assessing the strength of dependency that a receiver node N_j has on its feeder node N_i. Next, a discussion is presented on ways to access the criticality constraint in FDNA (α,β) WLR.

From Definition 6.7, criticality of dependency is the operability level β_{kj} such that the operability level of receiver node N_j with feeder nodes $N_1, N_2, N_3, \ldots, N_h$ can never be more than $P_k + \beta_{kj}$ for all $k = 1, 2, 3, \ldots, h$, where $0 \le \beta_{kj} \le 100$ and P_k is the operability level of feeder node N_k.

TABLE 6.1

A Strength of Dependency (SOD) Constructed Scale

Ordinal Scale	Receiver Node Baseline Operability Level (BOL)	Interval Scale BOL Range
5 **RED**	A receiver node with this rating has a **very low** operational capability prior to receiving its feeder node's contribution. Receiver nodes with this rating are those that operate in the range 0% to < 20% of their full operational capability (100 utils), *prior to receiving their feeder node contributions.* Receiver nodes with this rating are those with a **very high dependency** on their feeder nodes.	0 to < 20 utils
4 **ORANGE**	A receiver node with this rating has a **low** operational capability prior to receiving its feeder node's contribution. Receiver nodes with this rating are those that operate in the range 20% to < 40% of their full operational capability (100 utils), *prior to receiving their feeder node contributions.* Receiver nodes with this rating are those with a **high dependency** on their feeder nodes.	20 to < 40 utils
3 **YELLOW**	A receiver node with this rating has a **modest** operational capability prior to receiving its feeder node's contribution. Receiver nodes with this rating are those that operate in the range 40% to < 60% of their full operational capability (100 utils), *prior to receiving their feeder node contributions.* Receiver nodes with this rating are those with a **modest dependency** on their feeder nodes.	40 to < 60 utils
2 **GREEN**	A receiver node with this rating has a **high** operational capability prior to receiving its feeder node's contribution. Receiver nodes with this rating are those that operate in the range 60% to < 80% of their full operational capability (100 utils), *prior to receiving their feeder node contributions.* Receiver nodes with this rating are those with a **low dependency** on their feeder nodes.	60 to < 80 utils
1 **BLUE**	A receiver node with this rating has a **very high** operational capability prior to receiving its feeder node's contribution. Receiver nodes with this rating are those that operate in the range 80% to 100% of their full operational capability (100 utils), *prior to receiving their feeder node contributions.* Receiver nodes with this rating are those with a **very low dependency** on their feeder nodes.	80 to < 100 utils

From Definition 6.8, criticality of dependency can be viewed from a degradation perspective. In Definition 6.8, for a single feeder–receiver node pair, shown in Figure 6.11, the criticality of dependency constraint is the operability level β_{kj} (utils) that receiver node N_j degrades to from a reference point operability level $RPOLP_j$, when its feeder node is fully inoperable (wholly ineffective).

As stated earlier, degradation is measured from an operability level that has meaning with respect to the receiver node's performance goals or requirements. This is the receiver node's RPOL, a value in the interval $0 < RPOLP_j \leq 100$. A convenient reference point is a receiver node's BOL. The BOL offers an anchoring property. It is a receiver node's operability level prior to (or without) receiving its feeder node contribution.

COD Questions Protocol

If we use a receiver node's BOL as its RPOL, then the criticality of dependency constraint can be assessed from the following questions: *If the feeder node's performance is equal to zero in operational utility (the value or worth of its contribution) to its receiver node, then (1) will the receiver node degrade from its baseline operability level? (2)* ***If yes****, then to what operability level will the receiver node decline? (3)* ***If no****, then no criticality of dependency exists between the receiver node and its feeder node.*

If the answer to (3) is no, then no degradation occurs in the baseline operability level of the receiver node. A criticality of dependency is not present between the receiver and its feeder node from this reference perspective.

If the answers to (1) and (2) are **yes,** then the criticality of dependency constraint is set to the level of operability to which receiver node N_j will (or is anticipated to) decline. For instance, in Figure 6.25 suppose an event occurs such that N_i's performance level has 0 operational utility (value or worth) to N_j. If, because of this, the operability level of N_j degrades to 0 utils then $\beta_{ij} = 0$ utils. With this, the FDF for the single feeder–receiver node pair in Figure 6.25 is

$$\begin{aligned} P_j &= \text{Min}(\alpha_{ij}P_i + BOLP_j, CODP_j) = \text{Min}(\alpha_{ij}P_i + BOLP_j, P_i + \beta_{ij}) \\ &= \text{Min}(\alpha_{ij}P_i + 100(1-\alpha_{ij}), P_i + 0) = P_i \end{aligned} \tag{6.41}$$

The result in Equation 6.41 reflects Property 6.6, which states if receiver node N_j has dependency relationships on h feeder nodes (where $h = 1$ in Figure 6.25) with $\beta_{kj} = 0$ utils, for all $k = 1, 2, 3, \ldots, h$, then

$$P_j = \text{Min}(P_1, P_2, P_3, \ldots, P_h)$$

Thus, if $\beta_{kj} = 0$ utils for all $k = 1, 2, 3, \ldots, h$ feeder nodes then the operability level N_j is equal to the operability level of its weakest performing feeder

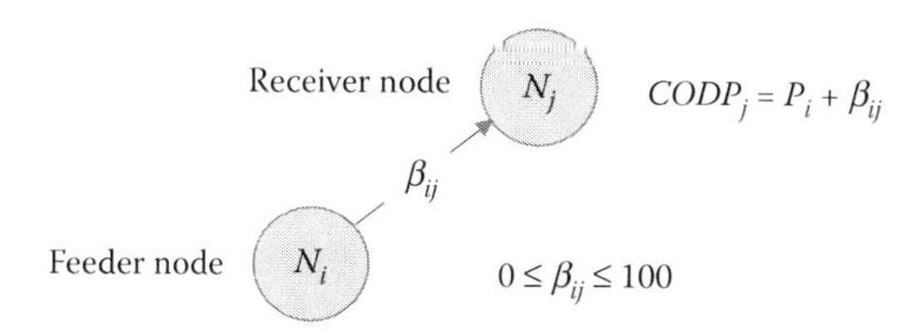

FIGURE 6.25
A feeder–receiver node criticality of dependency view.

node. From Definition 6.4, this condition is known as *maximum criticality*. Instead, in Figure 6.25 suppose the operability level of N_j degrades to 20 utils. In this case, $\beta_{ij} = 20$ utils. With this, the FDF for the single feeder–receiver node pair in Figure 6.25 is

$$P_j = \text{Min}(\alpha_{ij}P_i + BOLP_j, P_i + \beta_{ij}) = \text{Min}(\alpha_{ij}P_i + 100(1 - \alpha_{ij}), P_i + 20)$$

In summary, the criticality of dependency constraint is such that the operability level of a receiver node N_j with h feeder nodes can never be more than $P_k + \beta_{kj}$ for all $k = 1, 2, 3, \ldots, h$, where $0 \le \beta_{kj} \le 100$ and P_k is the operability level of feeder node N_k.

6.5 Network Operability and Tolerance Analyses

As mentioned earlier, critical considerations in assessing a mission's operational effectiveness are identifying, representing, and measuring dependencies between entities (e.g., programs, functions, technologies) necessary for its successful execution. The importance of understanding entity relationships is many-fold. Primary, is to study the ripple effects that degraded performance in one entity has on the performance of other dependent entities across a relationship network that characterizes a mission or capability. The ability of a mission or network to absorb the effects of nodal degradation while maintaining an acceptable level of operational effectiveness is known in FDNA as network tolerance. Modeling and measuring the tolerance of a network to nodal degradations is a major consideration in engineering system planning, design risk analysis, and investment decisions.

As seen from the preceding discussions, graph theory offers a visual representation of complex dependencies between entities and enables the design of formalisms that measure and trace the effectiveness of these relationships as they affect many parts and paths in a graph. Likewise, a mission or capability can be represented as a network of systems, infrastructures, or organizational entities expressed as nodes on a graph that depict direction, strength, and criticality of feeder–receiver dependency relationships. With this, algorithms can be designed to address questions such as the following:

> What is the effect on the ability of a mission to operate effectively if one or more entities or feeder–receiver chains degrade, or fail due to events or situations? How much operational degradation occurs and does it breach the mission's minimum effective level of performance?

The following illustrates an operability analysis from a network perspective using the FDNA calculus developed thus far. The analysis will include a perspective where the degradation tolerance of a receiver node to losses in feeder node performance is measured and discussed. First, we introduce the concept of a node's minimum effective operability level (MEOL).

Definition 6.9

The minimum effective operability level (MEOL) of a node is the utility associated with the minimum level of performance the node must achieve for its outputs to be minimally acceptable to stakeholders.

The MEOL is to recognize that not all nodes need to be fully operable for their outputs to have meaningful utility to stakeholders.

PROBLEM 6.10

Conduct an operability analysis of the FDNA graph in Figure 6.26. Use FDNA (α,β) WLR given the following: $BOLP_j = 50$ utils, $\beta_{ij} = 10$ utils, and $MEOLP_j = 80$ utils.

Solution

From Equation 6.40, if $BOLP_j = 50$ utils then receiver node N_j has a strength of dependency fraction $\alpha_{ij} = 0.50$. A criticality constraint of $\beta_{ij} = 10$ utils is given between receiver node N_j and feeder node N_i. From Definition 6.7, this means the operability level of N_j can never be more than $P_i + 10$, where P_i is the operability level of feeder node N_i. From this, the operability function of receiver node N_j is

$$P_j = \text{Min}(SODP_j, CODP_j) = \text{Min}(\alpha_{ij}P_i + BOLP_j, P_i + \beta_{ij})$$
$$= \text{Min}\left(\frac{1}{2}P_i + 50, P_i + 10\right)$$

where $BOLP_j = 100(1-\alpha_{ij})$, $0 \le \alpha_{ij} \le 1$, $0 \le \beta_{ij} \le 100$ and $0 \le P_i, P_j \le 100$. Table 6.2 presents an operability analysis of the FDNA graph in Figure 6.26. The analysis shows how the operability level of receiver node N_j improves with increasing levels of operability in feeder node N_i, subject to the conditions given.

TABLE 6.2

Problem 6.10 Operability Analysis with $\beta_{ij} = 10$ utils

N_i Operability Level	$SODP_j$	$CODP_j$	N_j Operability Level
0	50	10	10
5	52.5	15	15
10	55	20	20
15	57.5	25	25
20	60	30	30
25	62.5	35	35
30	65	40	40
35	67.5	45	45
40	70	50	50
45	72.5	55	55
50	75	60	60
55	77.5	65	65
60	80	70	70
65	82.5	75	75
70	85	80	80
75	87.5	85	85
80	90	90	90
85	92.5	95	92.5
90	95	100	95
95	97.5	105	97.5
100	100	110	100

Note: An operability level of $N_i = 80$ utils is when $SODP_j = CODP_j$.

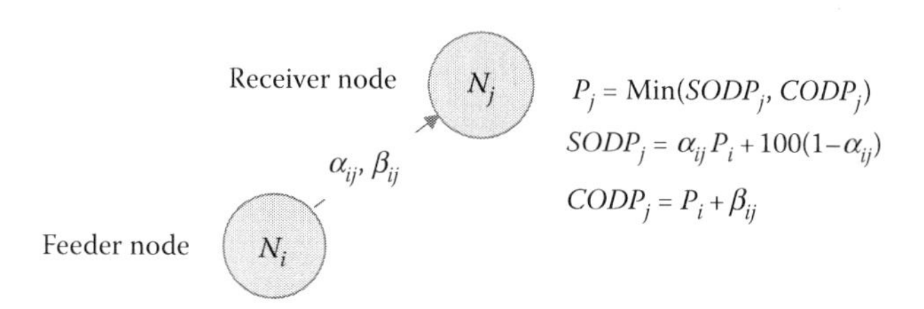

FIGURE 6.26
FDNA graph for Problem 6.10.

The minimum effective operability level of N_j is given to be 80 utils. This is the value or worth associated with the minimum level of performance N_j must achieve for its outputs to be minimally acceptable to stakeholders. In this example, receiver node N_j achieves its minimum effective operability level when feeder node N_i reaches 70 utils (in Table 6.2).

In Problem 6.10, we can also interpret the criticality constraint in terms of degradation in a receiver node's operability level. From Definition 6.8, recall that degradation can be measured from an operability level that has meaning with respect to the receiver node's performance goals or requirements, called the reference point operability level. In this problem, if the receiver node's reference point is its baseline operability level then N_j degrades to 10 utils if its feeder node's performance level has 0 operational utility (no value or worth) to N_j. Thus, N_j maintains an acceptable level of operational effectiveness as long as its feeder node operates at 70 utils or higher. Below that, for every one util decline in its feeder node's operability receiver node N_j degrades by one util of operability—until it bottoms out at its criticality constraint of 10 utils.

A number of other interesting results can be derived from Table 6.2. A few are as follows: the point where the operability level of N_j stages its transition from being determined by $CODP_j$ to being determined by $SODP_j$ occurs when N_j's feeder node N_i reaches

$$P_i = 100 - \frac{\beta_{ij}}{1-\alpha_{ij}} = 100 - \frac{10}{1-0.50} = 80 \text{ utils}$$

From this point forward, N_j improves in operability with increasing levels of operability in N_i, but it does so at half the rate of improvement produced by $CODP_j$. Specifically, for every one util increase in the operability level of N_i (from $P_i = 80$) there is an $\alpha_{ij} = 0.50$ util increase in the operability level of N_j.

If feeder node N_i's contribution is valued at 0 in operational utility to receiver node N_j then N_j degrades from its baseline operability level of 50 to 10 utils, rendering N_j nearly inoperable. In Problem 6.10, instead of $\beta_{ij} = 10$ utils suppose we have $\beta_{ij} = 30$ utils. Table 6.3 shows the resultant operability analysis with this criticality constraint. In this case, receiver node N_j achieves its minimum effective operability level when feeder node N_i reaches 60 utils instead of 70 utils (in Table 6.2). Thus, N_j maintains an acceptable level of operational effectiveness as long as N_i operates at 60 utils or higher.

In Table 6.3, the point where the operability level of N_j stages its transition from being determined by $CODP_j$ to being determined by $SODP_j$ occurs when N_j's feeder node N_i reaches

$$P_i = 100 - \frac{\beta_{ij}}{1-\alpha_{ij}} = 100 - \frac{30}{1-0.50} = 40 \text{ utils}$$

From this point forward, N_j improves in operability with increasing levels of operability in N_i, at a rate governed by $SODP_j$. Specifically, for every one util increase in the operability level of N_i (from $P_i = 40$) there is an $\alpha_{ij} = 0.50$ util increase in the operability level of N_j. Figure 6.27 compares the operability analysis in Table 6.2 with Table 6.3.

In Figure 6.27, the operability function for receiver node N_j is given by

$$P_j = \text{Min}\left(\frac{1}{2}P_i + 50, P_i + \beta_{ij}\right)$$

TABLE 6.3

Problem 6.10 Operability Analysis with $\beta_{ij} = 30$ utils

N_i Operability Level	$SODP_j$	$CODP_j$	N_j Operability Level
0	50	30	30
5	52.5	35	35
10	55	40	40
15	57.5	45	45
20	60	50	50
25	62.5	55	55
30	65	60	60
35	67.5	65	65
40	70	70	70
45	72.5	75	72.5
50	75	80	75
55	77.5	85	77.5
60	80	90	80
65	82.5	95	82.5
70	85	100	85
75	87.5	105	87.5
80	90	110	90
85	92.5	115	92.5
90	95	120	95
95	97.5	125	97.5
100	100	130	100

Note: An operability level of $N_i = 40$ utils is when $SODP_j = CODP_j$.

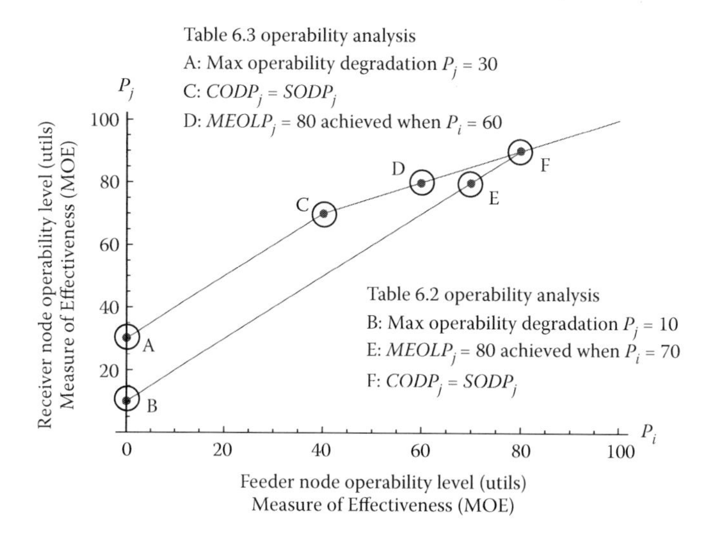

FIGURE 6.27
Tables 6.2 and 6.3 operability analysis comparisons.

where $\beta_{ij} = 30$ utils (top function) and $\beta_{ij} = 10$ utils (bottom function). Figure 6.28 is a further analysis of the dependency relationship in Figure 6.26 for varying criticality constraints β_{ij}. In Figure 6.28, the functions from top to bottom are given by, respectively,

$$P_j = \text{Min}\left(\frac{1}{2}P_i + 50, P_i + 50\right) = \frac{1}{2}P_i + 50 = SODP_j \tag{6.42}$$

$$P_j = \text{Min}\left(\frac{1}{2}P_i + 50, P_i + 40\right) \tag{6.43}$$

$$P_j = \text{Min}\left(\frac{1}{2}P_i + 50, P_i + 30\right) \tag{6.44}$$

$$P_j = \text{Min}\left(\frac{1}{2}P_i + 50, P_i + 20\right) \tag{6.45}$$

$$P_j = \text{Min}\left(\frac{1}{2}P_i + 50, P_i + 10\right) \tag{6.46}$$

$$P_j = \text{Min}\left(\frac{1}{2}P_i + 50, P_i + 0\right) = P_i = CODP_j \tag{6.47}$$

The top function in Figure 6.28 illustrates Property 6.2, which states that in a single feeder–receiver node pair, if $BOLP_j \leq \beta_{kj} \leq 100$, then $P_j = SODP_j$. In Equation 6.42, we have $BOLP_j = 50 = \beta_{kj}$. Hence, the operability level of N_j is strictly determined by $SODP_j$ in accordance with Property 6.2. Also, in Equation 6.42, observe there is no degradation in the baseline operability level of N_j if the operability level of N_i is 0. The bottom function in Figure

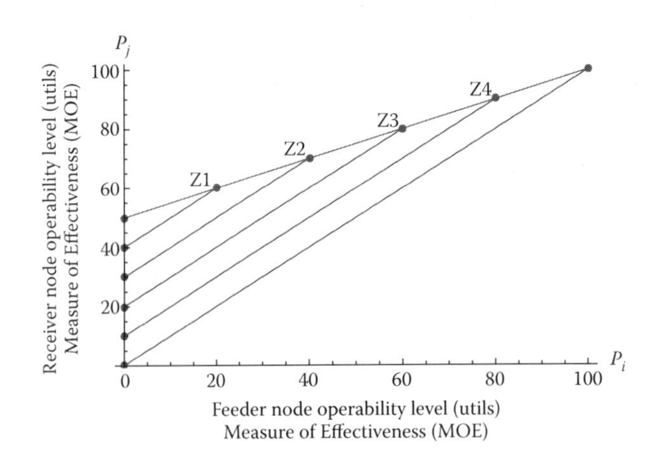

FIGURE 6.28
Table 6.2 operability analysis with $\beta_{ij} = 50, 40, 30, 20, 10, 0$.

6.28 illustrates Property 6.6, which states if receiver node N_j has dependency relationships on h feeder nodes with $\beta_{ij}=0$ for all $i=1,2,3,\ldots,h$, then

$$P_j = \text{Min}(P_1, P_2, P_3, \ldots, P_h)$$

In Equation 6.47, we have $\beta_{ij}=0$. Hence, the operability level of N_j is strictly determined by $CODP_j$. In Equation 6.47, observe there is complete degradation in the operability level of N_j if the operability level of N_i is 0. As discussed earlier, this occurs when a receiver node has maximum criticality of dependency on all of its feeder nodes. In such cases, the operability level of N_j is equal to the operability level of its weakest performing feeder node (refer to Definition 6.4). In Figure 6.28, Z1, Z2, Z3, and Z4 mark transition points where the operability level of N_j stages its transition from being determined by $CODP_j$ to being determined by $SODP_j$ for $\beta_{ij}=40,30,20,$ and 10, respectively. Why are no such transitions seen in the top or bottom functions in Figure 6.28?

PROBLEM 6.11

Conduct an operability analysis of the FDNA graph in Figure 6.22. Use FDNA (α,β) WLR given the following parameter values.

$$\begin{array}{ll}
\eta_1 \equiv \alpha_{1Cap_1}, \beta_{1Cap_1} = 0.90, 10 & \eta_2 \equiv \alpha_{3Cap_1}, \beta_{3Cap_1} = 0.45, 55 \\
\eta_3 \equiv \alpha_{3Cap_2}, \beta_{3Cap_2} = 0.65, 35 & \eta_4 \equiv \alpha_{6Cap_3}, \beta_{6Cap_3} = 0.90, 10 \\
\eta_5 \equiv \alpha_{4Cap_3}, \beta_{4Cap_3} = 0.85, 15 & \eta_6 \equiv \alpha_{53}, \beta_{53} = 0.30, 70 \\
\eta_7 \equiv \alpha_{31}, \beta_{31} = 0.15, 85 & \eta_8 \equiv \alpha_{21}, \beta_{21} = 0.28, 72
\end{array}$$

Solution

From the solution to Problem 6.9 we have the following FDFs:

$$P_1 = \text{Min}\left(\frac{\alpha_{21}P_2}{2} + \frac{\alpha_{31}P_3}{2} + 100\left(1 - \left(\frac{\alpha_{21} + \alpha_{31}}{2}\right)\right), P_2 + \beta_{21}, P_3 + \beta_{31}\right)$$

$$P_3 = \text{Min}(\alpha_{53}P_5 + 100(1 - \alpha_{53}), P_5 + \beta_{53})$$

$$P_{Cap_1} = \text{Min}\left(\frac{\alpha_{1Cap_1}P_1}{2} + \frac{\alpha_{3Cap_1}P_3}{2} + 100\left(1 - \left(\frac{\alpha_{1Cap_1} + \alpha_{3Cap_1}}{2}\right)\right), P_1 + \beta_{1Cap_1}, P_3 + \beta_{3Cap_1}\right)$$

$$P_{Cap_2} = \text{Min}(\alpha_{3Cap_2}P_3 + 100(1 - \alpha_{3Cap_2}), P_3 + \beta_{3Cap_2})$$

$$P_{Cap_3} = \text{Min}\left(\frac{\alpha_{4Cap_3}P_4}{2} + \frac{\alpha_{6Cap_3}P_6}{2} + 100\left(1 - \left(\frac{\alpha_{4Cap_3} + \alpha_{6Cap_3}}{2}\right)\right), P_4 + \beta_{4Cap_3}, P_6 + \beta_{6Cap_3}\right)$$

TABLE 6.4

Problem 6.11 Operability Analysis

Functional Dependency Network Analysis (FDNA): A Capability Portfolio							
FDNA Parameters							
α_{1Cap_1}	0.90	α_{4Cap_3}	0.85	β_{1Cap_1}	10	β_{4Cap_3}	15
α_{3Cap_1}	0.45	α_{53}	0.30	β_{3Cap_1}	55	β_{53}	70
α_{3Cap_2}	0.65	α_{31}	0.15	β_{3Cap_2}	35	β_{31}	85
α_{6Cap_3}	0.90	α_{21}	0.28	β_{6Cap_3}	10	β_{21}	72

If these leaf node feeders are functioning at these operability levels ...

At time t_1		**At time t_2**		**At time t_3**	
P_2	100	P_2	75	P_2	50
P_5	100	P_5	75	P_5	50
P_4	100	P_4	75	P_4	50
P_6	100	P_6	100	P_6	100

Then these receiver nodes are functioning at these operability levels ...

At time t_1		**At time t_2**		**At time t_3**	
P_3	100	P_3	92.50	P_3	85.00
P_1	100	P_1	95.94	P_1	91.88
P_{Cap_1}	100	P_{Cap_1}	96.48	P_{Cap_1}	92.97
P_{Cap_2}	100	P_{Cap_2}	95.13	P_{Cap_2}	90.25
P_{Cap_3}	100	P_{Cap_3}	89.38	P_{Cap_3}	65.00

Applying the FDNA parameter values to these equations yields the results in Table 6.4. The table presents the operability analysis that results from the four leaf node feeders in Figure 6.22 losing operability over time periods t_1, t_2, and t_3. Next, we explore the tolerance of each capability node in Figure 6.29 to degradations in feeder node performance. Specifically, we discuss ways to address the following:

> What is the effect on the ability of each capability node, shown in Figure 6.29, to operate effectively if its feeder nodes degrade or fail due to unwanted events or situations? How much operational degradation occurs in each capability node and is its minimum effective level of performance breached?

6.5.1 Critical Node Analysis and Degradation Index

A critical node analysis involves identifying which nodes in a graph are among the most influential to the operational effectiveness of other dependent nodes. One way to identify these nodes is to measure the ripple effects

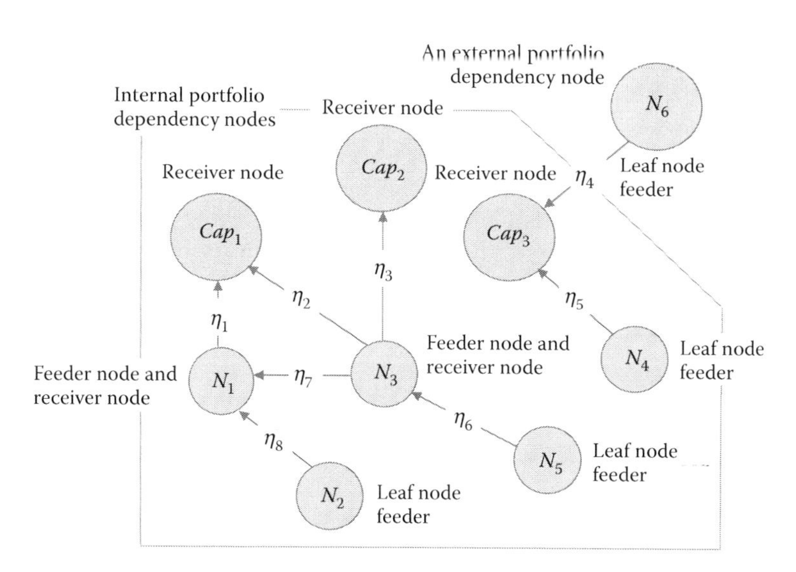

FIGURE 6.29
FDNA graph for Problem 6.12.

that a change in the operability level of one node has on the operability levels of other dependent nodes across a network. The following introduces a rule designed to measure these effects. The rule is called the *degradation index* (DI).

Definition 6.10: *Degradation Index**

The degradation index of receiver node N_j that is dependent on feeder node N_i is the instantaneous rate of change of N_j with respect to N_i and is denoted by $DIN_j|N_i$. When the dependency relationships between nodes are expressed by linear operability functions, then $DIN_j|N_i$ is the amount N_j decreases in operability level with every unit decrease in the operability level of N_i. In this case, the degradation index reflects a constant rate of change.

Definition 6.11 presents how to compute the degradation index. The computation involves applying the multivariate chain rule by forming the

* The degradation index is a rate of change measure. It can measure losses or gains in the operability levels of nodes, according to the situation of interest. In this section, the loss perspective is taken to illustrate how degraded performance in feeder nodes affects other dependent nodes in a network. Performance degradations can occur from circumstances that lead to operability levels being lower than planned (e.g., losses from the occurrence of unwanted events). The index can measure gains in the performance of nodes from circumstances that lead to operability levels being higher than planned (e.g., gains realized from successful investment decisions). Accordingly, Definition 6.10 can be interpreted to reflect the appropriate context.

correct "chaining" of derivatives as a function of the dependency relationships in the FDNA graph, assuming differentiability conditions exist.

Definition 6.11: ***Computing the Degradation Index***

(A) If the operability level of N_j is a function of the operability level of N_1 and the operability level of N_1 is a function of the operability level of N_2 then

$$DIN_j | N_1 = \frac{\partial P_j}{\partial P_1} \quad (6.48)$$

$$DIN_j | N_2 = \frac{\partial P_j}{\partial P_2} = \frac{\partial P_j}{\partial P_1} \cdot \frac{\partial P_1}{\partial P_2} \quad (6.48a)$$

where P_j, P_1, and P_2 are the operability functions of nodes N_j, N_1, and N_2, respectively.

(B) If the operability level of N_j is a function of the operability levels of N_1 and N_2, and the operability levels of N_1 and N_2 are each a function of the operability levels of N_3 and N_4, then

$$DIN_j | N_3 = \frac{\partial P_j}{\partial P_1} \cdot \frac{\partial P_1}{\partial P_3} + \frac{\partial P_j}{\partial P_2} \cdot \frac{\partial P_2}{\partial P_3} \quad (6.48b)$$

$$DIN_j | N_4 = \frac{\partial P_j}{\partial P_1} \cdot \frac{\partial P_1}{\partial P_4} + \frac{\partial P_j}{\partial P_2} \cdot \frac{\partial P_2}{\partial P_4} \quad (6.48c)$$

where P_j, P_1, P_2, P_3, and P_4 are the operability functions of nodes N_j, N_1, N_2, N_3, and N_4, respectively.

Since the operability levels of all nodes in an FDNA graph fall along the same 0 to 100 util scale, nodes with high degradation indices can be considered more influential to the network compared with those with low degradation indices. Interpreting the meaning of this influence must consider whether the dependency relationships between nodes are expressed by linear or nonlinear rates of change (i.e., by constant or non-constant rates of change).

PROBLEM 6.12

Compute the degradation index of each node in Figure 6.29, given the FDNA parameters in Problem 6.11 and the feeder node operability levels at time period t_3.

Solution

From Problem 6.11, we have the following operability functions.

$$P_1 = \text{Min}\left(\frac{\alpha_{21}P_2}{2} + \frac{\alpha_{31}P_3}{2} + 100\left(1 - \left(\frac{\alpha_{21} + \alpha_{31}}{2}\right)\right), P_2 + \beta_{21}, P_3 + \beta_{31}\right)$$

$$P_3 = \text{Min}(\alpha_{53}P_5 + 100(1 - \alpha_{53}), P_5 + \beta_{53})$$

$$P_{Cap_1} = \text{Min}\left(\frac{\alpha_{1Cap_1}P_1}{2} + \frac{\alpha_{3Cap_1}P_3}{2} + 100\left(1 - \left(\frac{\alpha_{1Cap_1} + \alpha_{3Cap_1}}{2}\right)\right), P_1 + \beta_{1Cap_1}, P_3 + \beta_{3Cap_1}\right)$$

$$P_{Cap_2} = \text{Min}(\alpha_{3Cap_2}P_3 + 100(1 - \alpha_{3Cap_2}), P_3 + \beta_{3Cap_2})$$

$$P_{Cap_3} = \text{Min}\left(\frac{\alpha_{4Cap_3}P_4}{2} + \frac{\alpha_{6Cap_3}P_6}{2} + 100\left(1 - \left(\frac{\alpha_{4Cap_3} + \alpha_{6Cap_3}}{2}\right)\right), P_4 + \beta_{4Cap_3}, P_6 + \beta_{6Cap_3}\right)$$

From Problem 6.11 (Table 6.4), we know that feeder nodes N_2, N_5, N_4, and N_6 at time period t_3 have operability levels 50, 50, 50, and 100, respectively. Given these levels, the operability functions for the receiver nodes are as follows:

$$P_1 = \frac{\alpha_{21}P_2}{2} + \frac{\alpha_{31}P_3}{2} + 100\left(1 - \left(\frac{\alpha_{21} + \alpha_{31}}{2}\right)\right)$$

$$P_3 = \alpha_{53}P_5 + 100(1 - \alpha_{53})$$

$$P_{Cap_1} = \frac{\alpha_{1Cap_1}P_1}{2} + \frac{\alpha_{3Cap_1}P_3}{2} + 100\left(1 - \left(\frac{\alpha_{1Cap_1} + \alpha_{3Cap_1}}{2}\right)\right)$$

$$P_{Cap_2} = \alpha_{3Cap_2}P_3 + 100(1 - \alpha_{3Cap_2})$$

$$P_{Cap_3} = P_4 + \beta_{4Cap_3}$$

To compute the degradation index of each node, we must formulate the correct "chaining" of derivatives as a function of the dependency relationships in Figure 6.29. One way to express this chaining is to use the "tree diagram technique." This involves labeling the segments between each pair of nodes in Figure 6.29, as shown in Figure 6.30.

Referring to Figure 6.30, and assuming that differentiability conditions exist, the degradation index of one node that depends on another node is the

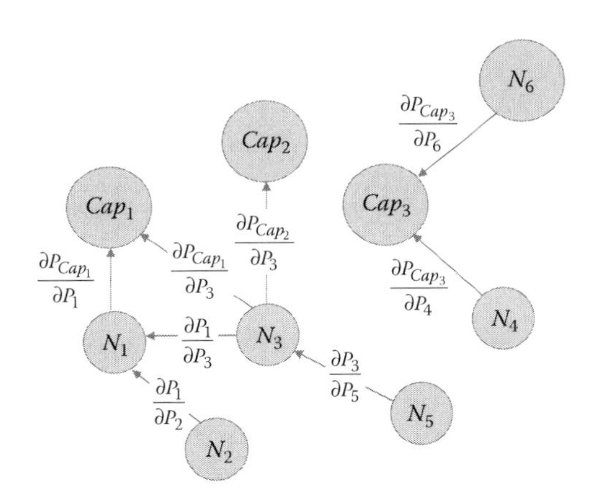

FIGURE 6.30
Chain rule formulation: The tree diagram technique.

sum over all paths between them of the product of the derivatives on each segment (or branch) along that path. With this, we have the following chain rule formulations for computing each node's degradation index.

$$DICap_1|N_2 = \frac{\partial P_{Cap_1}}{\partial P_1} \cdot \frac{\partial P_1}{\partial P_2} = \frac{\alpha_{1Cap_1}}{2} \cdot \frac{\alpha_{21}}{2} = \frac{0.90}{2} \cdot \frac{0.28}{2} = 0.063$$

$$DICap_1|N_1 = \frac{\partial P_{Cap_1}}{\partial P_1} = \frac{\alpha_{1Cap_1}}{2} = \frac{0.90}{2} = 0.45$$

$$DICap_1|N_5 = \frac{\partial P_{Cap_1}}{\partial P_1} \cdot \frac{\partial P_1}{\partial P_3} \cdot \frac{\partial P_3}{\partial P_5} + \frac{\partial P_{Cap_1}}{\partial P_3} \cdot \frac{\partial P_3}{\partial P_5} = \frac{\alpha_{1Cap_1}}{2} \cdot \frac{\alpha_{31}}{2} \cdot \alpha_{53} + \frac{\alpha_{3Cap_1}}{2} \cdot \alpha_{53}$$

$$= \frac{0.90}{2} \cdot \frac{0.15}{2}(0.30) + \frac{0.45}{2}(0.30) = 0.077625$$

$$DICap_1|N_3 = \frac{\partial P_{Cap_1}}{\partial P_1} \cdot \frac{\partial P_1}{\partial P_3} + \frac{\partial P_{Cap_1}}{\partial P_3} = \frac{\alpha_{1Cap_1}}{2} \cdot \frac{\alpha_{31}}{2} + \frac{\alpha_{3Cap_1}}{2}$$

$$= \frac{0.90}{2} \cdot \frac{0.15}{2} + \frac{0.45}{2} = 0.25875$$

$$DICap_2|N_5 = \frac{\partial P_{Cap_2}}{\partial P_3} \cdot \frac{\partial P_3}{\partial P_5} = \alpha_{3Cap_2} \cdot \alpha_{53} = 0.65(0.30) = 0.195$$

$$DICap_2|N_3 = \frac{\partial P_{Cap_2}}{\partial P_3} = \alpha_{3Cap_2} = 0.65$$

$$DICap_3|N_4 = \frac{\partial P_{Cap_3}}{\partial P_4} = \frac{\partial(P_4 + \beta_{4Cap_3})}{\partial P_4} = 1$$

$$DICap_3|N_6 = \frac{\partial P_{Cap_3}}{\partial P_6} = \frac{\partial(P_4 + \beta_{4Cap_3})}{\partial P_6} = 0$$

$$DIN_1|N_2 = \frac{\partial P_1}{\partial P_2} = \frac{\alpha_{21}}{2} = \frac{0.28}{2} = 0.14$$

$$DIN_1|N_3 = \frac{\partial P_1}{\partial P_3} = \frac{\alpha_{31}}{2} = \frac{0.15}{2} = 0.075$$

$$DIN_1|N_5 = \frac{\partial P_1}{\partial P_3} \cdot \frac{\partial P_3}{\partial P_5} = \frac{\alpha_{31}}{2} \cdot \alpha_{53} = \frac{0.15}{2}(0.30) = 0.0225$$

$$DIN_3|N_5 = \frac{\partial P_3}{\partial P_5} = \alpha_{53} = 0.30$$

These degradation indices reflect constant rates of change. With this, we can summarize these results in various ways. For example, relative to the three capability nodes Cap_1, Cap_2, and Cap_3 in Figure 6.29, we have

$$DICap_1|N_2 = 0.063 \text{ utils}, \; DICap_1|N_1 = 0.45 \text{ utils}$$

$$DICap_3|N_4 = 1 \text{ utils}, \; DICap_3|N_6 = 0 \text{ utils}$$

$$DICap_1|N_5 + DICap_2|N_5 = 0.077625 + 0.195 = 0.272625 \text{ utils}$$

$$DICap_1|N_3 + DICap_2|N_3 = 0.25875 + 0.65 = 0.90875 \text{ utils}$$

Thus, for every 1 util decrease in the operability level of node N_3 (a receiver and feeder node) there is a 0.90875 util decrease in the total operability level of receiver nodes Cap_1 and Cap_2. From this, the operability level of N_3 has a large influence on the operability levels of nodes Cap_1 and Cap_2. The index introduced in this section provides visibility into which nodes are among the most sensitive in influencing the operational effectiveness of the network as a whole.

6.5.2 Degradation Tolerance Level

Merriam-Webster defines tolerance as the allowable deviation from a standard or a range of variation permitted. Modeling and measuring network tolerance is an active area of research in the network science, engineering systems, and system sciences fields. A robust treatment of network tolerance is beyond the scope of this section. The aim of this discussion is to introduce the concept and present one of many perspectives on its measurement. For this, our focus is addressing the question:

> Given a set of leaf node feeders to a receiver node, what single operability level must each leaf node mutually equal or exceed to ensure the receiver node does not breach its minimum effective operability level?

Definition 6.12: Tolerance

Tolerance is the ability of an FDNA network to operate effectively within an allowable range of performance if, due to events or circumstances, one or more nodes or feeder-receiver chains degrade of fail.

Definition 6.13: Leaf Node Degradation Tolerance Level (LNDTL)

The degradation tolerance of a set of leaf nodes $N_1, N_2, N_3, \ldots, N_h$ to receiver node N_j is the operability level x that each leaf node must equal or exceed to ensure that N_j does not breach its minimum effective operability level ($MEOLP_j$).

From Definition 6.13, receiver node N_j breaches its minimum effective operability level if its set of leaf node feeders ***each*** fall below x, where $0 \leq x \leq 100$ utils. The value of x is the *common threshold operability level* that each leaf node feeder, in the set, must be greater than or equal to for the receiver node N_j to avoid an MEOL breach. A breach in receiver node N_j's MEOL can occur from combinations of leaf node operability levels that are different than x. The value derived for x is not affected by whether the operability levels of intermediate nodes along the paths between N_j and its leaf node feeders fall below their MEOLs, if these thresholds are identified. Noting these breaches is an important finding for management considerations.

PROBLEM 6.13

Consider the FDNA graph in Figure 6.20. Determine the degradation tolerance level of leaf node feeder N_2 if the MEOL of receiver node N_j is 82.5 utils, given the FDNA parameters

$$\alpha_{1j}, \beta_{1j} \equiv 0.25, 100, \quad \alpha_{2j}, \beta_{2j} \equiv 0.75, 100, \quad \alpha_{21}, \beta_{21} \equiv 0.50, 100$$

Solution

From Problem 6.7, the operability functions for the FDNA graph in Figure 6.20 are as follows:

$$P_j = \text{Min}\left(\frac{\alpha_{1j}}{2}P_1 + \frac{\alpha_{2j}}{2}P_2 + 100\left(1 - \frac{\alpha_{1j} + \alpha_{2j}}{2}\right), P_1 + \beta_{1j}, P_2 + \beta_{2j}\right)$$

$$P_1 = \text{Min}(\alpha_{21}P_2 + 100(1 - \alpha_{21}), P_2 + \beta_{21})$$

Given $\alpha_{1j}, \beta_{1j} = 0.25, 100$, $\alpha_{2j}, \beta_{2j} = 0.75, 100$, $\alpha_{21}, \beta_{21} = 0.50, 100$, and $0 < P_1, P_2, P_j \leq 100$ it follows that

$$\begin{aligned} P_j &= \text{Min}(0.125P_1 + 0.375P_2 + 50, P_1 + 100, P_2 + 100) \\ &= SODP_j = 0.125P_1 + 0.375P_2 + 50 \\ P_1 &= \text{Min}(0.50P_2 + 50, P_2 + 100) = SODP_1 = 0.50P_2 + 50 \end{aligned}$$

Given $MEOLP_j = 82.5$, it follows that $0.125P_1 + 0.375P_2 + 50 = 82.5$. Since the operability function of node N_1 is $P_1 = 0.50P_2 + 50$ it follows that

$$\begin{aligned} &0.125(0.50P_2 + 50) + 0.375P_2 + 50 = 82.5 \\ &\Rightarrow 0.4375P_2 + 56.25 = 82.5 \\ &\Rightarrow P_2 = 60 \text{ utils} \end{aligned}$$

In accordance with Definition 6.13, receiver node N_j breaches its minimum effective operability level of 82.5 utils if its leaf node feeder falls below $x = 60$ utils. This value is the threshold operability level that leaf node feeder N_2 must be greater than or equal to for receiver node N_j to avoid an MEOL breach.

From the above, the operability function for N_j can be expressed in terms of its leaf node feeder N_2 as $P_j = 0.4375P_2 + 56.25$. This is a linear function with a rate of change equal to 0.4375. This is the degradation index of N_j that can also be derived by the chain rule, presented in Section 6.5.1. With respect to Figure 6.20, the chain rule is

$$\frac{dP_j}{dP_2} = \frac{\partial P_j}{\partial P_1} \cdot \frac{\partial P_1}{\partial P_2} + \frac{\partial P_j}{\partial P_2} = 0.125(0.50) + 0.375 = 0.4375$$

Thus, there is a degradation of 0.4375 utils in the operability level of receiver node N_j for every unit decrease in the operability level of leaf node feeder N_2. Furthermore, as long as the operability level of leaf node feeder N_2 remains in the allowable range $60 \leq P_2 \leq 100$ utils then receiver node N_j will not breach its stated minimum effective operability level.

In Problem 6.13, criticality constraints did not affect receiver node operability levels. The MEOL of receiver node N_j fell at an operability level derived by the function composition of $SODP_j$ and $SODP_1$; in this case,

$$P_j = SODP_j = f(P_1, P_2),\ P_1 = g(P_2) = SODP_1 \Rightarrow P_j = 0.4375P_2 + 56.25$$

for all $0 \leq P_j, P_1, P_2 \leq 100$. In FDNA WLRs, operability function compositions may change as nodes in the network take values along the interval

0 to 100 utils. Identifying these changes and where they occur in this interval is necessary when computing LNDTLs, as defined herein.

PROBLEM 6.14

Consider the graph and FDNA parameters in Figure 6.19. Determine the LNDTL for the set of leaf node feeders if the MEOL of receiver node N_p is 86 utils.

Solution

From Problem 6.6, the operability functions for the FDNA graph in Figure 6.19 are as follows:

$$P_p = \text{Min}\left(\alpha_{jp}P_j + 100(1-\alpha_{jp}), P_j + \beta_{jp}\right)$$
$$P_j = \text{Min}(SODP_j, P_1 + \beta_{1j}, P_2 + \beta_{2j}, P_3 + \beta_{3j}, P_4 + \beta_{4j}, P_5 + \beta_{5j})$$

where

$$SODP_j = \frac{\alpha_{1j}}{5}P_1 + \frac{\alpha_{2j}}{5}P_2 + \frac{\alpha_{3j}}{5}P_3 + \frac{\alpha_{4j}}{5}P_4 + \frac{\alpha_{5j}}{5}P_5 + 100\left(1 - \frac{\alpha_{1j} + \alpha_{2j} + \alpha_{3j} + \alpha_{4j} + \alpha_{5j}}{5}\right)$$

Given the FDNA parameters in Figure 6.19, these equations become

$$P_p = \text{Min}\left(0.78P_j + 22, P_j + 55\right) = 0.78P_j + 22 \tag{6.49}$$

$$P_j = \text{Min}(SODP_j, P_1 + 30, P_2 + 45, P_3 + 50, P_4 + 100, P_5 + 100)$$

where

$$SODP_j = 0.07P_1 + 0.03P_2 + 0.18P_3 + 0.09P_4 + 0.13P_5 + 50$$

for $0 \le P_p, P_j, P_1, P_2, P_3, P_4, P_5 \le 100$.

From Definition 6.13, the degradation tolerance of leaf nodes N_1, N_2, N_3, N_4, and N_5 to receiver node N_p is the operability level x that each leaf node feeder must equal or exceed to ensure that N_p does not breach its minimum effective operability level. Thus, we must find x such that

$$P_j = \text{Min}(SODP_j, x + 30, x + 45, x + 50, x + 100, x + 100)$$

where

$$SODP_j = 0.07x + 0.03x + 0.18x + 0.09x + 0.13x + 50 = 0.50x + 50$$

Since $0 \le x \le 100$, it follows that

$$P_j = \text{Min}(SODP_j, x+30) = \text{Min}(0.50x+50, x+30) = \begin{cases} 0.50x+50 & \text{if } 40 \le x \le 100 \\ x+30 & \text{if } 0 \le x \le 40 \end{cases}$$

From this, we can write

$$P_j = \begin{cases} 0.50x+50 & \text{if } 40 \le x \le 100 \Rightarrow 70 \le P_j \le 100 \\ x+30 & \text{if } 0 \le x \le 40 \Rightarrow 0 \le P_j \le 70 \end{cases} \quad (6.50)$$

Given the MEOL of receiver node N_p is 86 utils, from Equation 6.49, it follows that

$$MEOLP_p = 86 = 0.78P_j + 22 \Rightarrow P_j = 82.0513$$

Since $P_j = 82.0513$ falls in the interval $70 \le P_j \le 100$, it follows that

$$82.0513 = 0.50x + 50 \Rightarrow x = 64.103 \text{ utils}$$

This solution is shown in Figure 6.31 with plots of Equations 6.49 and 6.50. Thus, receiver node N_p breaches its minimum effective operability level of 86 utils if each leaf node feeder falls below $x = 64.103$ utils. This value is the common threshold operability level that all five leaf node feeders must be greater than or equal to for receiver node N_p to avoid an MEOL breach. This does not preclude a breach in the receiver node's MEOL from combinations of leaf node operability levels that are different than x.

PROBLEM 6.15

Consider the graph and FDNA parameters in Figure 6.32. Determine the LNDTL for the set of leaf node feeders if the MEOL of receiver node N_j is 75 utils.

Solution

The operability functions for the FDNA graph in Figure 6.32 are as follows:

$$P_1 = \text{Min}\left(\frac{\alpha_{31}}{2}P_3 + \frac{\alpha_{21}}{2}P_2 + 100\left(1 - \frac{\alpha_{31}+\alpha_{21}}{2}\right), P_3 + \beta_{31}, P_2 + \beta_{21}\right) \quad (6.51)$$

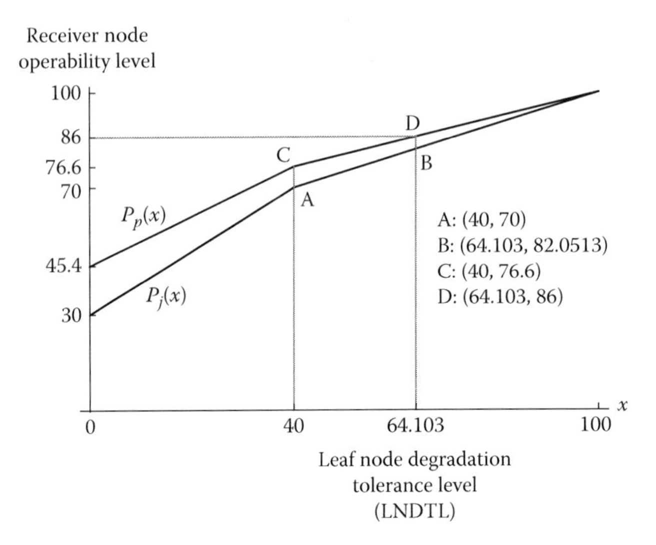

FIGURE 6.31
Problem 6.14: Leaf node degradation tolerance level.

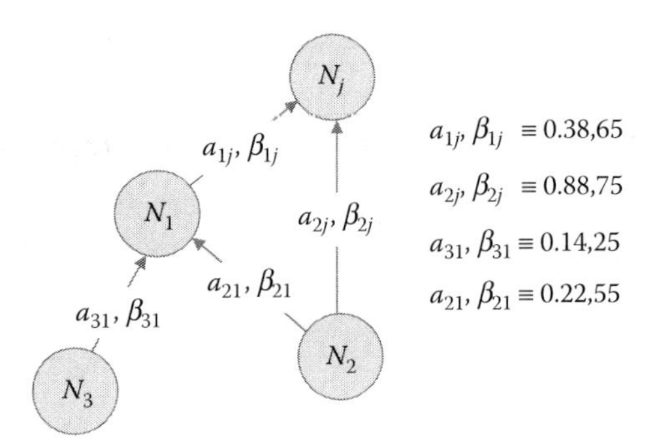

FIGURE 6.32
FDNA Graph for Problem 6.15.

$$P_j = \text{Min}\left(\frac{\alpha_{1j}}{2}P_1 + \frac{\alpha_{2j}}{2}P_2 + 100\left(1 - \frac{\alpha_{1j} + \alpha_{2j}}{2}\right), P_1 + \beta_{1j}, P_2 + \beta_{2j}\right) \quad (6.52)$$

for $0 \leq P_2, P_3, P_1, P_j \leq 100$. Given the FDNA parameters specified in Figure 6.32, these functions are as follows:

$$P_1 = \text{Min}(0.07P_3 + 0.11P_2 + 82, P_3 + 25, P_2 + 55)$$
$$P_j = \text{Min}(0.19P_1 + 0.44P_2 + 37, P_1 + 65, P_2 + 75)$$

From Definition 6.13, the degradation tolerance of leaf node feeders N_3 and N_2 to receiver node N_j is the operability level x that each leaf node must equal or exceed to ensure that N_j does not breach its minimum effective operability level. Thus, we must find x such that

$$\begin{aligned} P_1 &= \text{Min}(0.07x + 0.11x + 82, x + 25, x + 55) \\ &= \text{Min}(0.18x + 82, x + 25, x + 55) \end{aligned} \tag{6.53}$$

$$P_j = \text{Min}(0.19P_1 + 0.44x + 37, P_1 + 65, x + 75) \tag{6.54}$$

where $0 \le x \le 100$. From this, it follows that the operability function P_1 can be written as

$$P_1 = \text{Min}(0.18x + 82, x + 25, x + 55) = \text{Min}(0.18x + 82, x + 25) \tag{6.55}$$

Figure 6.33 depicts the behavior of P_1, given by Equation 6.55, along the interval $0 \le x \le 100$. From Figure 6.33, the operability function P_1 can be seen as

$$P_1 = \text{Min}(0.18x + 82, x + 25) = \begin{cases} 0.18x + 82 \text{ if } 69.5122 \le x \le 100 \\ x + 25 \text{ if } 0 \le x \le 69.5122 \end{cases} \tag{6.56}$$

Substituting Equation 6.56 for P_1 into Equation 6.54 we have the following:

$$P_j = \text{Min}(0.19(0.18x + 82) + 0.44x + 37, (0.18x + 82) + 65, x + 75), 69.5122 \le x \le 100$$

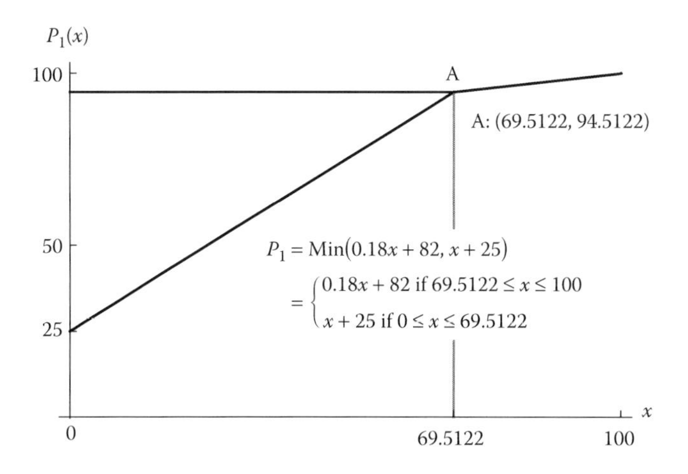

FIGURE 6.33
Behavior of operability function P_1.

$$P_j = \text{Min}\left(0.19(x+25)+0.44x+37,(x+25)+65,x+75\right), 0 \le x \le 69.5122$$

This simplifies to

$$P_j = \begin{cases} \text{Min}(0.4742x+52.58, 0.18x+147, x+75) \text{ if } 69.5122 \le x \le 100 \\ \text{Min}(0.63x+41.75, x+90, x+75) \text{ if } 0 \le x \le 69.5122 \end{cases} \quad (6.57)$$

Equation 6.57 simplifies to

$$P_j = \begin{cases} 0.4742x+52.58 \text{ if } 69.5122 \le x \le 100 \\ 0.63x+41.75 \text{ if } 0 \le x \le 69.5122 \end{cases} \quad (6.58)$$

Figure 6.34 depicts the behavior of P_j, given by Equation 6.58, along the interval $0 \le x \le 100$. Thus, receiver node N_j breaches its minimum effective operability level of 75 utils if each leaf node feeder falls below $x = 52.777$ utils. This value is the common threshold operability level that leaf node feeders N_3 and N_2 must be greater than or equal to for receiver node N_j to avoid an MEOL breach. This does not preclude a breach in the receiver node's MEOL from combinations of leaf node operability levels that are different than x.

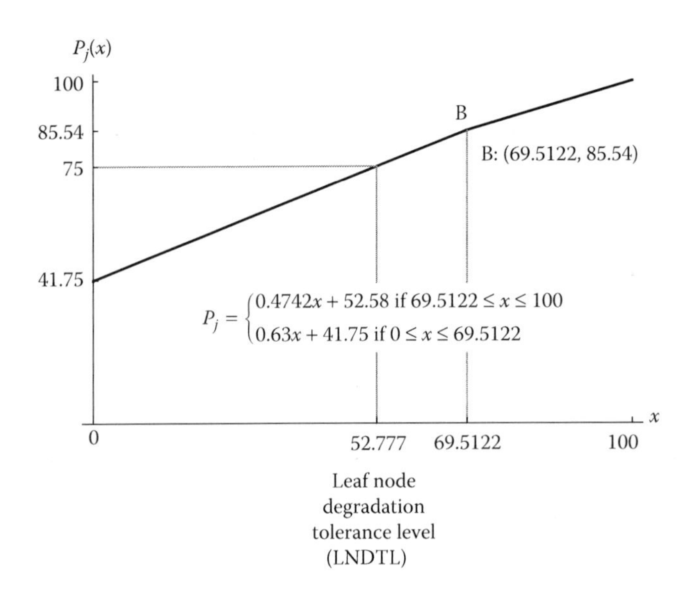

FIGURE 6.34
Problem 6.15: leaf node degradation tolerance level.

Various degradation indices can be seen in the solution to Problem 6.15. For example, the degradation index of receiver node N_j with respect to its leaf node feeders N_2 and N_3 is given by Equations 6.57 and 6.58, respectively

$$DIN_j \mid N_2 = \frac{\partial P_j}{\partial P_1} \cdot \frac{\partial P_1}{\partial P_2} + \frac{\partial P_j}{\partial P_2} \quad (6.57)$$

$$DIN_j \mid N_3 = \frac{\partial P_j}{\partial P_1} \cdot \frac{\partial P_1}{\partial P_3} \quad (6.58)$$

As mentioned earlier, operability function compositions in FDNA WLRs may change with the operability levels of nodes in a network. This can affect their degradation indices, which can affect leaf node degradation tolerance levels. Table 6.5 presents the degradation indices $DIN_j \mid N_2$ and $DIN_j \mid N_3$ for Problem 6.15 and how their values are affected by the operability function formed by the row-column composition.

Understanding the relationship between degradation indices and degradation tolerance levels in an FDNA graph is important. It provides insights into the rates with which critically important nodes may approach their minimum effective operability levels. With this, risk management options might be implemented that avoids performance breaches and ensures the network's operational levels remain within allowable ranges.

TABLE 6.5

Degradation Indices of $DIN_j|N_2$ and $DIN_j|N_3$ for Problem 6.15

DI Formula	$P_1 = SODP_1$	$P_1 = P_3 + \beta_{31}$	$P_1 = P_2 + \beta_{21}$
$P_j = SODP_j$	$DIN_j \mid N_2 = \frac{\alpha_{1j}}{2} \cdot \frac{\alpha_{21}}{2} + \frac{\alpha_{2j}}{2}$	$DIN_j \mid N_2 = \frac{\alpha_{2j}}{2}$	$DIN_j \mid N_2 = \frac{\alpha_{1j}}{2} + \frac{\alpha_{2j}}{2}$
	$DIN_j \mid N_3 = \frac{\alpha_{1j}}{2} \cdot \frac{\alpha_{31}}{2}$	$DIN_j \mid N_3 = \frac{\alpha_{1j}}{2}$	$DIN_j \mid N_3 = 0$
$P_j = P_1 + \beta_{1j}$	$DIN_j \mid N_2 = \frac{\alpha_{21}}{2}$	$DIN_j \mid N_2 = 0$	$DIN_j \mid N_2 = 1$
	$DIN_j \mid N_3 = \frac{\alpha_{31}}{2}$	$DIN_j \mid N_3 = 1$	$DIN_j \mid N_3 = 0$
DI Values	$P_1 = SODP_1$	$P_1 = P_3 + \beta_{31}$	$P_1 = P_2 + \beta_{21}$
$P_j = SODP_j$	$DIN_j \mid N_2 = 0.4609$	$DIN_j \mid N_2 = 0.44$	$DIN_j \mid N_2 = 0.63$
	$DIN_j \mid N_3 = 0.0133$	$DIN_j \mid N_3 = 0.19$	$DIN_j \mid N_3 = 0$
$P_j = P_1 + \beta_{1j}$	$DIN_j \mid N_2 = 0.11$	$DIN_j \mid N_2 = 0$	$DIN_j \mid N_2 = 1$
	$DIN_j \mid N_3 = 0.07$	$DIN_j \mid N_3 = 1$	$DIN_j \mid N_3 = 0$

The preceding discussion focused on addressing the question: *Given a set of leaf node feeders to a receiver node, what single operability level x must each leaf node mutually equal or exceed to ensure the receiver node does not breach its minimum effective operability level?* The value of x was the *common threshold operability level* that each leaf node feeder, in the set, must be greater than or equal to for the receiver node to avoid an MEOL breach. As mentioned earlier, this does not preclude a breach in the receiver node's MEOL from combinations of leaf node operability levels that are different than x.

For example, in Problem 6.15 if leaf node feeder N_3 becomes fully inoperable then leaf node feeder N_2 must have an operability level greater than or equal to 75.57 utils for the operability level of N_j to be greater than or equal to its MEOL of 75 utils. In this situation, the operability level of N_j reaches a maximum of 85.75 utils—it can never reach full operability ($P_j = 100$) if leaf node feeder N_3 is fully inoperable; that is,

$$\text{If } P_3 = 0 \text{ then } \begin{cases} P_2 \geq 75.57 \text{ for } P_j \geq MEOP_j = 75 \\ P_j \leq 85.75 \end{cases}$$

This result is shown in Figure 6.35 by the bottom-most line. Figure 6.35 presents a family of lines with each identifying the tolerance interval within which the operability level of N_2 must fall for receiver node N_j to avoid an MEOL breach—given the fixed operability level of N_3. Figure 6.35 also shows the maximum operability level of receiver node N_j, if leaf node feeder N_2 is fully operable and the operability level of N_3 is as shown.

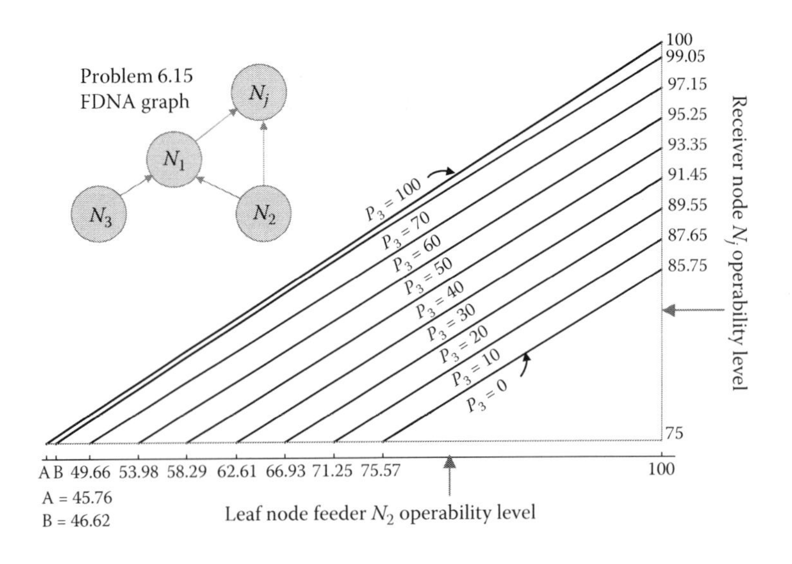

FIGURE 6.35
A family of tolerance intervals for Problem 6.15.

The analysis that went into constructing Figure 6.35 highlights one of many ways to structure and examine the tolerance of an FDNA graph. As mentioned earlier, modeling and measuring tolerance is an active area of research in the network science, engineering systems, and system sciences fields.

6.6 Special Topics

As stated earlier, FDNA is a new approach created to model and measure dependency relationships between suppliers of technologies and providers of services these technologies enable the enterprise to deliver. With any new approach, there are a number of extensions, special topics, and research areas to explore. This section discusses three areas. They are (1) regulating the operability function of dependent nodes, (2) an FDNA calculus to address nodes whose functionality are defined by constituent elements, and (3) ways to address cycle dependencies in an FDNA graph.

6.6.1 Operability Function Regulation

The flexibility in formulating the operability function between nodes is a desirable and important aspect of FDNA. Doing so requires understanding the context of the FDNA graph, the nature of its nodes, and the meaning of dependency relationships between them. For example, an operability function can be customized to regulate the rate that value (utils) from a feeder node flows into a receiver node.

Consider the single feeder–receiver node pair in Figure 6.36. What if circumstances are such that a receiver node is fully operable *before* its feeder node is fully operable? Can we regulate α_{ij} to address this circumstance?

Suppose we have a single feeder–receiver node pair as shown in Figure 6.36. In Figure 6.36, suppose there is no criticality of dependency constraint between receiver node N_j and feeder node N_i. Let α'_{ij} denote the regulated α_{ij} with

$$P_j = SODP_j = \alpha'_{ij}P_i + 100(1-\alpha_{ij}) \equiv \alpha'_{ij}P_i + BOLP_j$$

From this, it follows that $P_j = 100$ when

$$P_i = \frac{100 - BOLP_j}{\alpha'_{ij}} = 100\frac{\alpha_{ij}}{\alpha'_{ij}}$$

Furthermore, only when $\alpha_{ij} \le \alpha'_{ij}$ will $P_j = 100$ with $P_i \le 100$. Therefore, if $P_j = SODP_j = \alpha'_{ij}P_i + 100(1-\alpha_{ij})$ and $0 \le \alpha_{ij} \le \alpha'_{ij} \le 1$ then $P_j = 100$ when $P_i \le 100$. Under this $SODP_j$ regulation, feeder node N_i need not reach

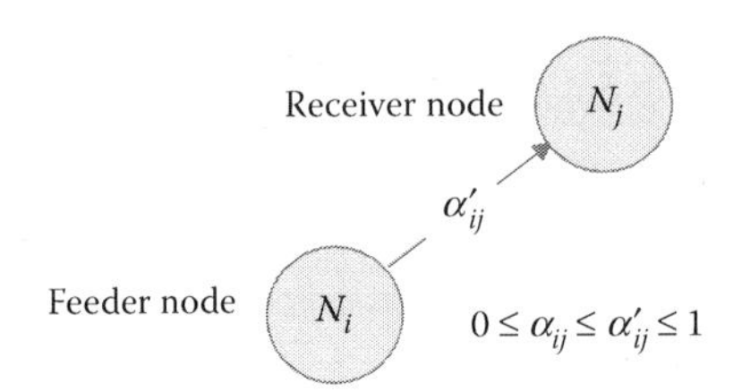

FIGURE 6.36
Regulated strength of dependency fraction α'_{ij}.

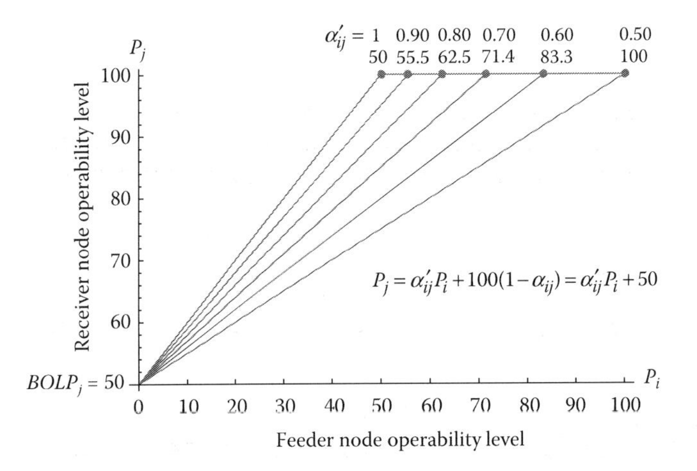

FIGURE 6.37
Families of $SODP_j$ functions for regulated α_{ij}.

an operability level of 100 utils for its receiver node N_j to become fully operable. This is illustrated in Figure 6.37. In Figure 6.37, observe that $BOLP_j = 100(1-\alpha_{ij}) = 50 \Rightarrow \alpha_{ij} = 0.50$. With this, a family of operability functions can be formed as shown in Figure 6.37. From left-to-right, they are:

$$P_j = P_i + 50 \Rightarrow P_j = 100 \text{ utils when } P_i = 50 \text{ utils}$$
$$P_j = 0.90P_i + 50 \Rightarrow P_j = 100 \text{ utils when } P_i = 55.5 \text{ utils}$$
$$P_j = 0.80P_i + 50 \Rightarrow P_j = 100 \text{ utils when } P_i = 62.5 \text{ utils}$$
$$P_j = 0.70P_i + 50 \Rightarrow P_j = 100 \text{ utils when } P_i = 71.4 \text{ utils}$$
$$P_j = 0.60P_i + 50 \Rightarrow P_j = 100 \text{ utils when } P_i = 83.3 \text{ utils}$$
$$P_j = 0.50P_i + 50 \Rightarrow P_j = 100 \text{ utils when } P_i = 100 \text{ utils}$$

This illustrates the flexibility FDNA offers in operability function formulation. FDNA enables operability functions to be designed to capture nuances in nodal dependencies and their interactions in very specific ways. Extending these ideas to FDNA graphs with multiple feeder nodes and complex nodal interactions is a rich area for further study.

6.6.2 Constituent Nodes

In an FDNA graph, suppose we have a node defined by two or more distinct components. Suppose each component makes a unique product. Suppose the value (utility or worth) of this node is a function of the value (utility or worth) of its products. In FDNA, such a node is called a constituent node.

Definition 6.14

An FDNA node characterized by two or more distinct components, where each component makes a unique product, is a *constituent node*. A node that is not a constituent node is a *single component node*. A single component node makes one and only one product.*

Definition 6.15

The operability level of a constituent node is a function of the operability levels of its distinct components.

Figure 6.38 illustrates an FDNA constituent node. In Figure 6.38, suppose N_i is a machine shop that manufactures five distinct components: cogwheels, gages, stamping dies, lathes, and rotor blades. In accordance with Definition 6.14, N_i is a constituent node. A constituent node is always separable into two or more distinct components. Nodes in the preceding discussions were not separable in this way. They delivered to, or received from, single component nodes. If the node in Figure 6.38 produced only cogwheels, then it would be a single component node.

A constituent node can be a feeder to, or a receiver from, other nodes in an FDNA graph. They may be other constituent nodes or other single component nodes. A component within a constituent node can be a feeder to, or a receiver

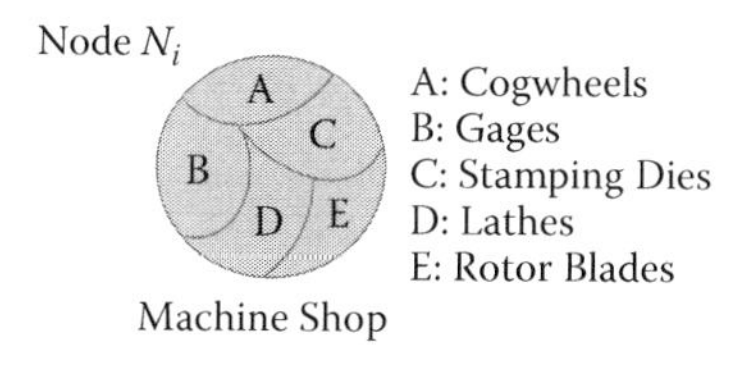

FIGURE 6.38
An FDNA constituent node.

* The nodes discussed in the preceding sections have all been single component node.

from, other components in the same node (an intracomponent dependency), or to other components in other constituent nodes, or to other constituent nodes (as a whole), or to other single component nodes in an FDNA graph.

If an FDNA node is a constituent node, then its operability level is defined as a function of the operability levels of its distinct components. The form of a component operability function is shaped by the particular dependency relationship present in the FDNA graph. For instance, the operability function of a component might be expressed by a single-dimensional value function. An example is shown in Figure 6.2. The operability function of a component might be expressed by a rule derived from one of the weakest link formulations, as discussed in the preceding sections. A combination of formulations might be appropriate for the components of a particular FDNA constituent node. These considerations are illustrated in the forthcoming examples.

If the value (worth or utility) of the output produced by each component in a constituent node meets certain independence conditions (Keeney and Raiffa, 1976), then from decision theory the overall operability function of the constituent node can be expressed as a linear additive sum of the component operability functions. Consider Figure 6.39.

In Figure 6.39, suppose A, B, and C are three distinct components that define constituent node N_i. Suppose the operability functions for A, B, and C are the value functions $P_A(x_A)$, $P_B(x_B)$, and $P_C(x_C)$, respectively. We can define the overall operability function of node N_i as

$$P_i = w_A P_A(x_A) + w_B P_B(x_B) + w_C P_C(x_C) \tag{6.79}$$

where w_A, w_B, and w_c, are nonnegative weights whose values range from 0 to 1, with $w_A + w_B + w_c = 1$, and $0 \leq P_i$, $P_A(x_A)$, $P_B(x_B)$, $P_c(x_c) \leq 100$. Equation 6.79 is a form of the additive value function in decision theory (Keeney and Raiffa, 1976; Kirkwood, 1997). From this, the operability level of an FDNA constituent node can be formally defined as follows.

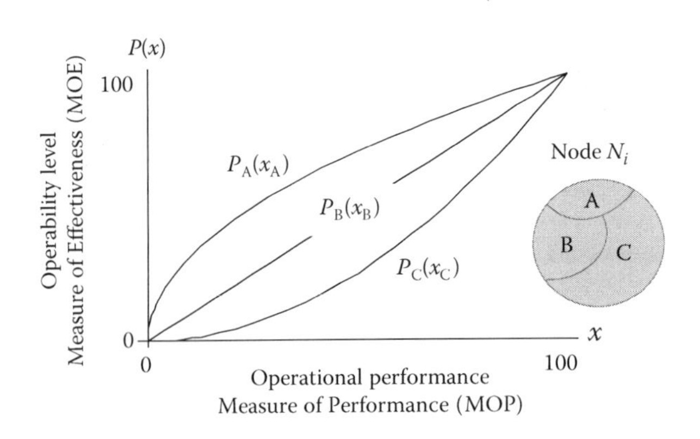

FIGURE 6.39
Component value functions for constituent node N_i.

Definition 6.16

The operability level P_y of an FDNA constituent node N_y with components $A_1, A_2, A_3, \ldots, A_s$ is given by

$$P_y = \sum_{i=1}^{s} w_i P_{A_i}(x_i) \tag{6.80}$$

where $w_1, w_2, w_3, \ldots, w_s$ are nonnegative weights whose values range from 0 to 1, $w_1 + w_2 + w_3 + \cdots + w_s = 1$, $P_{A_i}(x_i)$ is the operability function of A_i, and $0 \leq P_y, P_{A_i}(x_i) \leq 100$.

In Definition 6.16, if component A_i has a dependency relationship with another component (internal or external to N_y), then its operability function $P_{A_i}(x_i)$ is shaped by that particular relationship, as expressed in the FDNA graph. Equation 6.80 is a classical form of the Keeney-Raiffa additive value function. Figure 6.40 illustrates Definition 6.16.

PROBLEM 6.16

Formulate the FDF for the graph in Figure 6.41. Use FDNA (α, β) WLR.

Solution

Figure 6.41 consists of a constituent feeder node N_i and a receiver node N_j. In accordance with Definition 6.15, the operability level of constituent node N_i is a function of the operability levels of its three distinct components *A*, *B*, and *C*. In Figure 6.41, the arrow and the word *"All"* means receiver node N_j has a dependency relationship on each component contained in constituent feeder node N_i. With this, we can write the FDF for N_j as follows:

$$P_j = \text{Min}(\alpha_{ij} P_i + 100(1 - \alpha_{ij}), P_i + \beta_{ij})$$
$$P_i = w_A P_A + w_B P_B + w_C P_C$$

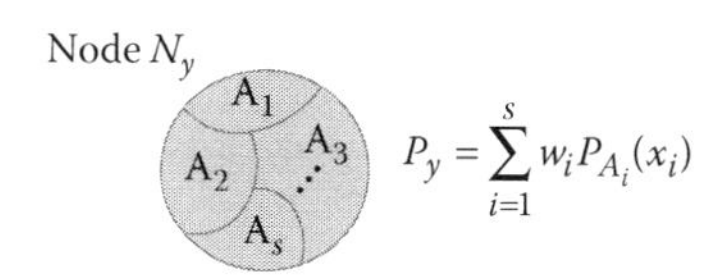

FIGURE 6.40
The operability function of constituent node N_y.

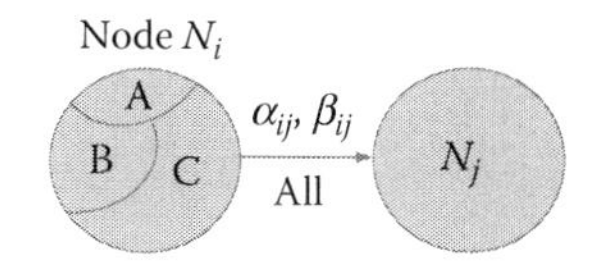

FIGURE 6.41
FDNA graph for Problem 6.16.

where w_A, w_B, and w_c are nonnegative weights whose values range from 0 to 1, $w_A + w_B + w_C = 1, P_A = P_A(x_A), P_B = P_B(x_B), P_C = P_C(x_C)$, and $0 \leq P_i, P_j, P_A, P_B, P_C \leq 100$. The terms P_A, P_B, and P_C are the operability levels of components A, B, and C, respectively. Their values derive from their value functions $P_A(x_A)$, $P_B(x_B)$, and $P_C(x_C)$, respectively. These value functions are the operability functions for A, B, and C.

PROBLEM 6.17

Formulate the FDF for the graph in Figure 6.42. Use FDNA (α, β) WLR.

Solution

Figure 6.42 consists of a feeder node N_i and a constituent receiver node N_j. In accordance with Definition 6.14, feeder node N_i is a single component node. The arrow pointing to the letters A, B means components A and B contained in N_j have a dependency relationship on feeder node N_i. With this, from Definition 6.16, the FDF for N_j is

$$P_j = w_A P_A + w_B P_B + w_C P_C$$

$$P_A = \mathrm{Min}(\alpha_{ij_A} P_i + 100(1 - \alpha_{ij_A}), P_i + \beta_{ij_A})$$

$$P_B = \mathrm{Min}(\alpha_{ij_B} P_i + 100(1 - \alpha_{ij_B}), P_i + \beta_{ij_B})$$

where w_A, w_B, w_C are nonnegative weights whose values range from 0 to 1, $w_A + w_B + w_C = 1, P_C = P_C(x_C)$ and $0 \leq P_i, P_j, P_A, P_B, P_C \leq 100$. The terms $\alpha_{ij_A}, \alpha_{ij_B}, \beta_{ij_A}$, and β_{ij_B} are strength and criticality of dependency parameters. The notation ij_A means component A contained in node N_j has a dependency relationship with node N_i. Likewise, the notation ij_B means component B contained in node N_j has a dependency relationship with node N_i. The term P_C is the operability level of component C. Its value derives from a value function $P_C(x_C)$. This value function represents the operability function for C.

PROBLEM 6.18

Formulate the FDF for the graph in Figure 6.43. Use FDNA (α,β) WLR.

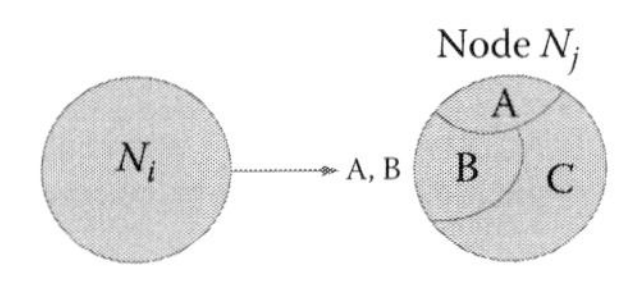

FIGURE 6.42
FDNA graph for Problem 6.17.

Solution

Figure 6.43 consists of a constituent feeder node N_i and a receiver node N_j. In accordance with Definition 6.14, receiver node N_j is a single component node. The two "puzzle" images and the "dot" on the left end of each arrow mean N_j has dependency relationships on components A and E contained in N_i. The point of each arrow touches the edge of N_j. This means the product created by N_j relies, to some degree, on the products created by components A and E contained in node N_i. With this, we can write the FDF for N_j as follows:

$$P_j = \text{Min}\left(\frac{\alpha_{i_A j} P_A}{2} + \frac{\alpha_{i_E j} P_E}{2} + 100\left(1 - \left(\frac{\alpha_{i_A j} + \alpha_{i_E j}}{2}\right)\right), P_A + \beta_{i_A j}, P_E + \beta_{i_E j}\right) \quad (6.81)$$

The terms $\alpha_{i_A j}, \alpha_{i_E j}, \beta_{i_A j}$, and $\beta_{i_E j}$ are strength and criticality of dependency parameters. The notation $i_A j$ means node N_j has a dependency on component A contained in node N_i. Likewise, the notation $i_E j$ means node N_j has a dependency on component E contained in node N_i. The terms P_A and P_E are the operability levels of components A and E, respectively. Their values derive from their value functions $P_A(x_A)$ and $P_E(x_E)$, which are the operability functions for A and E.

PROBLEM 6.19

Formulate the FDF for the graph in Figure 6.44. Use FDNA (α, β) WLR.

Solution

Figure 6.44 consists of two constituent nodes N_i and N_j. As shown in this figure, component G contained in N_j has a dependency relationship on

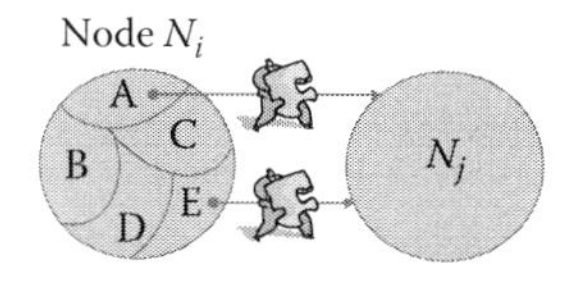

FIGURE 6.43
FDNA graph for Problem 6.18.

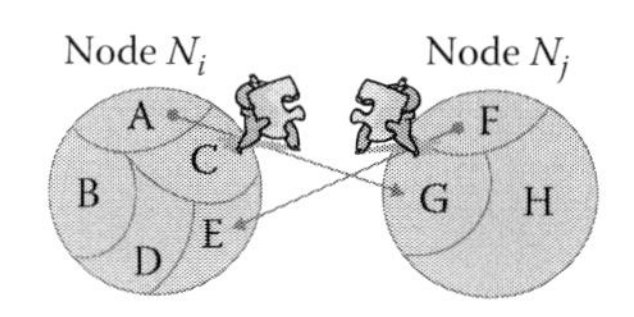

FIGURE 6.44
FDNA graph for Problem 6.19.

component A contained in N_i. Component E contained in N_i has a dependency relationship on component F contained in N_j. With this, we can write the FDFs for N_i and N_j as follows:

$$P_i = w_A P_A + w_B P_B + w_C P_C + w_D P_D + w_E P_E$$

$$P_j = w_F P_F + w_G P_G + w_H P_H$$

$$P_E = \text{Min}(\alpha_{j_F i_E} P_F + 100(1-\alpha_{j_F i_E}), P_F + \beta_{j_F i_E})$$

$$P_G = \text{Min}(\alpha_{i_A j_G} P_A + 100(1-\alpha_{i_A j_G}), P_A + \beta_{i_A j_G})$$

where $w_A, w_B, w_C, w_D, w_E, w_F, w_G$, and w_H, are nonnegative weights whose values range from 0 to 1, $w_A + w_B + w_C + w_D + w_E = 1$, $w_F + w_G + w_H = 1$, $P_A = P_A(x_A)$, $P_B = P_B(x_B)$, $P_C = P_C(x_C)$, $P_D = P_D(x_D)$, $P_F = P_F(x_F)$, and $P_H = P_H(x_H)$, and $0 \le P_i, P_j$, $P_A, P_B, P_C, P_D, P_E, P_F, P_G, P_H \le 100$.

The terms $\alpha_{j_F i_E}$, $\alpha_{i_A j_G}$, $\beta_{j_F i_E}$, and $\beta_{i_A j_G}$ are strength and criticality of dependency parameters. The notation $j_F i_E$ means component E contained in node N_i has a dependency on component F contained in node N_j. Likewise, the notation $i_A j_G$ means component G contained in node N_j has a dependency on component A contained in node N_i.

The terms P_A, P_B, P_C, P_D, P_F, and P_H are the operability levels of components A, B, C, D, F, and H, respectively. Their values derive from their value functions, which are the operability functions for these components contained in N_i and N_j.

PROBLEM 6.20

Formulate the FDF for the graph in Figure 6.45. Use FDNA (α, β) WLR.

Solution

The graph in Figure 6.45 consists of three nodes N_i, N_j, and Cap_1. Nodes N_i and N_j are constituent nodes. Component A in constituent node N_i is a feeder to component G in constituent node N_j. Component F in constituent node N_j is a feeder to component E in constituent node N_i. With this, we can write the FDFs for N_i, N_j, and Cap_1 as follows:

$$P_{Cap_1} = \text{Min}\left(\frac{\alpha_{iCap_1} P_i}{2} + \frac{\alpha_{jCap_1} P_j}{2} + 100\left(1 - \left(\frac{\alpha_{iCap_1} + \alpha_{jCap_1}}{2}\right)\right), P_i + \beta_{iCap_1}, P_j + \beta_{jCap_1}\right)$$

$$P_i = w_A P_A + w_B P_B + w_C P_C + w_D P_D + w_E P_E$$

$$P_j = w_F P_F + w_G P_G + w_H P_H$$

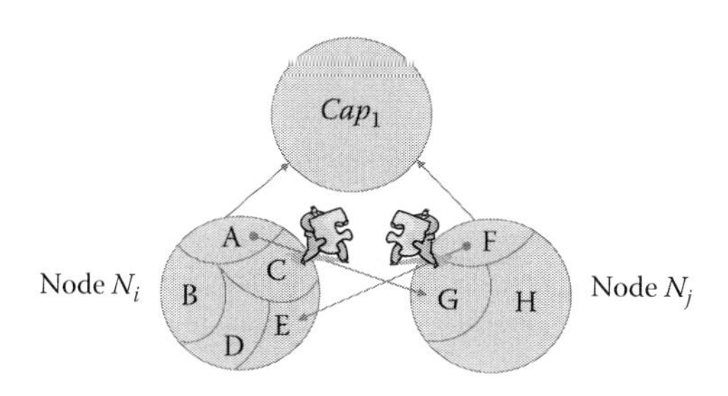

FIGURE 6.45
FDNA graph for Problem 6.20.

$$P_E = \text{Min}(\alpha_{j_F i_E} P_F + 100(1 - \alpha_{j_F i_E}), P_F + \beta_{j_F i_E})$$

$$P_G = \text{Min}(\alpha_{i_A j_G} P_A + 100(1 - \alpha_{i_A j_G}), P_A + \beta_{i_A j_G})$$

where $w_A, w_B, w_C, w_D, w_E, w_F, w_G$, and w_H, are nonnegative weights whose values range from 0 to 1, $w_A + w_B + w_C + w_D + w_E = 1$, $w_F + w_G + w_H = 1$, $P_A = P_A(x_A)$, $P_B = P_B(x_B)$, $P_C = P_C(x_C)$, $P_D = P_D(x_D)$, $P_F = P_F(x_F)$, $P_H = P_H(x_H)$, and $0 \le P_i, P_j, P_A, P_B, P_C, P_D, P_E, P_F, P_G, P_H \le 100$.

The terms $\alpha_{iCap_1}, \alpha_{jCap_1}, \beta_{iCap_1}$, and β_{jCap_1}, are strength and criticality of dependency parameters between N_i and Cap_1 and between N_j and Cap_1, respectively. The terms $\alpha_{j_F i_E}, \alpha_{i_A j_G}, \beta_{j_F i_E}$, and $\beta_{i_A j_G}$ are also strength and criticality of dependency parameters. The notation $j_F i_E$ means component E contained in node N_i has a dependency on component F contained in node N_j. Likewise, the notation $i_A j_G$ means component G contained in node N_j has a dependency on component A contained in node N_i.

The terms P_A, P_B, P_C, P_D, P_F, and P_H are the operability levels of components A, B, C, D, F, and H, respectively. Their values derive from their value functions, which are the operability functions for the components contained in N_i and N_j.

In Problem 6.20, suppose the FDNA parameters associated with the dependency relationships in Figure 6.45 are $\alpha_{iCap_1} = 0.90$, $\alpha_{jCap_1} = 0.45$, $\beta_{iCap_1} = 35$, $\beta_{jCap_1} = 60$, $\alpha_{j_F i_E} = 0.85$, $\alpha_{i_A j_G} = 0.30$, $\beta_{j_F i_E} = 25$, and $\beta_{i_A j_G} = 75$. If components A, B, C, and D contained in constituent node N_i and components F and H contained in constituent node N_j have operability levels shown in Table 6.6, then we can compute the operability levels for components E and G and nodes N_i, N_j, and Cap_1 as shown in Table 6.6. These computations assume equal weights among all components in constituent nodes N_i and N_j.

6.6.3 Addressing Cycle Dependencies

In general, a cycle dependency exists when a path connecting a set of nodes begins and ends with the same node. In graph theory, this is called a closed

TABLE 6.6

Computing Component and Constituent Node Operability Levels

Functional Dependency Network Analysis (FDNA): A Capability Portfolio			
	FDNA parameters		
α_{iCap_1}	0.90	β_{iCap_1}	35
α_{jCap_1}	0.45	β_{jCap_1}	60
$\alpha_{j_F i_E}$	0.85	$\beta_{j_F i_E}$	25
$\alpha_{i_A j_G}$	0.30	$\beta_{i_A j_G}$	75

If these components are functioning at these operability levels ...

At time t_1		At time t_2		At time t_3	
P_A	100	P_A	75	P_A	50
P_B	100	P_B	75	P_B	50
P_C	100	P_C	75	P_C	50
P_D	100	P_D	75	P_D	50
P_F	100	P_F	75	P_F	50
P_H	100	P_H	75	P_H	50

Then these receiver nodes are functioning at these operability levels ...

At time t_1		At time t_2		At time t_3	
P_E	100	P_E	78.75	P_E	57.50
P_G	100	P_G	92.50	P_G	85.00
P_i	100	P_i	75.75	P_i	51.50
P_j	100	P_j	80.83	P_j	61.67
P_{Cap_1}	100	P_{Cap_1}	84.78	P_{Cap_1}	69.55

path. In FDNA, Postulate 6.10 states that all FDNA graphs must be acyclic (without cycles). However, if a cycle dependency is identified then how might it be treated in an FDNA graph?

In FDNA, cycle dependencies can be evaluated by a procedure called *compartmentation*. Compartmentation is the process of assessing whether cycle dependencies can be resolved by replacing nodes along the closed path with a path that connects entities in acyclic relationships. In FDNA, these entities might be components in a constituent node. They would capture the basis for the original cycle but express the true nature of the cycle dependency in acyclic ways. Compartmentation is not guaranteed to resolve cycle dependencies. If a cycle dependency is found to be irresolvable, then this may signal the nodes in the path are truly not separable into acyclic relationships.

Figure 6.46 illustrates what a successful compartmentation of a cycle dependency might look like. Figure 6.46 is a modified FDNA graph from Problem 6.20. With the right-most graph in Figure 6.46, an operability analysis can

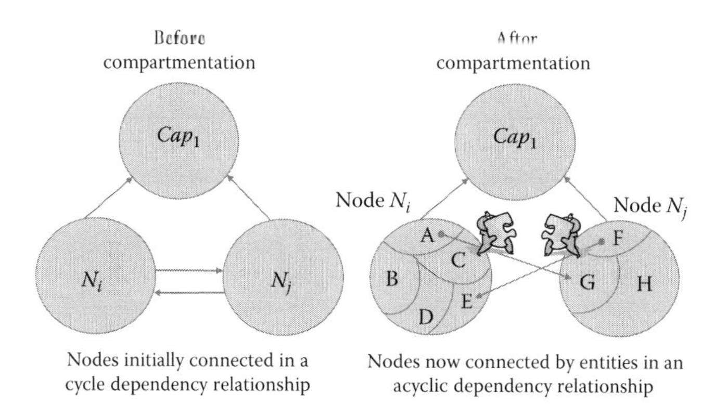

FIGURE 6.46
Compartmentation in an FDNA graph.

then be conducted as demonstrated with FDNA dependency functions provided in the solution to Problem 6.20.

6.7 Summary

Some of today's most critical and problematic areas in engineering management are identifying, representing, and measuring dependencies between entities involved in engineering an enterprise system. As discussed in this chapter, these entities can be suppliers of technologies and services and the users or receivers of these technologies and services.

The importance of understanding entity relationships is many-fold. Primary, is to study of ripple effects of failure of one entity on other dependent entities across an enterprise system. Providing mechanisms to anticipate these effects early in a system's design enables engineers to minimize dependency risks that, if realized, may have cascading negative effects on the ability of an enterprise to achieve capability objectives in the delivery of services to users.

This chapter presented FDNA, which enables designers, engineers, and managers to study and anticipate the ripple effects of losses in supplier-program contributions on an enterprise system's dependent capabilities, before risks that threaten these suppliers are realized. An FDNA analysis identifies whether the level of operability loss, if such risks occur, is tolerable in terms of acceptable levels of performance. This enables management to target risk resolution resources to those supplier programs that face high risk and are most critical to a system's operational capabilities.

FDNA is a new approach. Its calculus and use of graph theory to model complex dependency relationships enables addressing such problems in ways that can be difficult in matrix-based protocols, such as input–output (I/O) models in economic science (Leontief, 1966). FDNA has the potential to be a generalized approach for a variety of dependency problems, including those in input–output economics, critical infrastructure risk analysis, and nonstationary, temporal, dependency analysis problems.

There are a number of research areas to explore that will advance the theory and application of FDNA. These include the following:

Operability Function Formulations: This chapter presented weakest link rules originally designed for the FDNA operability function. Although these rules offer many desirable properties, other formulations are certainly possible. An area of further study is exploring other linear and non-linear forms for the FDNA operability function, discovering their properties, and identifying the conditions and contexts when they apply.

Analytical Scalability: Explore the analytical scalability of FDNA as a general dependency analysis methodology for large-scale logical topologies across a variety of problem spaces.

Nonstationary Considerations: Extend the FDNA calculus to include temporal features that address nonstationary dependency analysis problems. Research the integration of FDNA concepts within dynamic, time-varying modeling and simulation environments.

Degradation Tolerant, Resilient, and Adaptive Networks: Conduct research in network tolerance and resilience and ways to measure them in an enterprise system if, due to the realization of risks, one or more contributing programs or supplier–provider chains degrade or fail. Investigate the optimal design of adaptable supplier–provider networks. Explore ways a network can reconfigure surviving nodes to maintain operational effectiveness in the loss of critical nodes, from stationary and temporal perspectives.

Portfolio Optimization for Investment Decisions: Explore the integration of FDNA with portfolio optimization techniques. Build an analytic environment with optimization algorithms that will derive the best mix (or allocation) of resources needed to achieve an enterprise system's capability outcomes, subject to a variety of constraints such as cost, policy, and supplier–provider technology maturity considerations.

Probabilistic FDNA: Develop analytical protocols to conduct probabilistic analysis within the FDNA approach. This includes elicitation procedures for specifying uncertainty distributions around key FDNA parameters, such as α_{ij} and β_{ij}.

Alternative Dependency Analysis Approaches: Explore the relationship and contribution of FDNA to other dependency analysis approaches in the engineering systems and economics communities. These include design structure matrices (DSM), failure modes and effects analysis (FMEA), the Leontief input–output model (Leontief, 1966), the inoperability input–output model (IIM) (Jiang and Haimes, 2004, Santos, 2007), and system dynamics models.

Questions and Exercises

1. Answer the following given the FDNA graph in Figure 6.47.
   (A) Using the operability function given by Equation 6.10, compute the operability level of receiver node N_j if the operability levels of all feeder nodes are equally important to N_j.
   (B) Using the weakest link operability function given by Equation 6.11, what is the operability level of receiver node N_j?
   (C) Compare and contrast the results in (A) with the result in (B).
2. In Figure 6.47, suppose the operability function of node N_j is computed by the limit average weakest link rule. Suppose the operability levels of all three feeder nodes are equally important to N_j. If the operability level of receiver node N_j is 36 utils, then which feeder node is being constrained and what is the magnitude of its constraint?
3. Use FDNA (α, β) WLR to answer the following. Consider the dependency relationship between the widget production machine and its fluid supplier described in the narrative for the FDNA graph in Figure 6.12.
   (A) Determine α_{ij} and β_{ij} if $RPOLB_j = BOLP_j$.

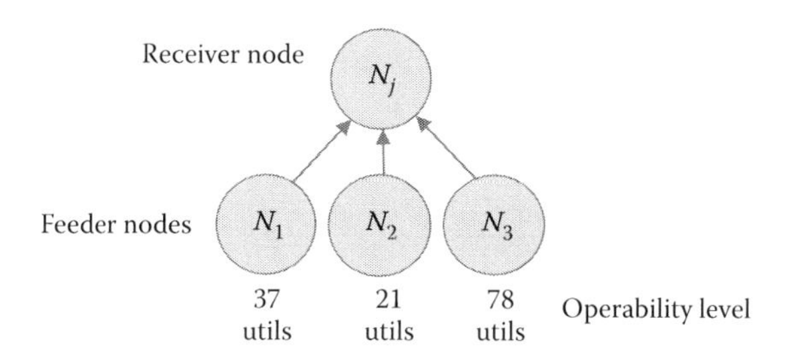

FIGURE 6.47
FDNA graph for Exercise 1.

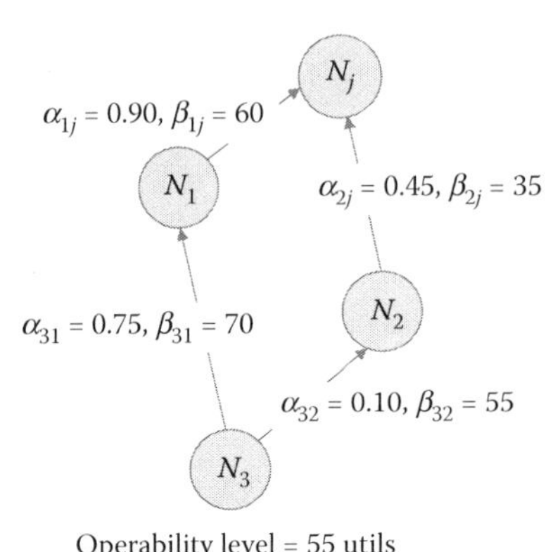

FIGURE 6.48
FDNA graph for Exercises 4 and 6.

(B) From (A) write the equations for $SODP_j$ and $CODP_j$.

(C) From (A) and (B) explain why $P_j = P_i$ for $0 \le P_i \le 100$, in this case.

4. Answer the following given the FDNA graph in Figure 6.48. Use the FDNA (α, β) WLR.
   (A) Compute the baseline operability levels of all nonleaf node feeders.
   (B) If the operability level of N_3 is 55 utils, determine the operability levels of the other nodes.
   (C) Determine the operability level of N_j, if N_2 has a maximum criticality of dependency on N_3.
   (D) Determine the operability level of N_j, if no criticality constraints are present in any dependency relationship.
5. In FDNA (α, β) WLR, if receiver node N_j has dependency relationships on h feeder nodes and $\beta_{kj} = 0$, for all $k = 1, 2, 3, \ldots, h$, then show that $P_j = \text{Min}(P_1, P_2, P_3, \ldots, P_h)$.
6. Answer the following, given the FDNA graphs in Figures 6.48 and 6.49.
   (A) Compute the degradation index of all receiver nodes in Figures 6.48 and 6.49.
   (B) Compute the degradation index of all receiver nodes in Figure 6.49 if $\beta_{jp} = 100$ and $\beta_{kj} = 100$ for all $k = 1, 2, 3, 4, 5$.
7. Answer the following given FDNA graph in Figure 6.50.
   (A) Show that the operability level of receiver node N_j is determined by $SODP_j$ when $100(1 - \alpha_{ij}) \le \beta_{ij} \le 100$ with $0 \le \alpha_{ij} \le 1$.
   (B) Show that the operability level of receiver node N_j crosses over from being determined by $CODP_j$ to being determined by $SODP_j$ when $P_i > 100 - \dfrac{\beta_{ij}}{1 - \alpha_{ij}}$, $0 \le \alpha_{ij} < 1$.

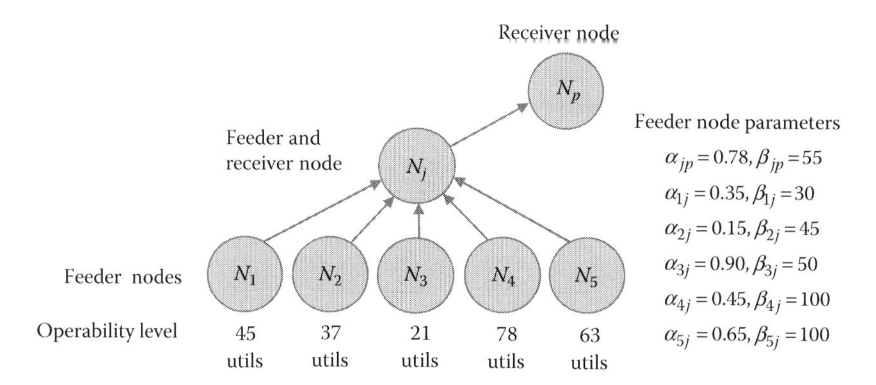

FIGURE 6.49
FDNA graph for Exercise 6.

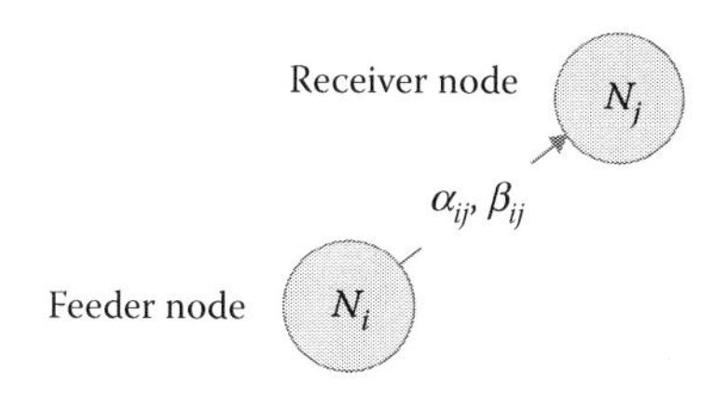

FIGURE 6.50
FDNA graph for Exercise 7.

TABLE 6.7
Table for Exercise 8

	$\beta_{ij} = 0$	$\beta_{ij} = 100$	$\beta_{ij} = BOLP_j$
$\alpha_{ij} = 0$	$P_j = P_i$	$P_j = 100$	$P_j = 100$
$\alpha_{ij} = 1$	$P_j = P_i$	$P_j = P_i$	$P_j = P_i$

8. Given the single feeder-receiver node pair in Figure 6.50, confirm the result for P_j shown in Table 6.7 for each row-column combination of α_{ij} and β_{ij}.
9. Formulate the FDNA dependency functions for the graph in Figure 6.51.
10. Formulate the FDNA dependency functions for the graph in Figure 6.52.
11. Formulate the FDNA dependency functions for the graph in Figure 6.53.
12. Conduct an operability analysis of the FDNA graph in Figure 6.26. Use FDNA (α, β) WLR given the following: $BOLP_j = 65$ utils, $\beta_{ij} = 50$ utils, and $MEOLP_j = 85$ utils. From this, answer the following:
    (A) Generate a new Table 6.2.

(B) Determine the feeder node operability level where the operability level of N_j stages its transition from being determined by $CODP_j$ to being determined by $SODP_j$.

13. Answer the following using the FDNA (α, β) WLR.

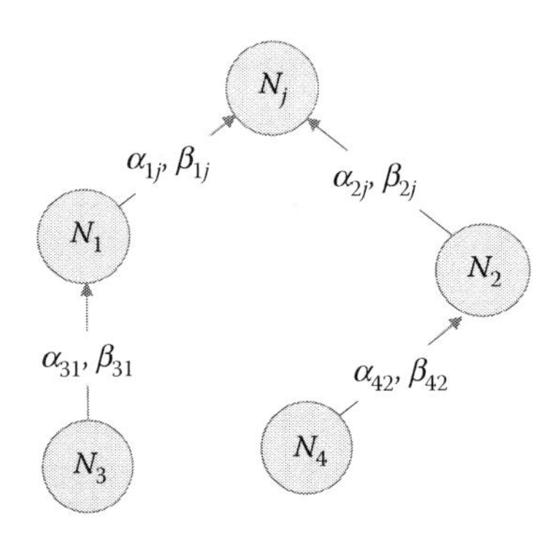

FIGURE 6.51
FDNA graph for Exercise 9.

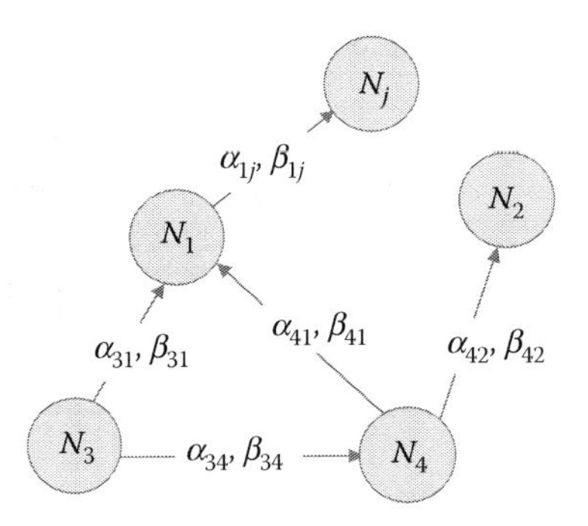

FIGURE 6.52
FDNA graph for Exercise 10.

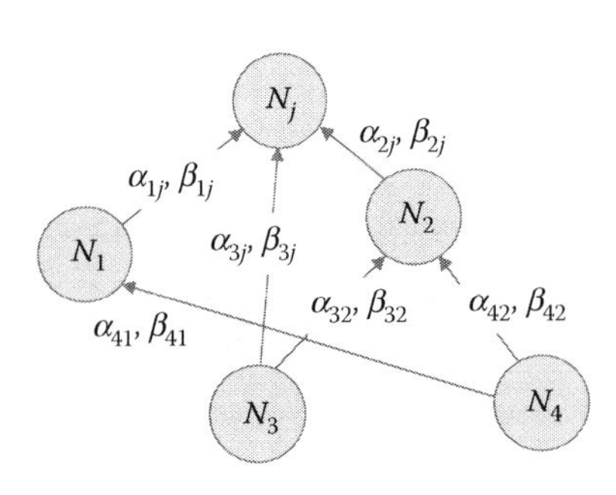

FIGURE 6.53
FDNA graph for Exercise 11.

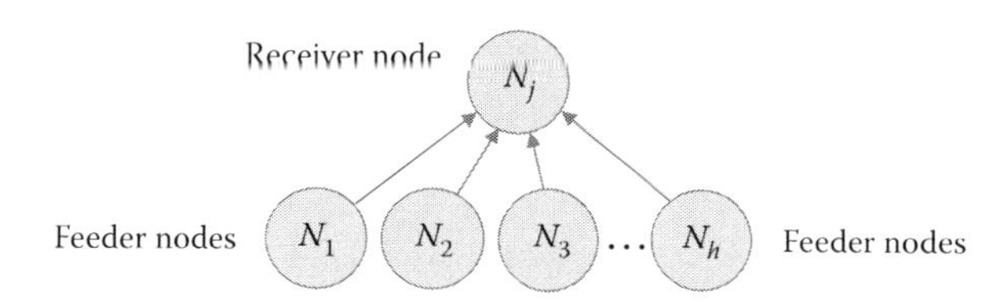

FIGURE 6.54
FDNA graph for Exercise 13.

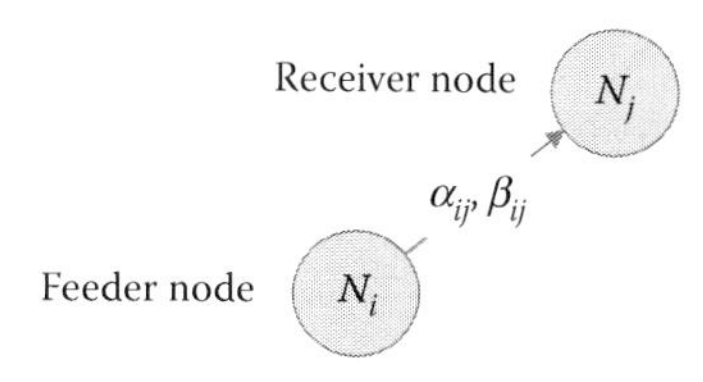

FIGURE 6.55
FDNA graph for Exercise 14.

(A) Suppose N_j is dependent on h feeder nodes as shown in Figure 6.54. If N_j becomes increasingly criticality unconstrained by the operability of its h feeder nodes, show that the operability function of N_j will increasingly approach $SODP_j$.

(B) Given (A), if N_j has no criticality constraints on these feeder nodes, show that $P_j = SODP_j$.

(C) Develop a graphic that illustrates the results from (A) and (B).

14. Suppose N_j is dependent on N_i as shown in Figure 6.55. If $\alpha_{ij} = 1$, prove that $P_j = SODP_j$ for all β_{ij}.
15. Derive the strength of dependency fractions for each dependency relationship in Figure 6.24 given the following: $BOLP_j = 55$ utils and α_{1j} and α_{2j} are each considered 50% more important than α_{3j} in improving the baseline operability level of receiver node N_j.
16. If the operability level of node N_3 is 70 utils for the FDNA graph in Problem 6.15, then

(A) Show that the degradation index of receiver node N_j with respect to N_2 is as follows:

$$\begin{cases} 0.63 & \text{for } 0 \le P_2 < 35.8427 \\ 0.4609 & \text{for } 35.8427 \le P_2 < 73.6364 \\ 0.44 & \text{for } 73.6364 \le P_2 \le 100 \end{cases}$$

(B) Find the operability level of node N_2 at the point where the operability levels of node N_1 and N_j are equal.

17. Verify the following, given the graph and FDNA parameters in Problem 6.15. Relate this to Figure 6.35.

$$\text{If } P_3 = 80 \text{ then} \begin{cases} P_2 \geq 46.335 & \text{for } P_j \geq MEOP_j = 75 \\ P_j \leq 99.734 \end{cases}$$

$$\text{If } P_3 = 90 \text{ then} \begin{cases} P_2 \geq 46.0468 & \text{for } P_j \geq MEOP_j = 75 \\ P_j \leq 99.867 \end{cases}$$

18. Consider the industry sectors oil and electricity. Suppose oil production requires electricity to run its facilities and electricity production requires oil for its generation and distribution. Given only this information, we have a cycle dependency between these sectors. This is shown by the FDNA graph in Figure 6.56 (the looped arrows imply a nodal intradependency is present).

    From Section 6.6, recall that compartmentation is the process of assessing whether cycle dependencies can be resolved by replacing nodes along the closed path with a path that connects entities in acyclic relationships. These entities might be components in a constituent node.

    Suppose compartmentation resolved the cycle dependency in Figure 6.56 into (1) the constituent nodes shown in Figure 6.57 and (2) into the acyclic relationships between its components, as shown in Figure 6.58.

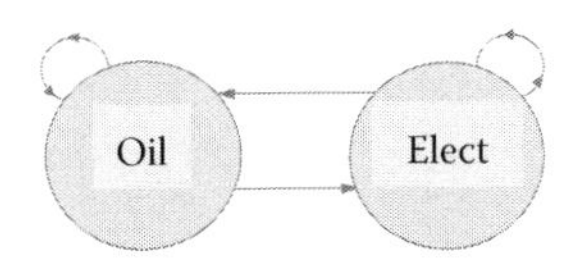

FIGURE 6.56
A cycle dependency between oil and electricity.

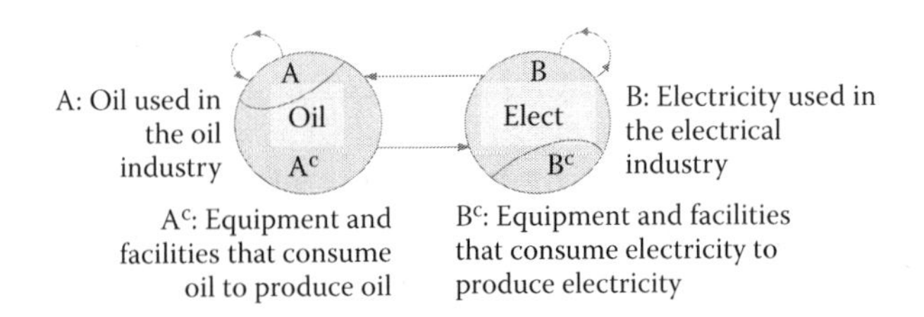

FIGURE 6.57
A constituent node representation of the cycle in Figure 6.56.

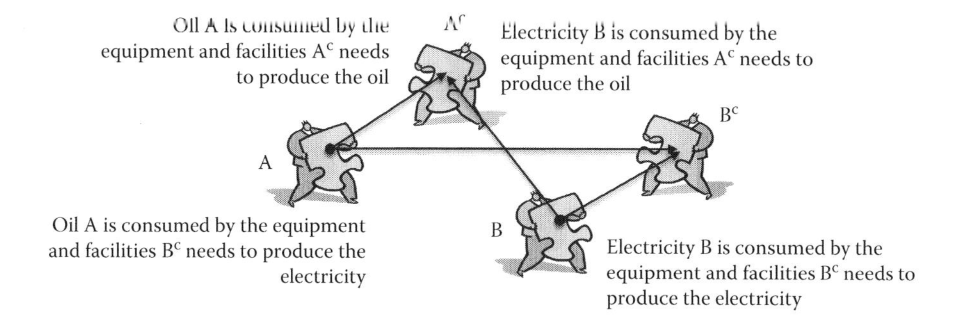

FIGURE 6.58
Acyclic component dependency relationships within Figure 6.56.

$$C = \begin{pmatrix} c_{11} & c_{12} \\ c_{21} & c_{22} \end{pmatrix} \begin{matrix} \text{Oil} \\ \text{Electricity} \end{matrix} \quad (\text{columns: Oil, Electricity})$$

Matrix row		Matrix column	
i = Input	$\longrightarrow$	j = Output	Equivalent
i = Producer	$\longrightarrow$	j = Consumer	Equivalent
i = Feeder	$\longrightarrow$	j = Receiver	Equivalent

FIGURE 6.59
An input–output matrix.

$$\alpha = \begin{pmatrix} A \rightarrow A^c & A \rightarrow B^c \\ B \rightarrow A^c & B \rightarrow B^c \end{pmatrix} = \begin{pmatrix} 0.35 & 0.15 \\ 0.25 & 0.10 \end{pmatrix}$$

Feeder $\longrightarrow$ Receiver

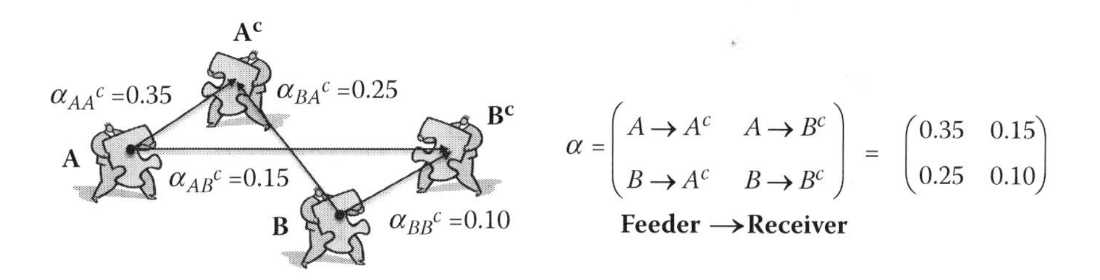

FIGURE 6.60
An input–output matrix of strength of dependency fractions.

From this, formulate the FDNA dependency functions for the graph in Figure 6.57 if the acyclic component relationships are as shown in Figure 6.58. Use the FDNA (α, β) WLR.

19. The two industry sectors in Exercise 18 can be represented by a protocol in economics known as the Leontief input–output matrix (Leontief, 1966). Such a matrix is shown in Figure 6.59.

From Exercise 18, suppose we populate Leontief's input–output matrix C with entries that reflect the strength of dependency fractions between the components in Figure 6.58. Suppose this is given by the alpha values on the FDNA component graph or equivalently by the entries in the α matrix in Figure 6.60.

Given the information in Figure 6.60, compute the operability levels of the Oil and Electricity nodes in Figure 6.57 if components A and B (in Figure 6.60) both have operability levels equal to 100 utils at time t_1, equal to 75 utils at time t_2, and equal to 50 utils at time t_3. Assume the FDNA (α, β) WLR applies. Assume no criticality constraints affect operability levels.

20. Questions for Research Projects or Papers

    (A) Explore the relationship between FDNA and Wassily Leontief's Nobel Prize winning work in the theory of input–output models, developed in the late 1950s, to measure economic consumption and demand across interdependent industries or sectors (Leontief, 1966).

    (B) In Section 6.7, areas for research in FDNA were identified. Select an area and review its description. Write a research paper on your selected area that discusses the issues, ways they might be addressed, and experiment with trial solutions that advance FDNA in the area being explored.

7

A Decision-Theoretic Algorithm for Ranking Risk Criticality

7.1 Introduction

This chapter presents a decision-theoretic approach for prioritizing risk management decisions as a function of risk measures and analysis procedures developed in the previous chapters. These measures are integrated into advanced ranking algorithms to isolate and prioritize which capabilities are most risk-threatened and qualify for deliberate management attention. This methodology enables decision-makers to target risk reduction resources in ways that optimally reduce threats posed by risks to critical capability outcome objectives.

7.2 A Prioritization Algorithm

Management decisions often involve choosing the "best" or the "most-preferred" option among a set of competing alternatives. Similar selection decisions exist in risk management. Instead of choosing the most-preferred alternative, risk management decisions involve choosing the most-preferred risks to reduce, or eliminate, because of their threats to capability. In either situation, the common question is "How to identify selecting the best option from a set of competing alternatives?" Addressing this question is the focus of rational decision-making, supported by a variety of analytical formalisms developed in the last 300 years.

In general, selection algorithms produce rankings of options from a finite set of competing alternatives as a function of how each alternative performs across multiple evaluation criteria. Algorithms that produce ordered rankings fall into two classes. These algorithms use either ordinal or cardinal methods to generate the rankings.

Ordinal methods apply scales to rate the performance of alternatives by numbers that represent order. Ordinal scales are common in the social sciences, where they are often used for attitude measurement. However, only the ordering of numbers on these scales is preserved. The distance between them is indeterminate (not meaningful). Arithmetic operations beyond "greater than," "less than," or "equal to" are impermissible. Thus, ordinal ranking algorithms isolate which alternative in a finite set of competing alternatives is more critical than the others. However, they cannot measure the distance between ranked alternatives.

Cardinal methods apply scales to rate the performance of alternatives by numbers that represent an ordered metric. This means the distance between numbers on a cardinal scale is determinate (meaningful). Examples of numbers on a cardinal scale include the probability measure or degrees centigrade. For ranking risk criticality, a cardinal-based approach is frequently employed. This approach provides the flexibility to join optimization protocols with ranking algorithms when optimal assignments of risk reduction resources, under a variety of constraints, need to be determined.

As mentioned above, there are many algorithms that can be tailored to address the problem of ranking risks (say) from most-to-least critical to an engineering system. Many have their origins in vNM expected utility theory (von Neumann and Morgenstern, 1944).

From utility theory, a well-established algorithm known as the linear additive model is a popular approach (Keeney and Raiffa, 1976). A form of the linear additive model is given by Equation 7.1 (refer to Chapter 3, Definition 3.6). Furthermore, it has been proved if the criteria in a selection problem are mutually preferentially independent, then the evaluator's preferences can be represented by an additive value function (Keeney and Raiffa, 1976).

7.2.1 Linear Additive Model

A value function $V_Y(y)$ is an additive value function if there exists n single-dimensional value functions $V_{X_1}(x_1), V_{X_2}(x_2), V_{X_3}(x_3), \ldots, V_{X_n}(x_n)$, satisfying

$$V_Y(y) = w_1 V_{X_1}(x_1) + w_2 V_{X_2}(x_2) + w_3 V_{X_3}(x_3) + \cdots + w_n V_{X_n}(x_n) \qquad (7.1)$$

where w_i for $i = 1, \ldots, n$ are nonnegative weights (importance weights), whose values range between 0 and 1 and where $w_1 + w_2 + w_3 + \cdots + w_n = 1$.

The linear additive model is representative of a class of decision rules known as compensatory models. Compensatory models allow trade-offs to compete between attributes (or criteria). For instance, an alternative with low scores on some attributes (or criteria) can improve in its attractiveness to a decision-maker if this is compensated by high values on other attributes; hence, the average or expected value effect results from compensatory decision models.

There is another class of decision rules known as compromise solution models. These rules assume that choice among alternatives depends on a reference point (e.g., an ideal set of outcomes on all attributes) and attempts to minimize the distance between alternatives and the reference point (Malczewski, 1999).

More commonly referred to as ideal point methods, these approaches generate a complete ranking of alternatives as a function of their relative distance from the hypothetical ideal (the alternative characterized by attributes with all ideal values). According to Malczewski (1999), "ideal point methods treat alternatives and their attributes as inseparable bundles, all competing for closeness in similarity to the ideal alternative."

The ideal point represents a hypothetical alternative that consists of the most desirable weighted normalized levels of each criterion across the set of competing alternatives. The alternative closest to the ideal point solution performs best in the set. Separation from the ideal point is measured geometrically by a Euclidean distance metric.

Ranking algorithms from vNM decision theory are rooted in maximizing the expected utility. In contrast, those that derive from ideal point methods are rooted in maximizing similarity to the ideal solution. The best alternative is the compromise solution relative to that reference point.

As seen in the previous chapters, complex dependency relationships are the norm and not the exception in engineering an enterprise system. Entities such as supplier–provider nodes play key roles in planning, engineering, and managing an enterprise. Their effects on the success or failure of delivering to users are such that trade-offs between them might not be realistic or even advisable. For this reason, we approach the ranking problem in this chapter by a compromise model. This does not preclude the use of compensatory models if and when trade-offs between entities are reasonable. This is a fruitful area of continued research.

7.2.2 Compromise Models

Figure 7.1 illustrates the motivation for the development of compromise solution models. Two alternatives A_1 and A_2 are shown in relation to two benefit criteria or attributes (Attribute 1 and Attribute 2). In Figure 7.1, observe that A_1 is closest to the ideal solution A^*, but A_2 is farthest from the negative ideal solution $A-$. Given this, which alternative do you choose?

An ideal point method that reconciles this question is the Technique for Order Preference by Similarity to Ideal Solution (TOPSIS) (Hwang and Yoon, 1995). TOPSIS is an ideal point method that ensures the chosen alternative is simultaneously closest to the ideal solution *and* farthest from the negative ideal solution. It chooses the alternative whose performance across all criteria maximally matches those that comprise the ideal solution.

TOPSIS assumes each attribute (or criterion) can be characterized by either monotonically increasing or decreasing utility. Here, we seek to maximize

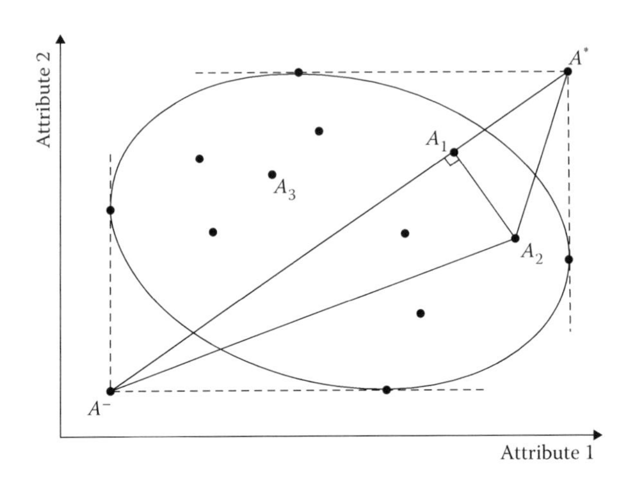

FIGURE 7.1
Euclidean distances to positive and negative ideal solutions. (Reprinted from Hwang and Paul, *Multiple Attribute Decision Making: An Introduction*, Sage University Paper Series in Quantitative Applications in the Social Sciences, 07–104, Sage Publications, Thousand Oaks, California, copyright 1995 by Sage Publications, Inc. With permission.)

attributes that offer a benefit and minimize those that incur a cost. TOPSIS generates an index that rank-orders competing alternatives from the most-to-least desired on the relative distance of each to the ideal solution.

The TOPSIS algorithms operate on a generalized decision matrix of alternatives as shown in Table 7.1. Here, the performance of an alternative is evaluated across competing criteria. The attractiveness of an alternative to a decision-maker is a function of the performance of each alternative across these criteria. Applying TOPSIS consists of the following steps and equations.

Step 1

Normalize the decision matrix of alternatives (Table 7.1). One way is to compute r_{ij} where

$$r_{ij} = \frac{x_{ij}}{\sqrt{\sum_i x_{ij}^2}} \quad i = 1,\ldots,m;\ j = 1,\ldots,n \tag{7.2}$$

Step 2

Compute a matrix of weighted normalized values according to v_{ij},

$$v_{ij} = w_j r_{ij} \quad i = 1,\ldots,m;\ j = 1,\ldots,n \tag{7.3}$$

where w_j is weight of the jth attribute (criterion).

TABLE 7.1

A Traditional Decision or Performance Matrix of Alternatives

Decision	**Criteria and Weights**				
	C_1	C_2	C_3	...	C_n
Alternative	w_1	w_2	w_3	...	w_n
A_1	x_{11}	x_{12}	x_{13}	...	x_{1n}
A_2	x_{21}	x_{22}	x_{23}	...	x_{2n}
A_3	x_{31}	x_{32}	x_{33}	...	x_{3n}
...	...	...	...	...	...
A_m	x_{m1}	x_{m2}	x_{m3}	...	x_{mn}

A Decision Matrix

Step 3

Derive the positive A^* and the negative A^- ideal solutions, where

$$A^* = \{v_1^*, v_2^*, \ldots, v_j^*, \ldots, v_n^*\} = \{(\max_i v_{ij} \mid j \in J_1), (\min_i v_{ij} \mid j \in J_2) \mid i = 1, \ldots, m\} \quad (7.4)$$

$$A^- = \{v_1^-, v_2^-, \ldots, v_j^-, \ldots, v_n^-\} = \{(\min_i v_{ij} \mid j \in J_1), (\max_i v_{ij} \mid j \in J_2) \mid i = 1, \ldots, m\} \quad (7.5)$$

where J_1 is the set of benefit attributes, and J_2 is the set of cost attributes.

Step 4

Calculate separation measures between alternatives as defined by the n-dimensional Euclidean distance metric. The separation from the positive-ideal solution A^* is given by

$$S_i^* = \sqrt{\sum_j (v_{ij} - v_j^*)^2} \quad i = 1, \ldots, m \quad (7.6)$$

The separation from the negative-ideal solution A^- is given by

$$S_i^- = \sqrt{\sum_j (v_{ij} - v_j^-)^2} \quad i = 1, \ldots, m \quad (7.7)$$

Step 5

Calculate similarities to positive-ideal solution as follows:

$$0 \le C_i^* = \frac{S_i^-}{(S_i^* + S_i^-)} \le 1 \quad i = 1, \ldots, m \quad (7.8)$$

Step 6

Choose the alternative in the decision matrix with the maximum C_i^*, or rank these alternatives from most-to-least preferred according to C_i^* in descending order. The closer C_i^* is to unity, the closer it is to the positive-ideal solution. The farther C_i^* is from unity, the farther it is from the positive-ideal solution. The optimal compromise solution produced from these steps is given by

$$Max\{C_1^*, C_2^*, C_3^*, \ldots, C_n^*\}$$

The alternative with the maximum C_i^* will be closest to the ideal solution and concurrently farthest from the least ideal solution.

7.2.3 Criteria Weights

Weighting the importance of each criterion in a decision matrix is a key consideration that influences the outcomes of a decision analysis. In the TOPSIS algorithm, criteria weights are entered at Step 2. How is weighting determined?

The literature presents many ways to derive criteria weights (Clemen, 1996; Kirkwood, 1997). These include subjective weighting methods, objective weighting methods, and a mix of these approaches. Subjective weights are primarily developed from judgment-driven assessments of criteria importance and can vary from person-to-person. Objective weights are primarily fact-driven and are developed by *quantifying the intrinsic information* (Diakoulaki, 1995) observed in each criterion. Objective weighting has the desirable feature of letting the "data" influence which criterion, in the set of criteria, is most important, which is next most important, and so forth.

The canonical approach to objective criteria weighting is the entropy method. Entropy* is a concept found in information theory that measures the uncertainty associated with the expected information content of a message. It is also used in decision theory to measure the amount of decision information contained and transmitted by a criterion.

The amount of decision information contained and transmitted by a criterion is driven by the extent the performance (i.e., "score") of each alternative is distinct and differentiated by that criterion. When alternatives (in a decision matrix) all have the same performance for a criterion, we say the

* In information theory, entropy measures the uncertainty associated with the expected information content of a message. The classical work in information entropy is available in Shannon, C. E., 1948. "A Mathematical Theory of Communication," *Bell System Technical Journal*.

criterion is unimportant. It can be dropped from the analysis because it is not transmitting distinct and differentiating information. The more distinct and differentiated the performance of competing alternatives on a criterion, the greater the amount of decision information contained and transmitted by that criterion; hence, the greater is its importance weight.

In decision theory, entropy is used to derive objective measures of the relative importance of each criterion (i.e., its weight) as it influences the performance of competing alternatives. If desired, prior subjective weights can be folded into objectively derived entropy weights. The following steps present the equations for computing entropy-derived objective weights used to derive the "most-preferred" alternative in a decision matrix.

Step 1

From the decision matrix in Table 7.1, compute p_{ij} where

$$p_{ij} = \frac{x_{ij}}{\sum_i x_{ij}} \quad i = 1, \ldots, m; \; j = 1, \ldots, n \tag{7.9}$$

Step 2

Compute the entropy of attribute (criterion) j as follows:

$$0 \le E_j = -\frac{1}{\ln(m)} \sum_i p_{ij} \ln p_{ij} \le 1 \quad i = 1, \ldots, m; \; j = 1, \ldots, n \tag{7.10}$$

Step 3

Compute the degree of diversification d_j of the information transmitted by attribute (criterion) j according to $d_j = 1 - E_j$.

Step 4

Compute the entropy-derived weight w_j as follows:

$$w_j = \frac{d_j}{\sum_j d_j} \quad j = 1, \ldots, n \tag{7.11}$$

If the decision maker has prior subjective importance weights λ_j for each attribute (criterion), then this can be adapted into w_j as follows:

$$w_j^{\bullet} = \frac{\lambda_j w_j}{\sum_j \lambda_j w_j} \quad j = 1, \ldots, n \tag{7.12}$$

In Step 2, observe that entropy weighting involves the use of the natural logarithm. Thus, weights derived from this approach require all elements in the

TABLE 7.2

A Decision Matrix that Yields an Undefined Entropy Weight

Capability Node	C-Node Risk Score	C-Node Risk Mitigation Dollars
C-Node 1	96.34	5.5
C-Node 2	54.67	7.8
C-Node 3	45.32	12.3
C-Node 4	77.78	0
C-Node 5	21.34	11.2
C-Node 6	66.89	9.3
C-Node 7	89.95	2.5

decision matrix be strictly greater than zero; otherwise, for some *ij* the term $p_{ij} \ln p_{ij}$ in the expression

$$0 \le E_j = -\frac{1}{\ln(m)} \sum_{i=1}^{m} p_{ij} \ln p_{ij} \le 1 \quad i = 1,\ldots,m;\ j = 1,\ldots,n$$

takes the indeterminate form $0 \cdot \infty$. In mathematics, this is one of seven indeterminate forms involving 0, 1, and ∞ whose overall limit is unknown. Table 7.2 shows a decision matrix in which this condition will arise, where *suppose* a true zero exists as shown in the last column of C-Node 4. However, it can be shown if $p \to 0$ (in the limit), then the expression $p \ln p$ approaches zero; hence, entropy calculations often use the convention $0 \cdot \ln 0 = 0$.

An alternative to objective weighting by the entropy measure is the variance-to-mean ratio (VMR). Like entropy, the VMR is a measure of the uncertainty or dispersion of a distribution. Distributions characterized by data with VMRs less than one are considered less random (more uniform) than those with VMRs greater than one. A probability distribution's VMR is often compared to the VMR of a Poisson distribution, whose VMR is exactly one. If alternatives in a decision matrix all have the same performance on a criterion, then the criterion's VMR is zero. Thus, the criterion can be dropped from the analysis because it is not transmitting distinct and differentiating information. Recall, this is also a property of the entropy measure.

Deriving objective weights for criteria by the VMR statistic* is done as follows:

$$w_j = \frac{\sigma_j^2/\mu_j}{\sum_j \sigma_j^2/\mu_j} \quad \mu_j \ne 0,\ j = 1,\ldots,n \tag{7.13}$$

* Note the VMR statistic requires the mean of a dataset be nonzero.

If a decision maker has prior subjective importance weights λ_j for each criterion, then this can be adapted into w_j by $w_j^{\bullet}$ in the same way shown in Step 4 of the entropy weighting method.

7.2.4 Illustration

This section illustrates how the TOPSIS algorithm can be used to prioritize risk management decisions. Suppose we have the following information in Table 7.3, which presents characteristics on seven capability nodes (C-Nodes) across five criteria. These are C-Node Risk Score, C-Node Functional dependency network analysis (FDNA), C-Node Risk Mitigation Dollars, C-Node Risk Reduction Benefit, and C-Node Criticality Level. They are defined as follows:

C-Node Risk Score

A value between 0 and 100 quantifies the C-Node's risk. The data for this criterion derives from the risk score equations, measures, and risk inheritance considerations developed in the preceding chapters.

C-Node FDNA

The quantified effect on a C-Node's operability if, due to the realization of risks, one or more contributing programs or supplier–provider chains degrade, fail, or are eliminated. The data for this criterion are expressed as a percentage below the minimum effective operational level (MEOL) defined in Chapter 6.

C-Node Risk Mitigation Dollars

The dollars (in millions) estimated to mitigate a C-Node's risks.

TABLE 7.3
Capability Node Risk Management Decision Matrix

Capability Node	C-Node Risk Score	C-Node FDNA	C-Node Risk Mitigation Dollars	C-Node Risk Reduction Benefit	C-Node Criticality Level
C-Node 1	96.34	5.0	5.5	85.0	3.0
C-Node 2	54.67	2.1	7.8	44.0	4.0
C-Node 3	45.32	3.2	12.3	78.0	3.0
C-Node 4	77.78	4.1	8.5	45.0	2.0
C-Node 5	21.34	0.1	11.2	56.0	1.0
C-Node 6	66.89	5.3	9.3	76.0	2.0
C-Node 7	89.95	3.3	2.5	25.0	5.0

C-Node Risk Reduction Benefit

A C-Node's risk-reduction benefit expected from expending its risk mitigation dollars. Here, benefit is expressed as the percent reduction in the C-Node's risk score.

C-Node Criticality Level

This criterion represents a C-Node's mission criticality with respect to the outcome objectives of the portfolio. The levels range from 1 (least critical) to 5 (most criticality). For convenience, assume these levels reflect numbers defined along a cardinal interval scale.

Given the above, suppose the management question is: "What is the most favorable ordering of C-Nodes in Table 7.3 that maximally reduces capability risk and minimizes the expense of risk mitigation dollars?" The TOPSIS algorithm can be applied to address this question. The algorithm's computational steps and results are shown and summarized in Tables 7.4a and 7.4b.

TABLE 7.4a

A TOPSIS-Derived Capability Risk Prioritization

Capability Node Risk Management Prioritization					
Capability Node	**C-Node Risk Score**	**C-Node FDNA**	**C-Node Risk Mitigation Dollars**	**C-Node Risk Reduction Benefit**	**C-Node Criticality Level**
C-Node 1	96.34	5.0	5.5	85.0	3.0
C-Node 2	54.67	2.1	7.8	44.0	4.0
C-Node 3	45.32	3.2	12.3	78.0	3.0
C-Node 4	77.78	4.1	8.5	45.0	2.0
C-Node 5	21.34	0.1	11.2	56.0	1.0
C-Node 6	66.89	5.3	9.3	76.0	2.0
C-Node 7	89.95	3.3	2.5	25.0	5.0
Sum	452.29	23.10	57.10	409.00	20.00
Root Sum of Squares	**182.742**	**9.770**	**23.083**	**163.728**	**8.246**
Normalized Matrix	**C-Node Risk Score**	**C-Node FDNA**	**C-Node Risk Mitigation Dollars**	**C-Node Risk Reduction Benefit**	**C-Node Criticality Level**
C-Node 1	0.5272	0.5118	0.2383	0.5192	0.3638
C-Node 2	0.2992	0.2149	0.3379	0.2687	0.4851
C-Node 3	0.2480	0.3275	0.5329	0.4764	0.3638
C-Node 4	0.4256	0.4197	0.3682	0.2748	0.2425
C-Node 5	0.1168	0.0102	0.4852	0.3420	0.1213
C-Node 6	0.3660	0.5425	0.4029	0.4642	0.2425
C-Node 7	0.4922	0.3378	0.1083	0.1527	0.6063
Norm of Normalized Cols	**1.000**	**1.000**	**1.000**	**1.000**	**1.000**

TABLE 7.4a (continued)

A TOPSIS-Derived Capability Risk Prioritization

Capability Node Risk Management Prioritization					
Entropy Matrix	**C-Node Risk Score**	**C-Node FDNA**	**C-Node Risk Mitigation Dollars**	**C-Node Risk Reduction Benefit**	**C-Node Criticality Level**
C-Node 1	0.2130	0.2165	0.0963	0.2078	0.1500
C-Node 2	0.1209	0.0909	0.1366	0.1076	0.2000
C-Node 3	0.1002	0.1385	0.2154	0.1907	0.1500
C-Node 4	0.1720	0.1775	0.1489	0.1100	0.1000
C-Node 5	0.0472	0.0043	0.1961	0.1369	0.0500
C-Node 6	0.1479	0.2294	0.1629	0.1858	0.1000
C-Node 7	0.1989	0.1429	0.0438	0.0611	0.2500
Sum	1.000	1.000	1.000	1.000	1.000
Entropy Measure	0.9589	0.9092	0.9577	0.9666	0.9496
Diversification Measure	0.0411	0.0908	0.0423	0.0334	0.0504
Entropy Weights	0.1592	0.3521	0.1640	0.1294	0.1953
Entropy Weight Sum	1.0000				

Entropy Weighted Normalized Matrix	**C-Node Risk Score**	**C-Node FDNA**	**C-Node Risk Mitigation Dollars**	**C-Node Risk Reduction Benefit**	**C-Node Criticality Level**
C-Node 1	0.0839	0.1802	0.0391	0.0672	0.0711
C-Node 2	0.0476	0.0757	0.0554	0.0348	0.0947
C-Node 3	0.0395	0.1153	0.0874	0.0616	0.0711
C-Node 4	0.0678	0.1478	0.0604	0.0356	0.0474
C-Node 5	0.0186	0.0036	0.0796	0.0442	0.0237
C-Node 6	0.0583	0.1910	0.0661	0.0600	0.0474
C-Node 7	0.0784	0.1189	0.0178	0.0198	0.1184

Tables 7.4a and 7.4b show the C-Node TOPSIS scores given the input data in Table 7.3. The C-Node with the largest TOPSIS score is most favorable with respect to the one that maximally reduces capability risk and minimizes the expense of risk mitigation dollars. The C-Node with the next largest TOPSIS score is the next most favorable, and so forth. The criterion with the largest weight has the most influence on the overall C-Node ranking. The criterion with the next largest weight has the next most influence on the overall C-Node ranking, and so forth. The ranking of C-Nodes by their TOPSIS scores is

C-Node 5 < C-Node 2 < C-Node 3 < C-Node 4 <
C-Node 7 < C-Node 6 < C-Node 1

TABLE 7.4b

A TOPSIS-Derived Risk Management Decision Prioritization

	C-Node Risk Score	C-Node FDNA	C-Node Risk Mitigation Dollars	C-Node Risk Reduction Benefit	C-Node Criticality Level		
A^* Positive-Ideal Soln (Ideal Vector)	0.0839	0.1910	0.0178	0.0672	0.1184		
A^- Negative-Ideal Soln (Nadir Vector)	0.0186	0.0036	0.0874	0.0198	0.0237		
S^* Euclidean Separation Distance	**C-Node 1** 0.0531	**C-Node 2** 0.1328	**C-Node 3** 0.1218	**C-Node 4** 0.1000	**C-Node 5** 0.2296	**C-Node 6** 0.0900	**C-Node 7** 0.0865
S^- Euclidean Separation Distance	**C-Node 1** 0.2056	**C-Node 2** 0.1111	**C-Node 3** 0.1301	**C-Node 4** 0.1573	**C-Node 5** 0.0257	**C-Node 6** 0.1983	**C-Node 7** 0.1752
Ranking Result Relative Closeness to Ideal Soln TOPSIS Score	**C-Node 1** 0.7949	**C-Node 2** 0.4554	**C-Node 3** 0.5165	**C-Node 4** 0.6114	**C-Node 5** 0.1007	**C-Node 6** 0.6880	**C-Node 7** 0.6696

Questions and Exercises

1. Using Table 7.5, find the ideal solution to which house to buy using the TOPSIS algorithm, where miles are to be minimized and quality ratings are to be maximized. Apply entropy-weighted criteria.
2. Solve Exercise 1 using the variance-to-mean ratio (VMR) as a basis for the criteria weights. How does the ideal solution found using VMR weighting compare to the ideal solution found using entropy weighting?
3. In Table 7.6, characteristics are given on seven C-Nodes across five criteria whose definitions are provided with Table 7.3. From these data, determine which C-Node is most favorable with respect to maximally reducing capability risk while minimizing the expense of risk mitigation dollars. Use the TOPSIS algorithm with entropy weighted criteria.

TABLE 7.5

Table for Exercise 1

Which House Is Best?	Distance to Work (miles)	Distance to Pizza Shop (miles)	Public Schools Quality (rating)
House A	20	2	4
House B	50	1	3
House C	34	3	1
House D	2	5	5
House E	18	3	2
House F	10	2	1
House G	27	2	5

TABLE 7.6

Table for Exercise 3

Capability Node	C-Node Risk Score	C-Node FDNA	C-Node Risk Mitigation Dollars	C-Node Risk Reduction Benefit	C-Node Criticality Level
C-Node 1	90.0	4.0	7.0	80.0	5.0
C-Node 2	45.0	3.2	9.0	50.0	3.0
C-Node 3	87.0	1.5	15.0	44.0	3.0
C-Node 4	55.0	5.5	4.0	45.0	4.0
C-Node 5	85.0	0.5	12.0	55.0	5.0
C-Node 6	35.0	6.7	8.8	70.0	2.0
C-Node 7	88.0	2.2	5.7	25.0	5.0

4. Solve Exercise 3 using the VMR as a basis for criteria weights. How does the ideal solution found under VMR weighting compare to the ideal solution found using entropy weighting?
5. Research the relationship between entropy weighting, VMR weighting, and standard deviation-to-mean weighting. Run empirical experiments to discover trends indicated in these weighting rules, when extreme ranges in values are present in a column of data (i.e., a criterion).

8

A Model for Measuring Risk in Engineering Enterprise Systems

8.1 A Unifying Risk Analytic Framework and Process

Previous chapters covered a great deal of ground. New risk analytic methods have been developed to address engineering an enterprise system from the perspective of a capability portfolio. This chapter describes how these methods relate and unify into a practical model for the analysis of risk in engineering today's enterprise systems.

8.1.1 A Traditional Process with Nontraditional Methods

In general, managing risk in engineering systems can be characterized by the process shown in Figure 8.1 (Blanchard and Fabrycky, 1990; Garvey, 2008). Although this process grew from engineering traditional systems,* its implementation involves nontraditional methods when applied to engineering an enterprise. Why is this?

As discussed earlier, today's information-age systems are characterized by their *ubiquity* and lack of specification. Systems such as the Internet are unbounded, present everywhere, and in places simultaneously. They are an *enterprise* of systems and systems of systems. By the use of advanced network and communications technologies, these systems continuously operate to meet the demands of globally distributed and uncountable users and communities.

Engineering enterprise systems is an emerging discipline that encompasses and extends "traditional" systems engineering to create and evolve "webs" of systems and systems of systems. They operate in a network-centric way to deliver capabilities via services, data, and applications through richly interconnected networks of information and communications technologies. More and more defense systems, transportation systems, and financial systems globally connect across boundaries and seamlessly interface with

* Traditional systems are generally regarded as systems characterized by well-defined requirements and technical specifications, predictable operational performance, adherence to engineering standards and manufacturing processes, and centralized management authority.

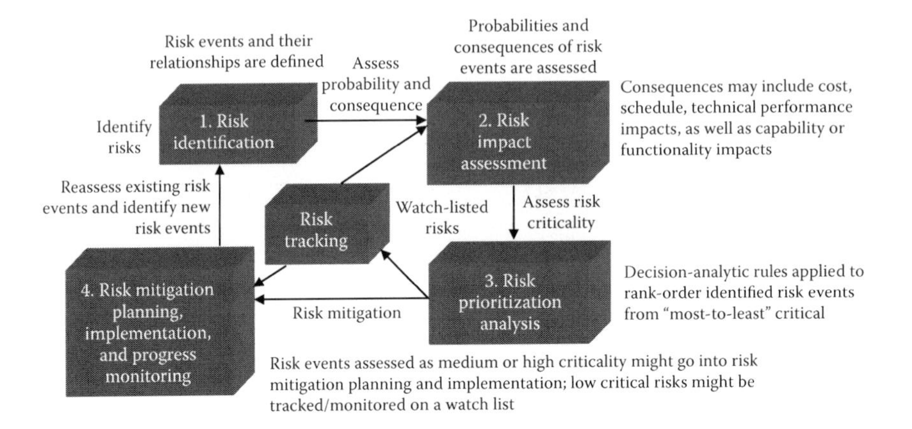

FIGURE 8.1
A traditional risk management process.

users, information repositories, applications, and services. These systems are an enterprise of people, processes, technologies, and organizations.

As discussed in Chapters 2 and 4, an enterprise system is often planned to deliver capabilities through portfolios of time-phased increments or evolutionary builds. Thus, risks can originate from many different sources (e.g., suppliers) and threaten enterprise capabilities at different points in time. Furthermore, these risks (and their sources) must align to the capabilities they potentially affect and the scope of their consequences understood. In addition, the extent enterprise risks may have unwanted collateral effects on other dependent capabilities must be captured and measured when planning where to allocate risk-reducing investments.

From a high-level perspective, the process for analyzing and managing risk in engineering enterprise systems is similar to that in engineering traditional systems. Scale, ubiquity, and decentralized authority in engineering enterprise systems drive the need for nontraditional risk analytic methods within the traditional process steps in Figure 8.1.

Recognizing and researching these distinctions have produced the formal methods herein. They aim to enable a holistic understanding of risks in engineering enterprise systems, their potential consequences, dependencies, and rippling effects across the enterprise space. When implemented, these methods provide engineering management a complete view of risks across an enterprise, so capabilities and performance objectives can be achieved via risk-informed resource and investment decisions.

8.1.2 A Model Formulation for Measuring Risk in Engineering Enterprise Systems

The following sections describe how the risk analytic methods in this book form a practical model for the analysis of risk in engineering today's enterprise systems. We will relate this model formulation to the fundamental process steps shown in Figure 8.1.

Step 1: Risk Identification

In engineering a traditional system, risk identification is the critical first step of the risk management process. Its objective is the early and continuous identification of risks to include those within and external to the engineering system project. As mentioned earlier, these risks are events that, if they occur, have negative impacts on the project's ability to achieve its outcome goals.

In engineering an enterprise system, risk identification needs to consider supplier risks. Supplier risks include unrealistic schedule demands placed on them by portfolio needs or placed by suppliers on their vendors. Supplier risks include premature use of technologies, including the deployment of technologies not adequately tested. Dependencies among suppliers can generate a host of risks, especially when a problem with one supplier generates a series of problems with others. Economic conditions can threaten the business stability or viability of suppliers and vendors. Unfavorable funding or political influences outside an enterprise can adversely affect its capability portfolios, its suppliers, or the supplier–vendor chains in ways that threaten enterprise goals and mission outcomes.

Risks that trace to "suppliers" are a major source of risk to the portfolio's ability to deliver capability to the enterprise. However, it is important to recognize that suppliers are not the only source of risk. Risks that threaten capabilities to be delivered by a portfolio can originate from sources other than those that affect only the portfolio's suppliers. These events can directly attack one or more capability nodes in a capability portfolio's hierarchy. For example, uncertainties in geopolitical landscapes may impact operational demands on capabilities that stress planned performance.

Risk identification also needs to consider dependencies. Dependencies between capability portfolios in families of portfolios, such as those that constitute an enterprise, are also potential risk sources. Here, outcome objectives for capabilities delivered by one capability portfolio may depend on the performance of capabilities delivered by another capability portfolio. Identifying risk events from non-supplier-related sources and capturing their contribution to a capability node's risk measure is an important consideration in a capability portfolio's risk assessment.

Figure 8.2 illustrates the risk analytic methods developed in this book that relate to this process step. Methods from Chapters 4 and 5 are applicable at this stage. Here, the capability portfolio is defined and expressed as a mathematical graph. This graph is used as a modeling "framework" within which risks can be assessed and capability risk measures derived. In this step, effort is spent defining capability in measurable contexts and aligning capability suppliers as they enable the portfolio to deliver capability.

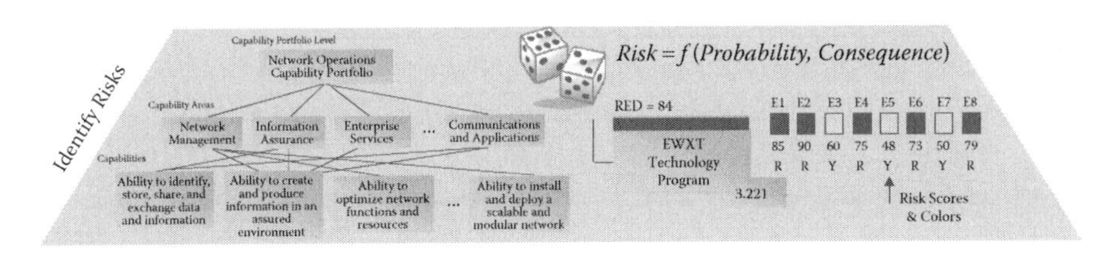

FIGURE 8.2
Risk analytic methods related to risk identification.

Step 2: Risk Impact (Consequence) Assessment

In engineering a traditional system, an assessment is made of the consequence each risk event could have on the engineering system project. Typically, this includes how the event could impact cost, schedule, or technical performance objectives. An assessment is also made of the probability each risk event will occur. This often involves subjective probability assessments, particularly if circumstances preclude a direct evaluation of probability by objective methods.

In engineering an enterprise system, findings from Step 1 feed into mathematical constructs that generate capability risk measures across an enterprise. These measures are determined by the calculus created in Chapter 4, where the enterprise problem space is represented by a supplier–provider metaphor in the form of a mathematical graph. This graph is a topology of nodes that depict supplier–provider–capability relationships unique to a capability portfolio.

Within this topology, mathematical rules can be developed which operate on these relationships to generate measures of capability risk. Here, a definition of capability risk is provided which considers the occurrence probabilities and consequences of risks that threaten capability. In this context, consequence is broadened beyond cost, schedule, and technical performance dimensions; risk consequence is evaluated according to a capability's *ability to achieve its outcome objectives* for the portfolio and ultimately for the enterprise.

Next, dependencies and risk correlationships that may exist in the enterprise are captured. Chapters 5 and 6 provide formalisms for analyzing dependency relationships and their effects on engineering and planning an enterprise system. Critical considerations in engineering enterprise systems are identifying, representing, and measuring dependencies between suppliers of technologies and providers of services to users.

The importance of dependency analysis in engineering an enterprise is manyfold. A primary concern is enabling the study of ripple effects of failure in one capability on other dependent capabilities across the enterprise. Providing mechanisms to anticipate these effects early in design enables engineers to minimize dependency risks that, if realized, can have cascading negative effects on the ability of an enterprise to deliver services to users.

One dependency is risk inheritance; that is, "How risk-dependent are capabilities so threats to them can be discovered before contributing programs (e.g., suppliers) degrade, fail, or are eliminated?" Chapter 5 provides a management metric, the risk correlationship (RCR) index, for capturing and measuring risk inheritance in an enterprise. The RCR index is a new management metric that measures risk inheritance between supplier programs and its ripple effects across a capability portfolio. The index identifies and captures horizontal and vertical impacts of risk inheritance, as it increases the threat that risks on one supplier program may adversely affect others and ultimately their contributions to portfolio capabilities.

The other dependency is operational dependence; that is, "What is the effect on the operability of capability if one or more contributing programs (e.g., suppliers) or supplier-provider chains degrade, fail, or are eliminated?" Chapter 6 presented an entirely new formalism called functional dependency network analysis (FDNA) for capturing and measuring operational dependence in an enterprise.

Factoring dependency considerations into the methods presented in this book enables the proper management of risk in an enterprise, specifically, investment decisions on where to target risk-reduction resources in ways that optimally reduce threats to capabilities posed by dependencies. Figure 8.3 illustrates the risk analytic methods in this book that relate to this process step.

Step 3: Risk Prioritization Analysis

At this step, the overall set of identified risk events, their impact assessments, and their occurrence probabilities are "processed" to derive a ranking of the most-to-least critical risks. Decision analytic techniques such as utility theory, value function theory, or ordinal methods are formalisms often used to derive this ranking.

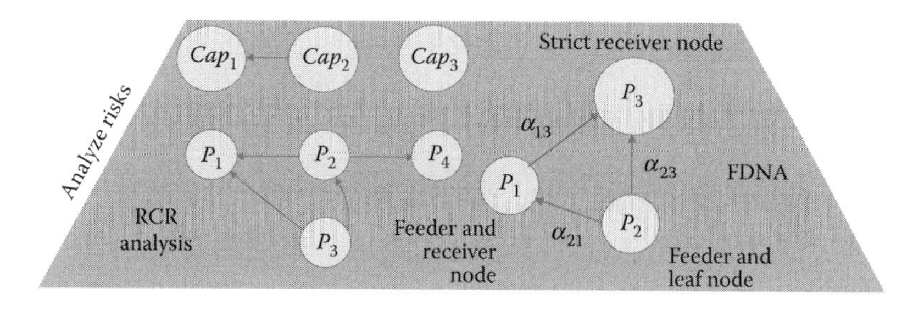

FIGURE 8.3
Risk analytic methods related to risk impact assessment.

Findings from Steps 2 and 3 feed into ranking algorithms that identify an optimal ordering of enterprise capabilities to risk manage. In Chapter 7, a decision-theoretic approach for prioritizing risk management decisions as a function of the risk measures and analysis procedures developed in the previous chapters is provided. These measures are integrated into advanced ranking algorithms to isolate and prioritize which capabilities are most risk threatened and qualify for deliberate management attention.

The outputs from these ranking algorithms enable decision-makers to target risk-reduction resources in ways that optimally reduce threats posed by risks to critical capability outcome objectives. Figure 8.4 illustrates the risk analytic methods developed in this book that relate to this process step.

Step 4: Risk Mitigation Planning and Progress Monitoring

This step involves the development of mitigation plans designed to manage, eliminate, or reduce risk to an acceptable level. Once a plan is implemented, it is continually monitored to assess its efficacy with the intent to revise its courses of action if needed.

Step 4 results from the integration of all the risk analytic methods developed in Chapter 4 through Chapter 7. With this, decision-makers have a logical and rational basis for addressing the choice problem of selecting which capability risks to mitigate or reduce as a function of their criticality to the portfolio and to the enterprise as a whole. Figure 8.5 illustrates this step.

Prioritize risks

Capability node	C-Node risk (score)	C-Node FDNA	C-Node mitigation (dollars)
C-Node 1	96.3	5.0	5.5
C-Node 2	54.7	2.1	7.8
C-Node 3	45.3	3.2	12.3
C-Node 4	77.8	4.1	8.5
C-Node 5	21.3	0.1	11.2

FIGURE 8.4
Risk analytic methods related to risk prioritization analysis.

FIGURE 8.5
Risk analytic methods related to risk management planning.

In summary, these four process steps bring the risk analysis approaches developed in this book into a coherent structure for representing, modeling, and measuring risk in engineering large-scale, complex systems designed to function in enterprise-wide environments.

With this work, the engineering management and systems engineering community has a generalized framework and computational model for the analysis of risk in engineering enterprise systems. This provides decision-makers formal ways to model and measure enterprise-wide risks, their potential multiconsequential impacts, dependencies, and their rippling effects within and beyond enterprise boundaries. Figure 8.6 visually summarizes the risk analytical framework and model formulation presented herein.

Making operational the model shown in Figure 8.6 has unique information needs. These can be grouped into two categories. The first group *addresses capability value*. The second one *addresses supplier contributions, criticality, and risks* as they relate to enabling the portfolio to deliver capability.

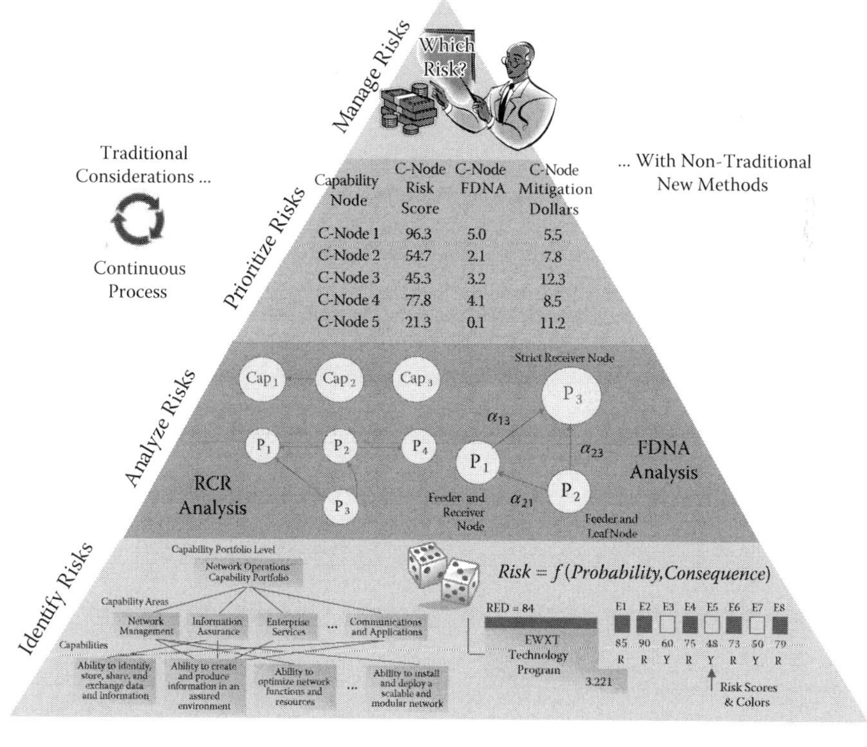

FIGURE 8.6
A risk analytical framework and model formulation.

Information needs to *address capability value* include the following:

- For each Tier 3 capability, shown in Figure 4.5, identify what performance standard (or outcome objective) each capability must meet by its scheduled delivery date.
- For each Tier 3 capability, identify the source basis for its performance standard (or outcome objective). Does it originate from user-driven needs, policy-driven needs, model-derived values, a combination of these, or from other sources?
- For each Tier 3 capability, identify to what extent the performance standard (or outcome objective) for one capability depends on others meeting their standards.

Information needs to *address supplier contributions, criticality, and risks* include the following:

- For each Tier 3 capability, identify which technology programs and technology initiatives are contributing to that capability.
- For each Tier 3 capability, identify what (specifically) are the contributions of its suppliers.
- For each Tier 3 capability, identify how supplier contributions enable the capability to achieve its performance standard (or outcome objective).
- For each Tier 3 capability, identify which technology programs and technology initiatives are critical contributors to enabling the capability to achieve its performance standard (or outcome objective).
- With the above, identify what risks originate from (or are associated with) suppliers that, if these events occur, negatively affect their contributions to capability.

Process tailoring, socialization, and establishing governance protocols are critical considerations in engineering risk management for enterprise systems. Overall, the risk analytic approaches described provide the following:

- Identification of risk events that threaten the delivery of capabilities needed to advance goals and capability outcome objectives of the enterprise.
- A measure of risk for each capability derived as a function of each risk event's occurrence probability and its consequence.
- An analytical framework and logical model within which to structure capability portfolio risk assessments—one where assessments can be combined to measure and trace their integrative influence on engineering the enterprise as a whole.

- Through the framework, ways to model and measure risk as capabilities are time-phased across incremental capability development approaches.
- Analytic transparency, where the methods herein provide decision-makers the trace basis and the event drivers behind all risk measures derived for any node at any level of the capability portfolio's hierarchy. With this, capability portfolio management has visibility and supporting rationales for identifying where resources are best allocated to reduce (or eliminate) events that threaten achieving the enterprise's capability goals.

8.2 Summary

Managing risk in engineering today's systems is more sophisticated and challenging than ever before. Lack of clearly defined boundaries, diminished hierarchical control, network-centric information exchange, and ubiquitous services are significant technical and managerial challenges faced in engineering enterprise systems. Increased complexity contributes to increased risks of system and management failures—particularly in systems designed to employ advanced, network-centric, information technologies (Daniels and LaMarsh, 2007). Few, if any, protocols exist for assessing and measuring risk in engineering enterprise systems from a capability portfolio perspective. Ways to address this problem has been the aim of the first eight chapters in this book. The advanced analytic methods presented provide a foundation on which to build solutions to a next set of hard risk analytic problems.

In summary, the material in this book falls at the interface between risk management methods for engineering traditional systems with those needed for engineering enterprise systems. Recognizing this is an essential first step toward addressing these challenges and discovering new methods and new practices in engineering risk management and its related disciplines.

Questions and Exercises

1. The following identifies research areas in risk analysis methods applicable to engineering enterprise systems. Select one or more of these areas and write an essay outlining your thoughts on ways to proceed. Identify what is important about each area and the challenges you see in addressing them. Identify work that may already

be underway in the engineering community that may be leveraged to address these areas.

Analytical Scalability: Research how to approach risk analysis in engineering enterprise systems that consist of dozens of capability portfolios with hundreds of supplier programs.

Explore representing large-scale enterprises by domain capability portfolio clusters and investigate a concept for portfolio cluster risk management—include a social-science perspective on the management of risk in engineering enterprise systems at this scale.

Explore the efficacies of alternative protocols for assessing and measuring risks from the supplier through the provider layers of a capability portfolio. Quality Function Deployment (QFD) and Design Structure Matrices (DSM) are protocols to consider.

Nonstationary Considerations: Extend the FDNA calculus presented in Chapter 6 to address nonstationary dependency analysis problems. Explore how FDNA can expand and integrate into time-varying modeling and simulation environments, such as those provided in systems dynamics methods and tools.

Optimal Adaptive Strategies: Research how to optimally adapt an engineering system's supplier–provider network to reconfigure its nodes to maintain operability if risks that threaten these nodes are realized. Consider this problem in stationary and nonstationary perspectives.

9

Random Processes and Queuing Theory

9.1 Introduction

Possibly one of the most important skills of a system engineer is to communicate in simple and readily comprehensible manner what in reality maybe a complex phenomenon in terms of "nomenclature and terminology that support clear, unambiguous communication and definition of the system and its functions, elements, operations, and associated processes" (INCOSE 2010, p. 4). Everyday, we hear the weather forecaster describe the day's expected high and low temperature, and rainfall or snow. When the weather forecaster describes a cold front moving into the region coupled with the vapor-rich breeze from the east will result into a high temperature of 45°C, you may ask yourself, how can he/she be so sure? No, the forecaster is never sure—but was able to convey a complex phenomenon with hundreds of factors into a compact and informative manner, albeit maybe inaccurate or imprecise. Also consider when a systems engineer is creating a testing plan to verify and validate system functionality, say the rate at which the new communication system drops calls coming from mobile antennas. The test may include a test parameter such as "dropped calls should not exceed 2-out-of-100." On both cases, a fairly complicated phenomena are expressed in such simple manner.

Systems engineer, similar to a weather forecaster, often needs to describe a phenomenon with the objective of helping himself or someone else arrive at a decision. Depending on the systems engineer's comprehension of the phenomenon and the needs of his audience, the engineer must be able to convey the appropriate information to project members tasked with developing or managing technical systems. Too much or too little information may negatively affect the decision-making process and the budget, schedule, or performance of the system being developed. Nonetheless, systems engineers must also be able to recognize that there will be situations in which describing phenomena in its simplest way may not be sufficient.

The following discussion compares and contrasts between deterministic and nondeterministic way of describing a phenomena, identifies advantages and disadvantages of using deterministic and nondeterministic models, and describes random process to be able to discern the two sources of uncertainty.

Furthermore, the importance of reducing uncertainty is emphasized and Markov process and its importance for systems engineers touched upon. Finally, the elements of a queuing system are identified, various performance parameters of queuing systems are described, and how to model common and specialized engineering problems as queuing systems are explained.

9.2 Deterministic Process

The simplest way to describe a phenomenon may be in a deterministic manner. Determinism is synonymous with definite, certain, sure, and can be quantitatively expressed as a point estimate such as when the day's forecasted high temperature is estimated as a single number of 45°C. Other deterministic expressions of phenomenon are the following:

- EPA describing the fuel consumption of a car to be, say 40 miles per gallon on the highway
- Structural engineers describing a bridge can hold up 60 tons of weight
- Pharmaceutical companies labeling a bottle to contain 8 ounces of medicine

There are countless examples of phenomenon described in a deterministic manner. Most of the times, these descriptions are sufficient and precise enough such that decision makers can appropriately choose among car models, type of vehicles to be allowed on a bridge, and the size of medicine bottle to buy at the pharmacy. As apparent by now, the beauty of determinism lies in its simplicity. A well thought-out deterministic description of phenomenon, with precision and accuracy appropriate for the decision maker and the decision scenario at hand, is one of the greatest tools and skill a systems engineer can have.

It may now be a good time to dwell more on a term used several times in the immediately preceding paragraphs—*describe*—as in describing a phenomenon. By using the term describe, the emphasis is less on the real phenomenon but more on how this phenomenon is construed in the minds of the engineers and how it is purposely communicated, compared to the term *define*. One can also think of the often-used term *model*, as in the high temperature is *modeled* (or *described*), to be a deterministic phenomenon. In the same way, an engineer can choose how to model a particular object (e.g., visual, schematic, etc.), or one can also choose to the same extent whether to describe a phenomenon as deterministic or otherwise. As with any type of model, a deterministic model of any phenomenon comes with many caveats.

These caveats of deterministic model of processes can be discussed both mathematically and philosophically. With this book aimed primarily at those with an engineering background, deterministic process will be aptly described from the mathematical perspective. Nonetheless, the philosophical perspective will also be discussed simply because engineers, managers, and any diligent problem solver will benefit greatly from knowing the philosophical basis of the deterministic process. This will also make it easier to comprehend the other side of the coin: random and probabilistic processes.

9.2.1 Mathematical Determinism

Let us look closer at forecasting the high temperature of a day. For brevity, let us express the forecasted high temperature as x, the low temperature as y, the humidity as z, and their relationship that results to the high temperature as g. Obviously, there is great simplification being made in grouping the rest of the meteorological phenomenon into a relationship g, but this will suffice for now. Assuming these are all the information that we know, then conventionally, the high temperature and the phenomenon can be expressed as

$$x = g(y, z) \tag{9.1}$$

Suppose that the forecaster obtained the high temperature estimate of 45°C based on a low temperature value of 33°C and humidity value of 85%. With the way the phenomenon is described by Equation 9.1, particular values of $y = 33$ and $z = 85\%$ results in the equation $x = g(33, 85) = 45$. To say that $x = 45$ every time, $y = 33$, and $z = 85$ is equivalent to saying that $x = g(y, z)$ is a deterministic process.

Definition 9.1: Deterministic Process

Deterministic process is when the same output is obtained every time the same set of inputs or starting conditions occur.

Generalizing the low-temperature example in the preceding paragraph, $g_t(Y, Z) = g(Y, Z) = X$ for all t, where t stands for time. There are many examples of situations when phenomena are described to be deterministic. Some examples are as follows:

- A computer program that provides the exact same output given the same input
- When an automobile tires is said to fail after 50,000 miles of road use
- When water level at a river is said to be 6 feet
- When the speed of the boat is said to be 10 nautical miles

9.2.2 Philosophical Determinism

To describe (or model) a process or phenomenon, such as a day's high temperature as a deterministic process carries with it a number of implications. The first implication is that there are some forms of causality between the inputs or starting conditions of the process and the outputs. The second implication is that the uncertainty regarding the true process is nonexistent or ignored. The concept of uncertainty itself deserves some discussion, but for now, the description of uncertainty as the *indefiniteness* about the outcome of a situation (Section 3.2) will suffice. The third implication and possibly the most significant for systems engineers is that either due to limitations on the knowledge about the true nature of the process or the context of the decision scenario and decision maker or both makes it appropriate to describe the process as a deterministic model.

Consider as an example the basic formula for force in classical physics, $F = m * a$, that is, force is equal to mass multiplied by acceleration. First, this particular model for the notion of force implies that there is causality between mass and acceleration, and force. That is, the absence of mass or acceleration (or both) results in the absence of force. Mathematically,

$$F = 0 \quad \text{if } m = 0 \text{ and/or } a = 0$$

Second, this model implies that the process expressed as $m * a$ describes for certain or determines the value of force without uncertainty. Finally, every time this notion of force is used by engineers in any of their activity implies that this description is appropriate for the scenario at hand.

9.3 Random Process

The previous section described the notion of determinism and deterministic process from both mathematical and philosophical perspectives. The discussion also presented many engineering problems expressed (or modeled) from the most basic mechanics and kinetics, and even physics and chemistry. In many situations, it is perfectly acceptable and appropriate to describe events and phenomenon in a deterministic manner. This is because many decisions can be made simply with deterministic statement and without great consequence.

Consider these example deterministic statements about projected costs:

- A haircut costs $10.
- I am thinking of buying a car that costs $2000.
- The federal government is going to spend $20 million to build a new bridge.

All these statements, if taken exactly the way they are stated, do not provide any information aside from a deterministic cost projection of a service, product, or infrastructure, that is, a haircut, a car, or a bridge, respectively. However, consider the situation where the real cost was found to vary from the projected cost by being 10% higher. That is, the actual cost of the haircut, the car, and the bridge turned to be $11, $2200, and $22 million, respectively. The consequence of being $1 short after a haircut may be only a slight inconvenience compared to a federal project that is $2 million short on budget. This shows that there may be cases, for one reason or another, that deterministic model of phenomenon may not suffice. It would have been more useful for the federal project planners to have known the range of uncertainties in the cost of the bridge, say a range of $20 ± $3 million. For one, this would have prompted the project planners to at least think of possible actions if indeed the bridge turned out to cost more than $20 million.

There is no denying that engineers have found themselves in many situations in which deterministic models are able to describe the system of interest only in a very limited way, whether in estimating the cost of a project or in the final dimension of a product. In this type of situations, deterministic models are often found lacking and insufficient.

Going back to the weather forecast, not many of us keep notes of the actual high temperature of days but experience tells us that on most days, the actual high will be few degrees higher or lower than the forecast. Nonetheless however unlikely, the actual high temperature may exactly be equal to the forecast on some days. However, there may be decision scenarios where more accurate and precise information is required. For example, launching of expensive and sensitive electronic instruments through high-altitude balloons may require more precise and accurate temperature forecast. What would normally be useful information as "high temperature of 45°C" may be lacking.

Definition 9.2: Random Process

A random process is a phenomenon that lacks predictability of its actual outcome (i.e., possesses the random property).

In general, the random property or simply *randomness* implies lack of predictability of the outcomes. Common examples are tossing a coin or rolling a die.

In Chapters 1 and 3, the concept of random event was briefly mentioned in discussing the axiomatic definition of probability. This concept is now defined more precisely.

Definition 9.3. Random Event

A random event is a set of outcomes resulting from a random process.

Consider the earlier examples of random processes and now augmented with their possible corresponding events. When we pertain to some real-world

TABLE 9.1

Random Process	Outcomes	Events
Tossing a coin	H or T Where H pertains to head and T pertains to Tail	For a single toss: H or T For two consecutive tosses: HH, TT, HT, or TH
Rolling a die	1, 2, 3, 4, 5, 6	For a single roll of a die: 1, 2, 3, 4, 5, or 6 Even-numbered outcomes of a single roll: (2, 4, 6)

random phenomena, most likely we are pertaining to random events that are actually outcomes of one or more random processes as shown in Table 9.1.

9.3.1 Concept of Uncertainty

In Chapter 3, uncertainty was described as the *indefiniteness* about the outcome of a situation. In everyday discussion, uncertainty commonly implies doubt, dubiety, ambiguity, lack of knowledge, and other similar terms. From the definition of a random process as a phenomenon that lacks predictability of its actual outcome, one will surmise that random process is characterized by uncertainty. It is beneficial to understand the two sources of uncertainty: aleatory and epistemic.

Definition 9.4: Epistemic Uncertainty

Epistemic uncertainty refers to uncertainty in our state of knowledge about certain phenomena. This is also known as reducible uncertainty, pertaining to its property to be reduced through investigation, reasoning, engineering interventions, and other forms of analyses.

Definition 9.5: Aleatory Uncertainty

Aleatory uncertainty, on the other hand, is due purely to the variation in outcomes of randomness. This is also known as irreducible uncertainty, pertaining to its property of not being affected by further investigation, reasoning, and other forms of interventions and analyses.

Many activities in systems engineering are meant to reduce uncertainty in one form or the another, such as in product and service quality control, logistics and supply chain optimizations, and others. In fact, the level of uncertainty is one of the factors a systems engineer needs to consider in designing a plan on how to develop a system (INCOSE 2010, p. iii). In reducing uncertainty, these two types of uncertainty are addressed by a large number of tools and techniques in a wide variety of disciplines and applications.

For example, quality control engineers in the manufacturing industry are used to looking into variations in the performance of certain machines. In particular, consider a machine that fills up 16-ounce boxes of breakfast cereals was observed to overfill or underfill boxes much too often than it is supposed to. Looking closely at the machine revealed a worn-out part that contributes to why the machine performs in such a way. After replacing the worn-out part, the machine now overfills and underfills boxes much less often. Further examination of the machine did not reveal any other mechanical reason that causes variability in the machine's performance. The variation in the machine's performance due to the worn-out part is an example of a variation with assignable cause (i.e., epistemic in nature) and was managed by analyzing the causes and implementing engineering intervention to eliminate these causes. Nonetheless, if there is still variability in the machine's performance, it can be simply attributed to randomness rather than on any assignable cause (i.e., aleatory type).

Previously, it was mentioned that simplification is achieved by describing or modeling a phenomenon as a function of another phenomenon. The example given earlier is the high temperature and how it can be expressed as a function in Equation 9.1. With the notion of uncertainty, one can say that there can be aleatory and epistemic uncertainty in this function. These uncertainties can be in our knowledge (i.e., epistemic) or in the actual values (i.e., aleatory) of the function g and variables y and z. That is, we can ask

Is the phenomena of a high temperature x really a function of $g(y,z)$ or can it be of another different form, say $x=f(y,z)$?

Are the values of the parameters actually y and z or could it be of other values?

These uncertainties in both the function and the variables of the model eventually translates to uncertainty in the phenomena x.

9.3.2 Uncertainty, Randomness, and Probability

For convenience, the definitions of uncertainty, randomness, probability, and random variable (from Chapter 3) are repeated here.

Uncertainty: indefiniteness about the outcome of a situation.

Randomness: a property of a process that describes lack of predictability of actual outcome.

Probability: a numerical measure that satisfies the Kolmogorov's axioms.

Random variable: a real-valued function defined over a sample space. The random variable is usually denoted by X and the sample space by Ω.

For convenience, the axiomatic definition of probability in Chapter 1 is repeated here using the notion of random variables and its sample space. Under this definition, it is assumed for each random event A, in a sample space Ω, there is a real number $P(A)$ that denotes the probability of A.

In accordance with Kolmogorov's axioms, probability is simply a numerical measure that satisfies the following:

Axiom 1 $0 \leq P(A) \leq 1$ for any event A in Ω.

Axiom 2 $P(\Omega) = 1$.

Axiom 3 For any sequence of mutually exclusive events $A_1, A_2, \ldots$ defined on Ω,

$$P(\bigcup_{i=1}^{\infty} A_i) = \sum_{i=1}^{\infty} P(A_i)$$

For any finite sequence of mutually exclusive events $A_1, A_2, \ldots, A_n$ defined on Ω,

$$P(\bigcup_{i=1}^{n} A_i) = \sum_{i=1}^{n} P(A_i)$$

Also described in Chapter 3 are the three most common interpretations of probability, namely,

1. Equally likely interpretation
2. Frequency interpretation
3. Measure of belief interpretation

It is important to point out that engineers often shift interpretation and hence use probabilities among these three interpretations. Intuitively, these three interpretations of probabilities go hand in hand in many engineering activities. For example, a quality engineer may conduct repeated trials of accelerated life tests to measure the failure of a part of a machine when subjected to 1000 cumulative hours of use. Based on these repeated trials, the frequency of failure of the particular part is expressed as probability, for example $P(A)=0.01$ where A is the event of part failure on or after 1000 cumulative hours of use. This probability can be interpreted as the frequency of failure of the part on or after 1000 cumulative hours of use is 1 in 100. Afterwards, this same information may be the basis for a preventive replacement plan for this part, in essence, shifting the interpretation of the probability to the belief that the chance of failure of the part after 1000 hours of use is 1%.

These three concepts of uncertainty, randomness, and probability are strongly related, and yet their relationship is not always clear to many engineers. To better place these concepts in context, consider the weather forecast example. If exactly knowing the high temperature for the day before it occurs is not possible, then this situation can be described as uncertain, that is, there is indefiniteness about the outcome of the situation. Now consider particularly the temperature readings for the day, that is, the value that can be read from the thermometer. Because the actual value of the thermal reading cannot be exactly predicted, then this reading can be described as a random process. For compactness, let us pertain to the thermal reading (i.e., event of a random process) as A, and consequently, A can be described as a random event. Furthermore, if A is described to be probabilistic, then it is assumed for each random event A in a sample space Ω there is a real number $P(A)$ that denotes the probability of A.

To further exemplify the concepts of uncertainty, randomness, and probability in engineering, assume that time has come when the science of weather forecasting has become so sophisticated that weather forecaster can accurately predict the high temperature for the day. That is, time has come that when a weather forecaster on TV says that "today's high temperature is a balmy 75°F," it is for certain that today's high temperature will actually be 75°F. However, this capability of meteorologists to perfectly predict the high temperature for the day does not change the fact that daily high temperature fluctuates from one day to the other. Another way to describe this is that weather forecaster may perfectly predict the weather but cannot control it. As such, daily high temperatures will still show the ups and downs expected as seasons change as we may know it today even if it is perfectly predictable.

At this point, we need to make a very significant shift in how we use probability. Previously, we used probability as a numerical measure to express random property of daily high temperature. But since we assumed that the time has come when daily high temperature is perfectly predictable, then one may say there is no use for the probability measure of daily high temperature. However in fact, there is another alternative use for probability measure aside from expressing lack of predictability—this is to use probability to express frequency of occurrence rather than measure of belief.

9.3.3 Causality and Uncertainty

An integral part of any engineering activity is the discovery of how things work and how to affect a desired result. Examples are as follows:

- A production engineer studies what causes a filling machine to overfill or underfill.
- A civil engineer studies the flow of water under various surfaces to design better irrigation systems.

In these examples, the engineers essentially infer causalities based on observations, for example, what triggers the machine to start and stop dispensing products or what causes a change in water flow directions and pressures. And as the engineers make inferred causalities, the uncertainty surrounding the phenomenon is reduced. This is in fact fundamental to any scientific activity, wherein conclusions are made based on rational and repeatable acts of discovery.

Definition 9.6: Causality

Causality is the relationship between two events, wherein the occurrence of one implies the occurrence of the other.

There are various ways in which engineers establish causal relationships among events, many of which are combinations of analyzing historical data, conducting experiments, and using statistical analyses. Nonetheless, the causal relationship among events can be meaningfully described for systems engineers in two ways.

Definition 9.7: Necessary Causes

A set of events B is described to be necessary to cause another set of events A if B is a required condition for the occurrence of A, not that A actually occurs.

Some implications of a situation where B is a necessary condition for the occurrence of A are as follows:

- Occurrence of A implies the occurrence of B
- Occurrence of B, however, does not imply that A will occur
- Nonoccurrence of B implies nonoccurrence of A

As an example, consider event B as being human and event A as being a systems engineer. By inductive reasoning of observation, one may safely infer that to be human is necessary to being a systems engineer: B is a necessary cause for A. Furthermore,

- Occurrence of A implies the occurrence of B; being a systems engineer implies being human;
- Occurrence of B, however, does not imply that A will occur; being human, however, does not imply being a systems engineer;
- Nonoccurrence of B implies nonoccurrence of A; not being human implies not being a systems engineer;

Definition 9.8: Sufficient Cause

A set of events B is described to be sufficient to cause another set of events A if the occurrence of B guarantees the occurrence of A. However, another set of events, say C, may alternatively cause A.

Some implications of a situation where B is a sufficient condition for the occurrence of A are as follows:

- Occurrence of A does not imply the occurrence of B.
- Occurrence of B may be not the only cause of occurrence of A.

As an example, consider as event B, a 100-year epic rainfall, and event A as the flooding of the Midtown Tunnel. It may be inferred that a 100-year epic rainfall is sufficient to flood the Midtown Tunnel: event A is sufficient to cause event B. However,

- Occurrence of A does not imply the occurrence B; flooding inside the Midtown Tunnel does not imply a 100-year epic rainfall;
- Occurrence of B may be not the only cause of A; aside from a 100-year epic rainfall, there may also be alternative cause to flooding the tunnel, such as broken water pipe in the tunnel, or leaks in the tunnel wall which let water in.

9.3.4 Necessary and Sufficient Causes

The notions of necessary and sufficient causes taken together form the foundation of judging causality in many fields including systems engineering and risk analysis. These concepts are the primary foundation in assuring the delivery of required functions during systems design and development, also known as goal operationalization, as well as the foundation of many risk management strategies.

Consider as an example the successful launching of rocket into space. During design and development, rocket engineers have a very clear list of functionalities and factors sufficient for successful launch, for example, enough thrust, right trajectory, weight, weather condition, and many others. All these functionalities and factors are sufficient to cause another event; in this case, the successful rocket launch. These being sufficient conditions, their presence are supposed to guarantee a successful launch.

Equivalently, necessary and sufficient causes are used to establish relationship between risk events and those that system engineers can affect to manage such risks. If the rocket engineers are tasked to prevent the risk of a failed launch, they will try to describe the potential causes of such risky event.

9.3.5 Causalities and Risk Scenario Identification

In the realm of risk management (to include risk assessment, analysis, and mitigation), the default preliminary step is the identification of risk scenarios. This step essentially determines what later on will be the focus of the rest of the risk management processes. "Risk identification is the process of recognizing potential risks and their root causes" (INCOSE, 2004, p. 62) and is essential in setting priorities for a more detailed risk assessment.

It is also evident that establishing causalities among events is at the very foundation of risk scenario identification. These causalities provide confidence that buildings we construct will stand a certain magnitude of earthquake, strong wind, and even the failure of some of its supporting structures. Nonetheless, risk scenarios must be expressed in a clear way to enable analysis and defensible management. Garvey (2008, p. 33) suggests the "condition-if-then construct" to express risk scenarios. In essence, this construct allows the undesirable consequence be stated conditioned on a contributing event or root cause.

As an example, consider the undesirable consequence, flooding of the Midtown Tunnel symbolized by A, and a known sufficient cause, 100-year epic rainfall symbolized by B. This risk scenario can be expressed using the condition-if-then construct as

$$A|B = \text{flooding of Midtown Tunnel conditioned on 100-year epic rainfall}$$

Risk scenarios expressed in this way facilitates the use of statistics and probabilities, that is, estimating $P(A|B)$ where P can be interpreted as either the chance of occurrence or the degree of belief. Furthermore, this construct also facilitates the search for other contributing events or causes, for example, $A|C$, $A|D$, which is a significant aspect of the entire risk management process.

Nonetheless, identifying risk scenarios, particularly the *unknown unknowns* is not a trivial process as shown by Parsons (2007), particularly for large and complex systems such as those in NASA's space exploration. This challenge applies to both identifying the root undesirable event A, as well as the various causes B, C, etc. Yet a complete set of risk scenarios is an ideal characteristic of an effective risk management process (Kaplan, 1997).

Again consider the flooding inside the Midtown Tunnel. Initially, the engineer may ask the question "What can cause the flooding inside the Midtown Tunnel?"

Possible responses may be as follows:

A. 100-year epic rainfall
B. Failure of water pump inside the tunnel
C. Failure of water gates at both ends of the tunnel
D. Leakage from the tunnel's walls and ceilings (Midtown Tunnel is an underwater tunnel)
E. Breakage in the city water pipes running inside the tunnel

$\{A\}, \{D\}$, and $\{E\}$ can be surmised as sufficient conditions for flooding in the tunnel. The curly brackets are used to mean separate and distinct set of conditions. That is, each of these events guarantees flooding inside the tunnel.

On the other hand, $\{B\}$ and $\{C\}$ by themselves will not cause flooding. Nonetheless, consider 100-year epic rainfall and failure of water pump inside the tunnel together, that is, $\{A,B\}$ sounds to be more compelling set of events to cause flooding, at least more compelling than just $\{A\}$ alone, the 100-year epic rainfall by itself.

In essence, sufficient conditions such as $\{A\}$ and $\{A,B\}$, though by definition of sufficiency can cause flooding in the tunnel by themselves, may provide different insights to someone who wants to analyze the risk of tunnel flooding. In particular, experience and common sense tell us the following:

The chances of occurrence of {A} may be different from {A, B}

The magnitude of floods resulting from {A} may be different from that of {A, B}.

The alternatives to manage flooding resulting from {A} may be different from that of {A, B}.

9.3.6 Probabilistic Causation

Lately, there has been an emphasis on expanding the traditional realm of risk scenarios to include those that would usually be seen as remote, unrelated, or are out-of-system bounds. Primarily, these have been the result of the observable but not well-understood transference of risks across system boundaries traditionally drawn by convention or convenience, that is, projects compared to programs as emphasized by Alali and Pinto (2009).

Furthermore, it has always been a challenge to assimilate the temporal domain of risk in the development of systems. As pointed out by Hofstetter et al. (2002) and more recently by Haimes (2009), actions meant to manage risks can create both further risks, as well as synergistic effects in the future, similar to a pebble dropped in the pond that creates ripples. These ripple effects, especially in the context of environmental risks, have proven to be a challenge from both the risk management, as well as the systems analysis perspective, as discussed by Hatfield and Hipel (2002).

From a systems analysis perspective, the two commonly held approaches to the identification of risk scenarios are bottom-up and top-down approaches. Bottom-up approach to risk identification is drawn from the systems analysis approach of the same name and relies on knowledge of what the elements of the systems are and how these elements are expected to work together. This approach is most commonly evident in reliability analysis and is embodied in tools or techniques such as failure mode and effect analysis (FMEA), fault trees, and alike. On the other hand, top-down approach to risk identification is drawn from the systems analysis approach of the same name and relies on knowledge of the objectives of the systems.

In practice, these two approaches of top-down and bottom-up applied together create synergy that provides risk analysts a more efficient identification of risk scenarios. The bottom-up approach, which relies heavily on empirical and historical data of previously known risks, coupled with knowledge of cause-and-effects, leads to a detailed set of risks with corresponding causes and effects. These risks are also termed as faults or failures in reliability analysis. The top-down approach, which relies on what is known or perceived to be objectives of the systems coupled with a process of logical elimination or exclusion, provides general set of risks. The distinction between these two approaches of identifying risk scenarios becomes more apparent in systems development for several reasons:

- The system being developed is not yet existing, and as such, all risk scenarios are in essence synthesized and results of informed conjecture.
- Usability of bottom-up approach to identifying risk events is limited and is dependent on the uniqueness of the system being developed and its comparability to existing systems.
- The mapping of systems development process with systems life cycle results to decisions in the systems development process to be predicated upon the perceived goals.
- The large number of possible risk scenarios coupled with the uncertainty in the potential consequences makes discerning the more important risk scenarios challenging.

It is evident that risk identification in the context of systems development is very much related to but not exactly the same in the traditional sense. The entire nature of systems development being primarily system goal driven places more emphasis on the top-down approach to risk identification.

Consider again the causal relationship of these two events:

A: being a systems engineer

B: being a human

From previous discussion of necessary causations, to be human is said to be necessary to being a systems engineer: B is a necessary cause for A. We can now couple this with the notion of conditional probability from Chapter 3. Since the occurrence of A necessarily implies the occurrence of B, that is, if someone is a systems engineer, then that necessarily implies that this someone is a human, and therefore, the associated conditional probability relating event B as a necessary cause of event A is

$$P(B|A) = 1$$

However as mentioned before, simply being human does not necessarily imply being a systems engineer, that is,

$$P(A|B) < 1$$

Consider another earlier example where it was said that a 100-year epic rainfall is sufficient cause of flooding inside the Midtown Tunnel:

A: flooding inside the Midtown Tunnel

B: 100-year epic rainfall,

Then,

$$P(A|B) = 1$$

However, being only a sufficient and not necessary condition, there can be other causes of flooding inside the Midtown Tunnel aside from a 100-year epic rainfall. As such

$$P(B|A) < 1$$

The notions of causation and conditional probability bring together the concept of probabilistic causation.

Definition 9.9: Probabilistic Causation

An evidence probabilistically causes the event of interest if the occurrence of the evidence increases the probability of occurrence of the event of interest.

The examples above show how events can be used as evidences in determining, possibly probabilistically, the occurrence of events of interest. In particular, being a systems engineer is a conclusive evidence of being human, in the same way as a 100-year epic rainfall is a conclusive evidence of flooding inside the Midtown Tunnel.

However, there are far more cases where evidences do not provide conclusions as strong as illustrated above. Consider a slight change to the first example such that the event of interest is now being a systems engineer and the evidence is being human.

A: being a systems engineer

B: being a human

As mentioned before, simply being human does not necessarily imply being a systems engineer, that is,

$$P(A|B) < 1$$

Occurrence of evidence B merely establishes the plausibility of event A. Let us extend this example and consider a third event, say,

C: having an engineering degree

If it was established that event C is true, then similar to the discussion of Bayes' rule in Chapter 3, the truthfulness of event A has to be revised. The occurrence of evidence C, similar to that of B, increases the chance that A is indeed true. Even more, the occurrence of both B and C provides stronger evidence to the truthfulness of A than B alone. That is,

$$P(A \mid B) < P(A \mid [B \text{ and } C]) < 1$$

Therefore, the occurrence of events that can be construed as evidences forms the collection of probabilistic causes for events of interests. As engineers and risk analysts, these events play an important role in the assessment of risks and the management strategies that may follow.

These examples show that evidences can be necessary or sufficient without being both.

Definition 9.10: Necessary and Sufficient Cause

An event (or evidence) B is necessary and sufficient condition for another event A if A occurs if and only if B occurs. That is,

$$P(A \mid B) = 1$$

and

$$P(A' \mid B') = 1$$

Ideally, decisions in engineering and risk management should be based on evidences that are both necessary and sufficient because this implies complete knowledge of the causality of a risk event. As a result of such knowledge, a risk manager can look at all possible causes of the risk event and accordingly institute risk management activities. Nonetheless, in less-than-ideal scenarios, risk analysts need to carefully and diligently consider as many evidences as possible to establish the plausibility of particular risk events and revise this plausibility as new evidences become available.

Consider the development of a new software system. System developers usually identify capabilities the new system ought to perform and identify functionalities that will deliver such capabilities. As an example, a financial software may be designed to provide capability to perform transactions over secure federal intranets. Some functionality that will enable this capability is as follows:

- Issue and accept security tokens
- Send and receive encrypted information
- Accept, verify, and issue encryption keys

In more general terms, consider a particular required capability A. Based on analysis of the components and associated functionalities of the new software system, the necessary and sufficient conditions for implementing capability A is given by the set of functionalities B. By definition of necessary and sufficient conditions, (A = true) if and only if (B = true) and

$$P(A \mid B) = 1$$

It can be noted that for this new software system (or any other system for that matter), the failure to deliver a required capability is undesirable and thus is a risk event. Since B is both necessary and sufficient, then

$$P(A' \mid B') = 1 \tag{9.2}$$

Furthermore, consider a case where the set of functionalities B can be decomposed to a finite number of system functionalities $\{b_1, b_2, b_3, \ldots, b_i\}$ such that B = true if ($b_i = true$) for all $\{b_1, b_2, b_3, \ldots, b_i\}$. That is,

$$P(A \mid b_1 \cap b_2 \cap b_3, \cap \ldots \cap b_i) = 1 \tag{9.3}$$

Equations 9.2 and 9.3 together provide conditions for failure of the new software system to deliver the capability as B = false if (b_i = *false*) for at least one *i* in $\{b_1, b_2, b_3, \ldots, b_i\}$:

$$P(A' \mid \exists\, i\, b_i') = 1 \tag{9.4}$$

Equations 9.3 and 9.4 emphasize two sets of important information, as described by Pinto et al. (2010):

1. Identifying necessary and sufficient functionalities, by definition, will lead to assuring a required capability.
2. Identifying sufficient functionalities, coupled by negation, will assure complete list of functional scenarios that can cause (the risk event) of not delivering a capability.

Going back to the case of capability to perform transactions over secure federal intranet, say that this is capability A that needs to be assured. Assume that the necessary and sufficient functionalities that will enable this capability are as follows:

- Issue and accept security tokens, b_1
- Send and receive encrypted information, b_2
- Accept, verify, and issue encryption keys, b_3

By definition, the following statements are true:

- Capability A will be possible only if functionalities b_1, b_2, and b_3 are all operational.
- Any failure in functionalities b_1, b_2, and b_3 will result in failure of capability A.

In summary, identifying necessary and sufficient conditions establishes functionalities that assure delivery of a required capability. Consequently, the same set of necessary and sufficient conditions establishes functionalities whose failure equally assures the nondelivery of the capability. As such, necessary and sufficient conditions are essential aspect of identifying causalities of failing to deliver a capability. In analyzing risks of not delivering a capability, this provides an important starting point toward more extensive risk management.

9.4 Markov Process

The preceding sections established the inescapable challenges presented by uncertainty surrounding risk events and established the importance of looking for evidence to reduce such uncertainty in the form of necessary and sufficient conditions. Engineers and risk analysts alike have adapted and learned to use various simplifying models and associated tools, and techniques to make reducing uncertainty more manageable. One such simplifying model is the Markov process.

A Markov process, named after the Russian mathematician Andrey Markov (1856–1922), describes that future state of the system depends only on its present states, and not on any past states. This is often associated with the phrase "memoryless." To be more precise, consider a probabilistic variable X that can be observed through time and thus can be compactly represented as a probabilistic process $X(t)$. This probabilistic process $X(t)$ is said to have a Markov property if for any instant along time $t_i, i = 1, 2, \ldots, n$

$$\begin{aligned} P(X(t_{n+1}) = x_{n+1} | X(t_n) = x_n, X(t_{n-1}) = x_{n-1}, \ldots, X(t_1) = x_1) \\ = P(X(t_{n+1}) = x_{n+1} | X(t_n) = x_n) \end{aligned}$$

This process, having the Markov property and termed as a Markov process, has states that depend only on the immediately previous instant and not on any other earlier instants.

More interestingly, if time t_n is now and t_{n+1} is the immediate future, then a Markov process future states depend only on the present and not on the past. This property is why this process is also termed to be memoryless, that is, it has no memory of the past.

Common examples of real system usually modeled as memoryless are as follows:

- Automated bank teller machines whose future rate of dispensing cash depends only on its current rate.

- A production machine whose rate of failure tomorrow depends only on its failure rate today.
- A telephone switch whose rate of dropping the next call depends only on the rate it drops the current call.

Markov process is a popular way to describe real events in engineering and risk management due to several reasons, especially the following.

Compactness of information. The memoryless property of Markov processes allows analysts the convenience of compactly summarizing the effects of past states of a system on its future states through its current state. Consider a production machine whose rate of failure tomorrow depends only on its failure rate today. If the engineer in charge of maintaining this machine has the failure rate of the machine today, only that single information will be the key in determining the failure rate for tomorrow. When tomorrow comes, the engineer needs only the tomorrow's failure rate to determine the failure rate for the next day and so on.

Convenience of analysis. The Markov property is a well-studied property of probabilistic processes that brings into consideration decades of research and hundreds of academic and application-based publications. Even though each engineer and risk analyst may be looking at different systems, there is a great chance that they will find an acceptable Markov process approximation of their system with accompanying sets of solution formulas for their fundamental needs by scouring the body of literature.

The Markov property of probabilistic processes is often summarized in terms of state transition probabilities or the chance that the system will be in a particular state given its current state. Consider the simple event of tossing a fair coin with the face of the coin landing up is the system state of interest. With only two states, head and tail, and with equal chance of occurrence for a fair coin, the probabilities of the coin's next state at t_{n+1} given the current state at t_n are

$$P(X(t_{n+1}) = head \mid X(t_n) = head) = 0.5$$

$$P(X(t_{n+1}) = head \mid X(t_n) = tail) = 0.5$$

$$P(X(t_{n+1}) = tail \mid X(t_n) = head) = 0.5$$

$$P(X(t_{n+1}) = tail \mid X(t_n) = tail) = 0.5$$

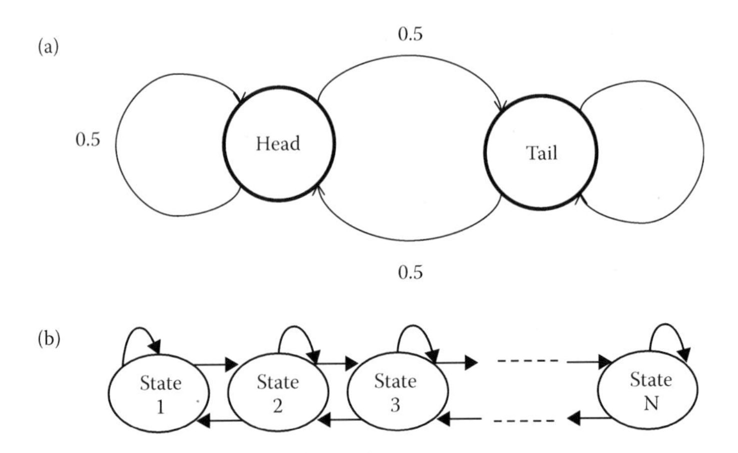

FIGURE 9.1
(a) State transition diagram for tossing a fair coin as a Markov process. (b) State transition diagram for a birth and death process.

These transition probabilities can also be expressed as a matrix

$$\begin{vmatrix} 0.5 & 0.5 \\ 0.5 & 0.5 \end{vmatrix}$$

9.4.1 Birth and Death Process

If the random variables denoting the states of a Markov process are discrete in nature, then these discrete random variables form a Markov chain. If a special type of Markov chain with the restriction that at each step of the chain, the state transition, if any, can occur only between immediately neighboring states, then a very useful process is defined.

Definition 9.11: Birth and Death Process

Birth and death process is a special type of Markov chain with the restriction that at each step of the chain, the state transition, if any, can occur only between immediately neighboring states (as shown in Figure 9.1a).

In a birth and death process with a current state i at time t_n, the state at the immediately succeeding time period t_{n+1} can either be unchanged at i or be changed to the immediately neighboring states $i-1$, or $i+1$. This is illustrated in Figure 9.1b.

9.5 Queuing Theory

Engineering problems often arise from scarcity of resources such as server time (man or machine time), monetary resources, material resources, and information. A broad category of engineering systems with scarce resources is system of flows and is defined as system wherein transactions flow through a server to go from one point to another (Kleinrock, 1976). Systems of flows abound and come in different forms and complexities. Systems of flows can be large and complex and engineers have always tried to use models for analysis. A system of flows is the basis of study termed as queuing theory or the study of queues.

Queues, based on its common usage, pertains to persons, objects, or any other entities that are lining up and waiting for something. This apparent simple nature of queues is one reason queuing theory and its applications are very popular. Consider Figure 9.2, which shows a number of entities, queuing and waiting for their turn to be served by a server. The entities and server illustrated in Figure 9.2 could represent many particular examples. Consider some examples in Table 9.2. The entities can be cars lined up in front of a red stop light, the server, waiting for the light to turn green. The entities

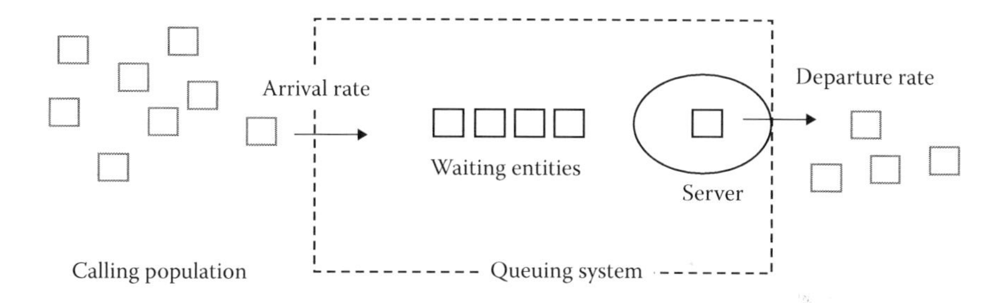

FIGURE 9.2
Schematic of a simple queuing system.

TABLE 9.2
Examples of Systems Being Modeled as a Queuing Systems and Possible Entities and Servers

System Being Modeled	Entity	Server	Characteristics
Barber shop	Humans	Barber	Number of barbers and barber's chairs How fast barbers cut hair Length and style of hair
Emergency room	Patients	Triage nurse	Time of the day Number of doctors on duty
Doctor's office	Patients	Doctors	Day of the week Type of practice
Internet at home	Packets of information	Internet router	Connection speed Routing protocol Volume of information
Street intersection	Cars	Stop light	Volume of cars Length of time light is red or green

can be humans waiting for their turn at a barber shop. Though they may not be literally lined up in front of the barber's chair as they may be sitting down, there is a particular order by which they will be served. The entities could also be packets of information waiting for their turn to be routed by an internet router. For this particular example, the packets of information are not physically lined up in front of the server or any of the service provided by the internet router is not directly observable by the naked eye. Nonetheless, to model what is occurring as a queuing system provides immeasurable benefits to engineers, resulting in the effectiveness and efficiency of the internet many of us are now benefiting from.

9.5.1 Characteristic of Queuing Systems

Inherent to any modeling effort is the selective choice on which among system characteristics will be included in the model. The choice is made based on the purpose of the modeling effort and the characteristics of the system being emphasized (Jain, 1991). The chosen characteristics are translated to variables or parameters that serve as building blocks of the model (Haimes, 1998). As a result, a model can be judged as appropriate for the analysis if the model has as building blocks the characteristics considered by the analyst as important and provide results that satisfy the analyst's requirements. Consider the examples in Table 9.2. Some of the characteristics of these systems are typical of how we will describe the system, for example, a barber shop can be typically described in terms of the number of barbers and barber chairs it has and by how fast they cut hair. However, for a particular patron waiting in line, it may also be relevant to know the length and styles of haircut of those ahead of the patron because these affect how long they will be served, hence the patron's waiting time, or the available seats inside the barber shop for waiting customers.

In more general terms, queuing systems can be described by a number of characteristics, including:

A. Description of how entities arrive into the queuing system (e.g., M stands for Markovian arrivals, GI for general (any distribution) arrivals, PH for phase-type arrivals).
B. Description of the service provided by the server (e.g., M stands for Markovian service, GI for general (any distribution), PH for phase-type service).
C. Number of servers in the system.
D. Capacity of the queue (i.e., the maximal number of entities that can be queued in the system).
E. System population (i.e., the maximal number of entities that can arrive in the queue).

F. Queuing discipline, which can be FIFO (first come first serve), LIFO (last come first serve), or any other queuing discipline. If this argument is missing, then by default the queuing discipline is FIFO. For Figure 9.2, it is implied that the queuing discipline is FIFO.

These characteristics are expressed primarily by the Kendall notation A/B/C/D/E/F made up of the preceding characterizations.

It is noticeable that several important characterizations—the arrival process of entities and service process—can be described using the Markov process. In essence, this implies that there are cases where the arrival of entities is memoryless. From the pioneering work by Jackson (1957), analyses of queuing models with Markovian arrival and service are done with steady-state probabilities. This also implies that the effects of initial conditions are nonexistent.

Of particular importance in characterizing queuing models with Markovian arrival and service times are two parametric distributions, namely Poisson and exponential distributions.

9.5.2 Poisson Process and Distribution

Poisson process is a collection of $N(t)$ random variables, where t stands for time $t \geq 0$, and $N(t)$ is the number of events that have occurred from time 0 to time t and has the following properties:

- Events do not occur simultaneously.
- The number of events occurring in any interval of time after time t is independent of the number of arrivals occurring before time t, that is, the process is memoryless.

Consider the arrival of cars at an intersection. If one marks on a timeline when a car arrives at an intersection, then this timeline may look like that of Figure 9.3.

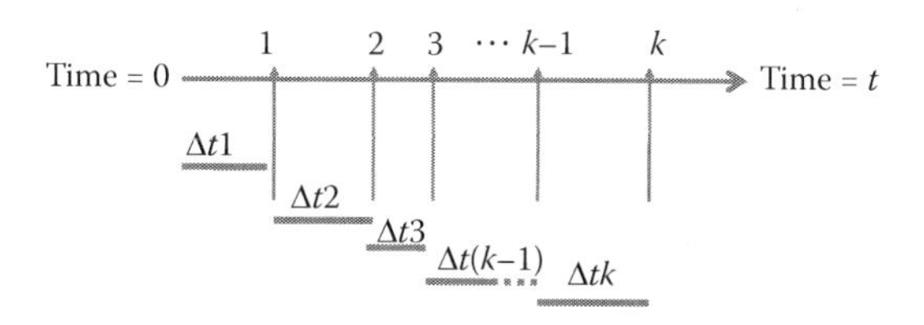

FIGURE 9.3
Timeline marked with the occurrence of events of interest, for example, arrival of a car in an intersection.

If the arrival of cars can be described as follows:

- No two or more cars arrive at the intersection at the same time, that is, events do not occur simultaneously
- The number of cars arriving after time t is independent of the number of arrivals occurring before t, that is the process is memoryless,

then this process can be described as a Poisson process, and the number of arrivals within a time period has a Poisson distribution.

Definition 9.12: Poisson distribution

The probability of k arrivals in t time periods in a Poisson process has a Poisson distribution expressed as

$$P_k(t) = \frac{(\lambda t)^k}{k!} e^{-\lambda t}$$

9.5.3 Exponential Distribution

Consider the time between arrivals of cars illustrated in Figure 9.2, the Δts. Essentially, these times between occurrences of events in a Poisson process are random variables whose probabilities have an exponential distribution.

Definition 9.13: Exponential Distribution

In a Poisson process, the distribution of time, say x, between occurrences of events have an exponential distribution and is given by

$$P(x) = \mu e^{-\mu x}$$

9.6 Basic Queuing Models

There are several well-established basic queuing models that are often used in engineering and other fields of applications. These basic models can be used to analyze, in particular, simple systems and also as building blocks for modeling more complex systems. Two basic queuing models are described in this section: single-server and multiple-server models.

9.6.1 Single-Server Model

A single-server model characteristically describes the presence of only one server to accommodate the entities as shown in Figure 9.2. We can analyze this type of queuing system by making some simplifying assumptions:

1. There are infinite number of spaces for entities waiting to be served.
2. The arrival of entities from the calling population is known and can be described by a constant rate λ, in terms of entities per time period. Furthermore, the probability of entities arriving into the queuing system within a very short time period $\Delta t \to 0$ is
   a. $P(\text{one arrival}) = \lambda t$
   b. $P(\text{no arrival}) = 1 - \lambda t$
   c. $P(\text{more than one arrival}) = 0$
3. The rate of service provided by the server can be described by a constant rate μ, in terms of entities per time period. Furthermore, the probability of entities departing from the queuing system (provided there are entities being served) within a very short time period $\Delta t \to 0$ is
   a. $P(\text{one departure}) = \mu t$
   b. $P(\text{no departure}) = 1 - \mu t$
   c. $P(\text{more than one departure}) = 0$

Consider that the state of the queuing system is the number of entities inside the system, composed of both those waiting and being served. The state of the systems are then nonnegative integer values $0, 1, 2, \ldots, N, \ldots, \infty$. Since it was assumed that there are infinite spaces for waiting entities, there can also be as many system states.

Let the state of the queuing system be a random variable and the probability that there are N entities inside the queuing system, and hence the state is N, at time t, has a function $p_N(t)$. By ignoring terms $(\Delta t)^Y$, with $Y = 2$ or higher since $\Delta t \to 0$, then

$$p_0(t + \Delta t) = p_0(t)[1 - \lambda\Delta t] + p_1(t)\mu\Delta t$$

$$p_N(t + \Delta t) = p_N(t)[1 - \lambda\Delta t - \mu\Delta t] + p_{N-1}(t)\lambda\Delta t + p_{N+1}(t)\mu\Delta t, \quad \text{for } N > 0.$$

Invoking the axiomatic definition of probabilities, $\Sigma_{\forall i}\, p_i(t) = 1$, for all $t \geq 0$ and taking the limits as $\Delta t \to 0$ gives a partial derivative

$$\frac{dp_0(t)}{dt} = -\lambda p_0(t) + \mu p_1(t)$$

$$\frac{dp_N(t)}{dt} = -(\lambda + \mu)p_0(t) + \lambda p_{N-1}(t) + \mu p_{N+1}(t), \quad \text{for } N > 0$$

These two differential equations describe the states of the queuing system illustrated in Figure 9.2.

Equilibrium condition is reached when

$$\frac{dp_0(t)}{dt} = 0 = -\lambda p_0(t) + \mu p_1(t)$$

$$\lambda p_0(t) = \mu p_1(t)$$

$$p_1(t) = \frac{\lambda}{\mu} p_0(t)$$

If the time argument (t) is dropped from the notation and $\rho = \lambda/\mu$, then

$$p_1 = \rho p_0$$

Similarly,

$$p_{N+1} = (1-\rho)p_N - \rho p_{N-1} = \rho p_N = \rho^{N+1} p_0, \quad \text{for } N \geq 1$$

Invoking the axiomatic definition of probabilities, $\Sigma_{\forall i}\, p_i = 1$, the system state probabilities of a queuing system illustrated in Figure 9.2 are

$$p_i = \rho^i (1-\rho), \quad \text{for all possible states } i = 0, 1, \ldots, \infty \tag{9.5}$$

However, a condition must be placed for a queuing system to be stable such that the number of entities inside any queuing system does not keep on continuously increasing. This condition is that the rate of arrival of entities must be less than the rate of departure, that is, $\lambda < \mu$ and therefore $\rho < 1$. Knowing the equilibrium state probabilities allow the calculation of various performance parameters of the queuing system described in the following.

9.6.2 Probability of an Empty Queuing System

There are situations when knowing the likelihood of the queuing system being empty can be helpful in designing a system. Consider the case of a bank manager who may need to know how often they can replenish the cash machine in a day that can only be done when no one is using the cash machine.

From Equation 9.5 and $i = 0$, that is, the state of the system when it has no entity,

$$p_0 = \rho^0 (1-\rho) = (1-\rho)$$

$$= \left(1 - \frac{\lambda}{\mu}\right)$$

PROBLEM 9.1

Consider an automated bank teller machine (ATM) whose operating system needs regular updates by uploading small pieces of codes through remote uploads. To minimize disruption of service, this remote uploads can only be done when the machine is not serving any client and that no client is waiting to be served. If the mean arrival rate of clients is 10 per hour and the mean service rate is 15 per hour, what is the chance that there are no clients being served or waiting to be served?

Solution

Modeling the ATM as a queuing system with $\lambda = 10$ per hour and $\mu = 15$, the chance that the system is empty is

$$P_0 = \left(1 - \frac{\lambda}{\mu}\right) = \left(1 - \frac{10}{15}\right) = 0.33$$

9.6.3 Probability That There Are Exactly *N* Entities Inside the Queuing System

Consider Equation 9.5 and expressing this in terms of the arrival and departure rates,

$$p_N = \rho^N (1 - \rho) = \left(\frac{\lambda}{\mu}\right)^N \left(1 - \frac{\lambda}{\mu}\right)$$

PROBLEM 9.2

Consider a walk-up ATM being installed in a place where winter can be bitterly cold. The engineer decided to install the ATM inside a climate-controlled area with enough space for 5 people, including that one using the ATM. If the mean arrival rate of clients is 10 per hour and the mean service rate is 15 per hour, what is the chance that there will be people needing to wait outside the climate-controlled area?

Solution

Essentially, what needs to be estimated is the chance that there are more than 5 people inside the queuing system. This is

$$\begin{aligned} P_{N>5} &= 1 - P_{N\le 5} \\ &= 1 - (P_0 + P_1 + P_2 + P_3 + P_4 + P_5) \\ &= 0.088 \end{aligned}$$

9.6.4 Mean Number of Entities in the Queuing System

One of the most insightful performance parameters of a queuing system is the number of entities inside the system. For engineering systems with limited space such as a factory, knowing how much work in process will be waiting for a machine may be important in designing the shop floor layout.

Consider Equation 9.5, which provides the probability that there are i entities inside the queuing system. Using the notion of expected value provides the calculation for expected number of entities inside the system to be

$$Q = \sum_{i=0}^{\infty} ip_i = \sum_{i=0}^{\infty} i\rho^i(1-\rho) = \frac{\rho}{1-\rho}$$
$$= \frac{\lambda}{\mu} \times \frac{1}{1-(\lambda/\mu)}$$
$$= \left(\frac{\lambda}{\mu-\lambda}\right)$$

9.6.5 Mean Number of Waiting Entities

The number of entities waiting to be served is essentially the number in the queue. Since we are looking at a single server that can serve only one entity at a time, the number of entities in the queue will be one less than the number inside the system:

$$Q_q = \sum_{i=0}^{\infty} (i-1)p_i = \frac{\rho}{1-\rho} - (1-p_0)$$
$$= \frac{\rho}{1-\rho} - \rho = \frac{\rho^2}{1-\rho}$$
$$= \left(\frac{\lambda^2}{\mu(\mu-\lambda)}\right)$$

9.6.6 Average Latency Time of Entities

The average latency time of an entity, that is, time inside the system waiting and being served can be important for systems where throughput is critical for system operation, for example, a well-balanced assembly line. Another consideration can be for systems where any delay can be critical to health and safety, for example, an emergency room triage service. However, one critical aspect that determines latency time of entities is the queue discipline or rule used to choose among the waiting entities to be served next. First come first served (FCFS), also known as first in first out (FIFO), is a common queue discipline. FCFS and FIFO make sense in many instances where there may be n discernable differences among entities, similar to a mass product assembly line. These rules can also be perceived as just and fair in common service systems such as banks, barber shops, or fast-food restaurants.

However, several more assumptions can simplify the calculation of the average latency time of entities in a queuing system. In particular, assume the following:

1. The arrival of entities into the queuing system is a Poisson process. That is, the distribution of times between arrivals is exponentially distributed with mean $1/\lambda$.
2. The service completion of entities, and thus their departure, is a Poisson process. That is, the distribution of times between departures is exponentially distributed with mean $1/\mu$.
3. Queue discipline is FCFS.

Then the average latency time of an entity is

$$L = \sum_{i=0}^{\infty} \frac{(k+1)}{u} p_k = \frac{1}{\mu(1-\rho)}$$
$$= \frac{1}{\mu - \lambda}$$

9.6.7 Average Time of an Entity Waiting to Be Served

The average time of an entity waiting to be served, that is, waiting in the queue prior to service will be one mean service time less than the average latency time. That is

$$L_q = L - \frac{1}{\mu} = \frac{\rho}{\mu(1-\rho)}$$
$$= \frac{1}{\mu(\mu - \lambda)}$$

In summary, the basic assumptions in a single-server model in Kendall notation are shown in Table 9.3. However, also shown in Table 9.3 are other possible assumptions and corresponding Kendall notations.

Significant studies have been done in estimating performance parameters of various queuing models. This provides systems engineers the convenience of not having to develop queuing models and derive performance measures for many types of systems. Nonetheless, the extensive research done in queuing systems also presents systems engineers the challenge of choosing the appropriate model for a particular system and for a specific

TABLE 9.3
Kendall Notation for Compactly Describing Queuing Models

Kendall Variable	Possible Characterization	Equivalent in Kendall Notation
A	Poisson arrival rate with λ mean arrival rate	M
	Constant arrival rate	C
	Arrival rate with general distribution	G
B	Exponential service times with μ mean service rate	M
	Constant arrival rate	C
	Arrival rate with general distribution	G
C	Only 1 server	1
	Multiple servers	S
D	Infinite queue length	∞
	Finite queue length	F
E	Infinite calling population	∞
	Finite queuing calling population	F
F	First-in, first-out queue discipline	FIFO
	Last-in, first-out queue discipline	LIFO
	Other priority queue discipline	Several notations

purpose. Presented in Tables 9.4 to 9.8 are the more common variations of queuing models that may be most useful for modeling many engineering systems. For more sophisticated and specialized queuing models, one may look at other sources specializing on this topic, including Bhat and Basawa (1992), Gorney (1981), and Daigle (1992).

Nonetheless, performance measures for the more commonly used queuing models presented in this chapter are as follows:

- Basic single server, $M/M/1/\infty/\infty/FIFO$ (see Table 9.4)
- Single server, constant service time $M/C/1/\infty/\infty/FIFO$ (see Table 9.5)
- Single server, finite queue length, $M/M/1/F/\infty/FIFO$ (see Table 9.6)
- Single server, finite calling population, $M/M/1/\infty/F/FIFO$ (see Table 9.7)
- Multiple server model, $M/M/S/\infty/\infty/FIFO$ (see Table 9.8)

9.7 Applications to Engineering Systems

Queuing models are often used in analyzing various aspects of engineered systems to describe the scenarios surrounding limitations of resources. Consider the problem of a systems engineer in the process of developing remote sensing equipment meant to relay data from remote places to satellites and eventually to a research laboratory. In particular, the engineer is

TABLE 9.4

Summary Performance Metrics for a Basic Single-Server Model, M/M/1/∞/∞/FIFO

Probability that no entities are in the queuing system, i.e., queuing system is empty	$P_0 = \left(1 - \frac{\lambda}{\mu}\right)$
Probability of exactly n entities in the queuing system	$P_n = \left(\frac{\lambda}{\mu}\right)^n \left(1 - \frac{\lambda}{\mu}\right)$
Average number of entities in the system, including those waiting and being served	$Q = \left(\frac{\lambda}{\mu - \lambda}\right)$
Average number of entities waiting to be served	$Q_q = \left(\frac{\lambda^2}{\mu(\mu - \lambda)}\right)$
Average latency time of an entity, i.e., time inside the system waiting and being served	$L = \frac{1}{\mu - \lambda}$
Average time of an entity waiting to be served, i.e., waiting in the queue	$L_q = \frac{1}{\mu(\mu - \lambda)}$
Chance (or proportion of time) the server is busy, i.e. the server's utilization factor	$\rho = \frac{\lambda}{\mu}$

TABLE 9.5

Performance Metrics for a Single Server with Constant Service Time, M/C/1/∞/∞/FIFO

Probability that no entities are in the queuing system, i.e., queuing system is empty	$P_0 = \left(1 - \frac{\lambda}{\mu}\right)$
Average number of entities waiting to be served	$Q_q = \left(\frac{\lambda^2\sigma^2 + (\lambda/\mu)^2}{2(1 - \lambda/\mu)}\right) = \left(\frac{\lambda^2}{2\mu(\mu - \lambda)}\right)$ Since $\sigma = 0$ when service time is constant,
Average number of entities in the system, including those waiting and being served	$Q = Q_q + \frac{\lambda}{\mu}$
Average latency time of an entity, i.e., time inside the system waiting and being served	$L = L_q + \frac{1}{\mu}$
Average time of an entity waiting to be served, i.e., waiting in the queue	$L_q = \frac{Q_q}{\lambda}$
Chance (or proportion of time) the server is busy, i.e., the server's utilization factor	$\rho = \frac{\lambda}{\mu}$

TABLE 9.6

Performance Metrics for a Single Server with Finite Queue Length, M/M/1/F/∞/FIFO

Probability that no entities are in the queuing system, i.e., queuing system is empty	$P_0 = \left(\dfrac{1-(\lambda/\mu)}{1-(\lambda/\mu)^{M+1}}\right)$
Probability of exactly n entities in the queuing system	$P^n = (P_0)\left(\dfrac{\lambda}{\mu}\right)^n$ for $n \le M$
Average number of entities in the system, including those waiting and being served	$Q_Q = \dfrac{\lambda/\mu}{1-(\lambda/\mu)} - \dfrac{(M+1)(\lambda/\mu)^{M+1}}{1-(\lambda/\mu)^{M+1}}$

TABLE 9.7

Performance Metrics for Single Server with Finite Calling Population, M/M/1/∞/F/FIFO

Probability that no entities are in the queuing system, i.e., queuing system is empty	$P_0 = \dfrac{1}{\sum_{n=0}^{N} \dfrac{N!}{(N-n)!}\left(\dfrac{\lambda}{\mu}\right)^n}$, where N = population size
Probability of exactly n entities in the queuing system	$P_n = \dfrac{N!}{(N-n)!}\left(\dfrac{\lambda}{\mu}\right)^n P_0$, where $n = 1, 2, \ldots, N$
Average number of entities in the system, including those waiting and being served	$Q = Q_q + (1 - P_0)$
Average number of entities waiting to be served	$Q_q = N - \left(\dfrac{\lambda+\mu}{\lambda}\right)(1-P_0)$
Average latency time of an entity, i.e., time inside the system waiting and being served	$L = L_q + \dfrac{1}{\mu}$
Average time of an entity waiting to be served, i.e., waiting in the queue	$L_q = \dfrac{Q_q}{(N-Q)\lambda}$

trying to determine which type of processor will be used to assure that data are processed and transmitted by the equipment in a timely manner.

If the engineer chooses to use queuing model to help in this decision scenario, then the problem can be abstracted by modeling the system primarily made up of information as entities or jobs and the processor as a server, similar to that shown in Figure 9.2. The modeling process can be summarized by specifying the characteristics of the model based on the Kendall notation as shown in Table 9.9.

TABLE 9.8

Performance Metrics for Multiple Server Model, M/M/S/∞/∞/FIFO

Probability that no entities are in the queuing system, i.e. queuing system is empty	$P_0 = 1\Big/\left(\left(\sum_{n=0}^{n=s-1}\frac{1}{n!}\left(\frac{\lambda}{\mu}\right)^n\right)+\frac{1}{s!}\left(\frac{\lambda}{\mu}\right)^s\left(\frac{s\mu}{s\mu-\lambda}\right)\right)$
Probability of exactly n entities in the queuing system	$P_n = \begin{cases}\frac{1}{s!s^{n-s}}\left(\frac{\lambda}{\mu}\right)^n P_0, & \text{for } n>s\\ \frac{1}{n!}\left(\frac{\lambda}{\mu}\right)^n P_0, & \text{for } n\le s\end{cases}$
Average number of entities in the system, including those waiting and being served	$Q=\frac{\lambda\mu(\lambda/\mu)^s}{(s-1)!(s\mu-\lambda)^2}P_0+\frac{\lambda}{\mu}$
Average latency time of an entity, i.e. time inside the system waiting and being served	$L=\frac{Q}{\lambda}$
Average number of entities waiting to be served	$Q_q = Q-\frac{\lambda}{\mu}$
Average time of an entity waiting to be served, i.e., waiting in the queue	$L_q = L-\frac{1}{\mu}$
Chance (or proportion of time) the server is busy, i.e., the server's utilization factor	$\rho=\frac{\lambda}{s\mu}$

However, similar to any modeling activity, the effectiveness and efficiency of using queuing models to approximate a real system will rely on multiple factors:

1. The purpose for the modeling activity (e.g., decision scenario, insights needed by decision makers, variables needed by the decision scenario, etc.)
2. The accuracy and precision of information required to support the decision scenario
3. The information known about the real system (e.g., the speed of a computer processor for a web server)
4. The cost of deducing, inferring, or estimating new information with the required accuracy (cost of market research to estimate potential traffic into a web server)
5. The resources available for the modeling activity
6. The benefits of the modeling activity
7. Other alternatives to using queuing models.

TABLE 9.9

Modeling a Remote Sensing Equipment as a Queuing System and Possible Kendall Notation

Kendall Variable	Characteristic of the Remote Sensing Equipment	Possible Characterization	Equivalent in Kendall Notation
A	Description of the arrival of data to be processed by the processor	Time between arrival of data is exponentially distributed, i.e., Markovian	M
B	Description of how the processor process the data	Time needed to process data is exponentially distributed, i.e., Markovian	M
C	Number of processors in the remote sensing equipment	1	1
D	Capacity of the remote-sensing equipment to store data prior to processing	Infinite storage capacity	∞
E	The maximal number of data needs to be processed	Infinite	∞
F	The order the data in storage will be processed	First in, first out	FIFO

These factors all contribute in varying degrees to the systems engineer's choice of using queuing models for the analysis and the particular type of model to use. Nonetheless, the availability of many variations and possible customization of queuing models may result in lower resources needed for the modeling activity.

Even though queuing model is popular in engineering application, a systems engineer needs to recognize that there are alternative models. Other approaches to modeling are the fluid flow model (Kulkarni, 1997) and the fractal queuing model (Ashok et al. 1997). More recently, a deterministic and statistical network calculus based on queuing models has been developed that puts bounds on the performance of queue in the form of effective service curves (Liebeherr et al., 2001) and bounds on the length of the waiting line and on latency have been estimated (Agrawal et al., 1999, Andrews, 2000, Kesidis and Konstantopoulos, 1998 all cited by Liebeherr et al., 2001).

Another field related to the modeling of risk and queue is the study of hard real-time systems. Hard real-time systems pertain to systems of flow wherein some of the transactions have guaranteed latency time. Buttazzo (1997) and Peng et al. (1997) approached the problem of assuring latency times by formulating scheduling algorithms. Such algorithms determine the feasibility for a certain sets of transactions with known arrival times and service times in the sense that all guaranteed latency times are satisfied. Some algorithms consider precedence constraints that essentially have priorities among transactions. There are also algorithms that consider servers with priorities similar to servers in queuing systems with priorities. However, the algorithms are basically on determining the feasibility of the system with respect to constraints in latency times. The algorithms do not facilitate the reconfiguration of the system to address infeasibility or the measurement of the risk of extreme latencies.

The product form, the fluid flow, and the fractal queuing models fail to answer a couple of needs for analysis of a realistic system. First, the assumption that the arrival and service processes have Markovian property cannot always be the case in real systems (Jain, 1991). Moreover, precise solutions are available when interarrival and processing time distributions have the Markovian property and not for other more general distributions, for example, Gumbel-type distributions. Second, the results of the analysis describe the performance of the system as expected value rather than as a distribution of possible values. Even though variance describes the spread of the values, it is still hinged on the central tendency value (recall that variance is $E[x-\overline{x}]$ where $\overline{x}$ is the mean). In cases when the distributions of the latency times are available, the solutions can be cumbersome and do not facilitate changes to configuration of the queuing systems. Such results do not provide the opportunity to perform analysis on the tails of the distributions representing the extreme latencies.

9.8 Summary

Knowing how and when to use deterministic and nondeterministic way of describing phenomena is an important skill of engineers and practitioners. Each way has its own advantages and disadvantages and primarily depends on the decision scenario at hand and available information that will appropriately support the use of either deterministic or nondeterministic models.

Important foundations for using nondeterministic models are the notions of random process and uncertainty. Furthermore, the importance of reducing uncertainty cannot be overemphasized in the field of engineering.

Finally, using queuing models can be an effective and efficient way for the analysis of many engineering systems. Nonetheless, applying queuing models requires that elements of the engineering system and their correspondence with elements of a queuing system must be identified, in addition to carefully choosing various performance parameters to be used in the analysis.

Questions and Exercises

True or False

1. One of the most important skills of a system engineer is to communicate in simple and readily comprehensible manner.
2. Systems engineer often needs to describe phenomenon with the objective of helping himself or someone else arrive at a decision.
3. Too much never negatively affect the decision-making process and affect the budget, schedule, or performance of the system being developed.
4. The simplest way to describe a phenomenon may be in a deterministic manner.
5. Determinism is synonymous with definite, certain, sure, and can be quantitatively expressed as a point estimate.
6. Phenomenon described in a deterministic manner is never sufficient and precise enough such that decision makers can appropriately choose among alternatives.
7. A well thought-out deterministic description of phenomenon, with precision and accuracy appropriate for the decision maker and the decision scenario at hand, is one of the greatest tools and skill a systems engineer can have.
8. Deterministic process is the one in which the same output is obtained every time the same set of inputs or starting conditions occur.
9. Deterministic process implies that there is some form of causality between the inputs or starting conditions of the process and the outputs.
10. A random process is a phenomenon that has predictability of its actual outcome (i.e., posses the random property).
11. A random event is a set of outcomes resulting from a random process.
12. Aleatory uncertainty refers to uncertainty in our state of knowledge about certain phenomena.
13. Aleatory uncertainty is also known as reducible uncertainty, pertaining to its property to be reduced through investigation, reasoning, engineering interventions, and other forms of analyses.
14. Epistemic uncertainty, on the other hand, is due purely to the variation in outcomes of randomness.
15. Aleatory is also known as irreducible uncertainty, pertaining to its property of not being affected by further investigation, reasoning, and other forms of interventions and analyses.

16. The concepts of uncertainty, randomness, and probability are weakly related, and their relationship is always clear to many engineers.
17. An integral part of any engineering activity is the discovery of how things work and how they affect a desired result.
18. Fundamental to any scientific activity is making conclusions based on rational and repeatable acts of discovery.
19. Correlation is the relationship between two events, wherein the occurrence of one implies the occurrence of the other.
20. A set of events B is described to be necessary to cause another set of events A if B is a required condition for the occurrence of A, not that A actually occurs.
21. A set of events B is described to be necessary to cause another set of events A if the occurrence of B guarantees the occurrence of A.
22. Establishing causalities among events is at the very foundation of risk scenario identification.
23. Two commonly held approaches to risk scenarios identification are bottom-up and top-down approaches.
24. Bottom-up approach to risk identification relies on the knowledge of what the elements of the systems are and how these elements are expected to work together.
25. Top-down approach to risk identification relies on the knowledge of the objectives of the systems.
26. An evidence probabilistically causes the event of interest if the occurrence of the evidence increases the probability of occurrence of the event of interest.
27. An event (or evidence) B is necessary and sufficient condition for another event A if B occurs if and only if A occurs.
28. Necessary and sufficient functionalities, by definition, will never assure a required capability.
29. Identifying sufficient functionalities, coupled by negation, will assure complete list of functional scenarios that can cause (the risk event) not delivering a capability.
30. A Markov process describes that future state of the system depends only on its present states, and not on any past states.
31. Memoryless property of Markov processes will not allow analysts to compactly summarizing the effects of past states of a system on its future states through its current state.
32. Birth and death process is a special type of Markov chain with the restriction that at each step of the chain, the state transition, if any, can occur only between immediately neighboring states.

Qualitative

1. Identify which of the following statements are deterministic and which are probabilistic.
   (A) It will rain tomorrow.
   (B) There is one-in-10-million chance of winning jackpot at the state lottery.
   (C) My car broke down after 10 years.
   (D) The butter I bought weighs one pound.
2. What is mathematical determinism?
3. Provide five examples of mathematical expressions that shows deterministic process, for example, $F = m^*a$.
4. Provide examples of situations where phenomena can be appropriately described deterministically.
5. Compare and contrast between deterministic and nondeterministic way of describing a phenomenon.
6. Identify the advantages and disadvantages of using deterministic and nondeterministic models.
7. Provide an example of a decision scenario when the statement *an automobile tire will fail after 50,000 miles of road use* is sufficient. When is it not?
8. Identify and describe a random process.
9. Compare and contrast two sources of uncertainty in predicting the outcome of tossing a coin.
10. In your own words, describe epistemic and aleatory uncertainty.
11. Describe the importance of reducing uncertainty.
12. Compare and contrast necessary causes from sufficient causes.
13. What are necessary and sufficient causes?
14. What is the importance of necessary and sufficient causes in systems engineering?
15. Describe a Markov process and compare with any other random process.
16. Identify the elements of a queuing system.
17. Identify and describe various performance parameters of a queuing system.
18. Describe what may happen in a queuing model under the following situations:
    (A) When $\lambda < \mu$
    (B) When $\lambda > \mu$
    (C) When $\lambda = \mu$

Case Studies

1. Consider the event of your identity being stolen by an identity thief.
   (A) Identify information that you would consider as evidence supporting the occurrence of this event. Identify as many as you can.
   (B) Rank the information from most supportive to least supportive.
   (C) Discuss the information on the top and at the bottom of your list.
2. Consider the event of being late going to work or school tomorrow.
   (A) Identify as many sufficient conditions as possible.
   (B) Identify as many necessary conditions as possible.
   (C) Identify as many necessary and sufficient conditions as possible.
   (D) What factors do you think determine the number of necessary and sufficient conditions for a given event?
3. A systems engineer is designing a new data processor to handle incoming request to access information stored in a multimedia server.
   (A) Describe this particular system in terms elements of a queuing model.
   (B) Describe assumptions about particular elements of the model if the system will be modeled as an $M/M/1/\infty/\infty/FIFO$ queuing system.
   (C) Closer investigation revealed that requests arrive at an average rate of 100 per second while average request processing rate is 150 per second. If the system is modeled as $M/M/1/\infty/\infty/FIFO$, what would be the model performance measures in terms of the following?
      i. Probability that no entities are in the queuing system, that is, queuing system is empty
      ii. Probability of exactly three entities in the queuing system
      iii. Average number of entities in the system, including those waiting and being served.
      iv. Average number of entities waiting to be served
      v. Average latency time of an entity, that is, time inside the system waiting and being served
      vi. Average time of an entity waiting to be served, that is, waiting in the queue
      vii. Chance (or proportion of time) that the server is busy, that is, the server's utilization factor
   (D) Identify and describe possible reasons that will compel an analyst to model this system as $M/C/1/\infty/\infty/FIFO$ instead of $M/M/1/\infty/\infty/FIFO$.

4. A new generation of bioauthentication technology is being deployed at the entrance of a library facility. At the heart of this new technology is a software program that can compare a person's bio profile with those on a database.
   (A) An engineer suggested modeling this system as M/C/1/∞/∞/FIFO queuing system. Identify particular information that will support such a suggestion.
   (B) Another engineer suggested modeling this system as M/M/1/∞/∞/FIFO queuing system. Identify particular information that will support such a suggestion.
   (C) Investigations revealed that requests to compare bio profile comes at a mean rate of 5 per minute, and each bio profile can be compared with those on a database in an approximately constant time of 0.5 seconds. Determine the following:
      i. There are no requests in the system
      ii. Average number of requests waiting to be served
      iii. Average number of requests in the system, including those waiting and being served
      iv. Average time of requests spent inside the system waiting and being served
      v. Average time of request waiting to be served
      vi. Chance (or proportion of time) the software is processing a request
5. A team of systems engineers is trying to improve the current design of a bank currently with three teller windows. One engineer suggested modeling each one of the teller as an M/M/1/F/∞/FIFO queuing system, whereas another engineer suggested modeling the entire bank as an M/M/S/∞/∞/FIFO queuing system.
   (A) Identify, describe, and discuss the advantages and disadvantages of the two suggestions.
   (B) Identify important information that will support using an M/M/1/F/∞/FIFO model.
   (C) Identify important information that will support using an M/M/S/∞/∞/FIFO model.
   (D) Closer investigation revealed that clients arrive to the bank at a mean rate of 0.1 per minute and each teller can served a client at a rate of 0.2 per minute. The bank currently has space for a total of 15 clients. Half of the engineering team decided to use M/M/1/F/∞/FIFO queuing system to model the bank. Determine the following:
      i. Probability that there are no clients in the bank

ii. Probability of exactly three clients in the bank
iii. Average number of clients in the system, including those waiting and being served

(E) The other half of the engineering team decided to use M/M/S/∞/∞/FIFO queuing system to model the bank.
   i. Identify necessary assumption for this model to be valid.
   ii. Determine the probability that no client is in bank.
   iii. Determine the probability of exactly three clients in the bank.
   iv. Determine the average number of clients in the bank.
   v. Determine the average time clients spends inside the bank.
   vi. Determine the average number of client waiting to be served.
   vii. Determine the average time of a client spends waiting to be served.
   viii. Determine the proportion of time the teller is busy serving clients.

6. Two material-request procedures are being considered for deployment in a school library by a team of systems engineers, librarians, and teachers. The team decided to model the procedures as M/M/1/∞/F/FIFO queuing systems. The following information is available: mean arrival rate of request for library materials is 6 per hour, procedure A can handle a request in an average of 2 minutes while procedure B can handle a request in an average of 3 minutes. There are approximately 200 students in the school and each one is not expected to make more than one request each time.

(A) Compare the two procedures in terms of the following performance measures:
   i. Probability that no entities are in the queuing system, that is, queuing system is empty
   ii. Probability of exactly two entities in the queuing system
   iii. Average number of entities in the system, including those waiting and being served
   iv. Average number of entities waiting to be served
   v. Average latency time of an entity, that is, time inside the system waiting and being served
   vi. Average time of an entity waiting to be served, that is, waiting in the queue

(B) Which procedure is the better choice?

10

Extreme Event Theory

10.1 Introduction to Extreme and Rare Events

Extreme events refer to phenomena that have relatively extreme high or low degree of magnitude while rare events refer to phenomena that have relatively very low frequency of occurrence. The use of the term *relative* in describing both extreme and rare events is important because, as will be discussed later, the relative difference between the magnitudes and frequency of occurrence of these events is the primary reason why they are difficult to analyze and manage. As such, events that are both extreme and rare can be defined as follows.

Definition 10.1: Extreme and Rare Events and Phenomena

Extreme and rare events and phenomena have relatively very low frequency of occurrence and at the same time have relatively extreme high or extreme low degree of magnitude.

There are some notable examples of extreme and rare events. One of the most notable is the recent global economic downturn. During the period of 2007–2009, the prices of stocks of many of the world's leading companies plummeted to very low levels. Another extreme and rare event is the magnitude 10 earthquake that occurred in 2011 off the coast of Japan.

It is important to note that events that are extreme are not necessarily rare. Take as an example a pendulum and the angle it makes relative to the vertical. As illustrated in Figure 10.1, the most extreme angles the pendulum makes are θ_{max} and θ_{min} because there will be no angles greater than θ_{max} or less than θ_{min} as illustrated in Figure 10.1. However, the figure also shows that for a frictionless pendulum, these extreme angles occur at every swing. That is, these extreme angles occur with the same frequency as any other measured angle of the pendulum and thus are extreme but not rare.

In a much similar way, rare events are not necessarily extreme. Consider as an example the number of wheels on vehicles in the streets. At the extreme low end are single-wheeled unicycles (if we include in consideration human-powered vehicles) and 18-wheeled tractor trailer trucks on the extreme high

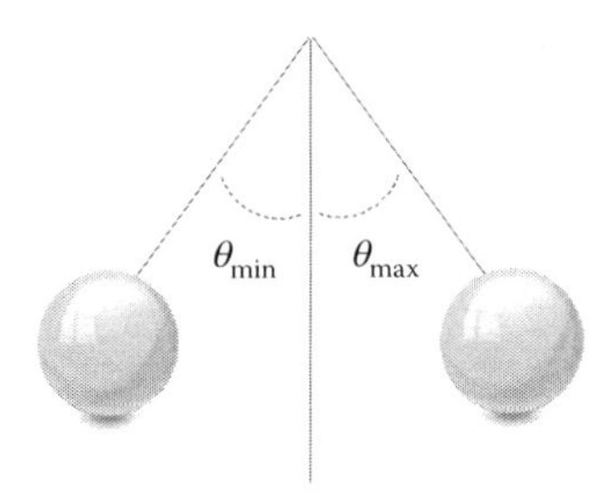

FIGURE 10.1
Illustration of extreme but common events.

end. Something of rarity are five-wheeled vehicles. Having five wheels on a vehicle may be rare indeed but is obviously not extreme.

10.2 Extreme and Rare Events and Engineering Systems

Extreme and rare events are important in the analysis of risks in engineering enterprise systems for two reasons: (1) if magnitude is construed to be directly related to consequence, then these events can be associated with relatively very high degree of monetary loss, loss of lives, property damage, physical and mental agony, loss of goodwill, and other negative consequences. As an example, extremely high rainfall may cause extremely high property damage, as well as extremely low rainfall can cause suffering due to food and water shortage and (2) the relatively very small chance of occurrence limits the use of traditional engineering analysis and design. This is brought about by the use of probability and statistics in many engineering activities and the reliance on analysis and design based on averages.

Consider the following examples.

1. *Quality of air in urban areas* is usually sampled during various times of the day, with special interest on the highest concentration of pollutants staying below a certain threshold. Extremely high concentration of pollutants may trigger a city-wide health advisory.
2. *Drinking water in a geographic area* may be sampled immediately after an earthquake to test for the presence of contaminants. Extremely high concentration of contaminants and bacteria may indicate damage in the water pipes.
3. *Destructive testing of structural materials* such as concrete blocks and steel beams are used to ensure sound bridges and buildings. An extremely low breaking point of a steel beam may be a sign of inferior alloy and may trigger more testing or stoppage of its usage.

4. *Mean time between failure* (MTBF) is a common measure of the reliability of equipment including those used in safety critical applications in hospitals. An extremely short observed time between failures for a particular type of equipment may be a sufficient reason to issue a recall on all equipment currently in use.

These examples all show scenarios in which systems engineers and risk analysts may need to pay more attention to the extreme events because of the possible consequences in that realm of possible events.

10.3 Traditional Data Analysis

Consider a hypothetical example of an engineer trying to design a new irrigation dam at a particular point in the James River in Virginia. An integral part of the design process is to analyze the likelihood of various flood levels at that particular point of the river. This will ensure that the new dam will be constructed appropriately high enough to avoid destructive flooding in the outlying areas.

The engineer conducted measurement of water level at a measuring point known as Kiln Bend in the James River. This was done in such a way that four measurements were taken throughout the day for 30 days. These measurements are shown in Table 10.1a.

Traditional data analysis would suggest that the four measurements for each day are averaged to obtain a daily average measurement. As an example, consider day 1 with readings of 29.8, 16.7, 24.3, and 9.6. These numbers are averaged to get an averaged measurement for day 1 such that

$$\textit{Average measurement for day 1} = \frac{29.8 + 16.7 + 24.3 + 9.6}{4} = 20.11.$$

The averaging process is done for all the 30 days, and the average values are shown in the right-most column of Table 10.1a.

The succeeding steps would be a sequence of steps meant to provide insights for the engineers:

1. Eyeballing the data: The simplest analysis could be to plot the daily averages with respect to their occurrence as shown in Figure 10.2a. At this step, an engineer may want to get insights as to whether there is perceivable trend or pattern in the occurrences of measured water level values. Another visual representation of the data is their frequency of occurrence by grouping the daily averages into bins regardless of when those measurements were recorded. This is also known as a histogram and is shown in Figure 10.2b. A histogram

TABLE 10.1a

Measurements of Water Level at Kiln Bend in the James River

	Reading 1	Reading 2	Reading 3	Reading 4	Average (Sample Mean)
Day 1	29.8	16.7	24.3	9.6	20.1
Day 2	25	6.9	10.7	3.7	11.6
Day 3	4.1	16.3	25	15.7	15.3
Day 4	24.1	10.1	14.6	11.9	15.2
Day 5	3.2	5.1	29.1	19.9	14.3
Day 6	12.4	14.1	6.8	10.7	11.0
Day 7	8.2	3.1	3	21.8	9.0
Day 8	20.1	13.7	10	10.3	13.5
Day 9	24.3	4.3	15.2	17.1	15.2
Day 10	7.7	29.2	7.4	26.8	17.8
Day 11	28.6	2.3	23.7	9	15.9
Day 12	28	20.1	29.9	13	22.8
Day 13	26.8	9.9	13.8	18.1	17.2
Day 14	26.9	17	4.8	29.3	19.5
Day 15	6.5	5.8	6.7	11.1	7.5
Day 16	13.9	9.8	14.1	21.6	14.9
Day 17	25.9	24.4	9.5	12.5	18.1
Day 18	4.7	2.5	29.1	22.2	14.6
Day 19	7.7	4.6	6.4	19.2	9.5
Day 20	23.3	6.5	10.6	15.1	13.9
Day 21	1.8	13.9	12.9	18	11.7
Day 22	21.6	8.1	18.1	3.9	12.9
Day 23	7.9	27.3	24.8	18.8	19.7
Day 24	26.2	23.1	13.7	24.6	21.9
Day 25	23.4	6.7	26	22.2	19.6
Day 26	10.2	2.4	4.2	20.9	9.4
Day 27	7	3.6	27.8	8.2	11.7
Day 28	8.3	19.8	14.1	7.7	12.5
Day 29	23.1	27.6	5.2	21.2	19.3
Day 30	4.5	4.1	2	20	7.7

can provide insights into how data are distributed over its range of values. For this case, one may observe what seems to be the two most frequent occurrences seen as peaks in the histogram: one on water level 17 and another on water level 20. Another visual analysis is to graph the frequencies in the histogram one on top of the other in succession, also known as cumulative frequency. As shown in Figure 10.2c, it is notable that cumulative frequency often has on its vertical axis % of total frequency instead of actual frequency as in the case of a histogram.

TABLE 10.1b

Basic Descriptive Statistics for Data in Table 10.1a

Basic Statistics for Daily Averages Water Level at Kiln Bend in the James River	
Mean of sample means	14.77
Standard error of sample means	0.76
Median of sample means	14.74
Standard Deviation of sample means	4.19
Sample Variance of sample means	17.53
Kurtosis of sample means	-0.82
Skewness of sample means	0.07
Range of sample means	15.23
Minimum of sample means	7.53
Maximum of sample means	22.75
Sum of sample means	442.95
Count of sample means	30.00

2. Basic statistics: After visually analyzing the data, the next step can be to generate basic statistics that will describe the data as shown in Table 10.1b. These statistics can be conveniently generated using computer applications for managing spreadsheets and statistical analysis.

However, descriptive statistics are "central tendencies" of the events and extreme events are averaged out with other events, for example, mean square error (MSE) in curve fitting. As such, traditional data analysis does not provide long-run description of extreme events that may be critically important in the design of the dam in the James River.

10.4 Extreme Value Analysis

The analysis of data presented in Section 10.3 is typical and can answer questions such as the following:

- What is the average water level?
- What have been the lowest and the highest water level on record?
- What is the most frequent water level?

In extreme value analysis, the interest is in the parametric model of the distribution of the maxima and the exceedances, which basically involves

(a)

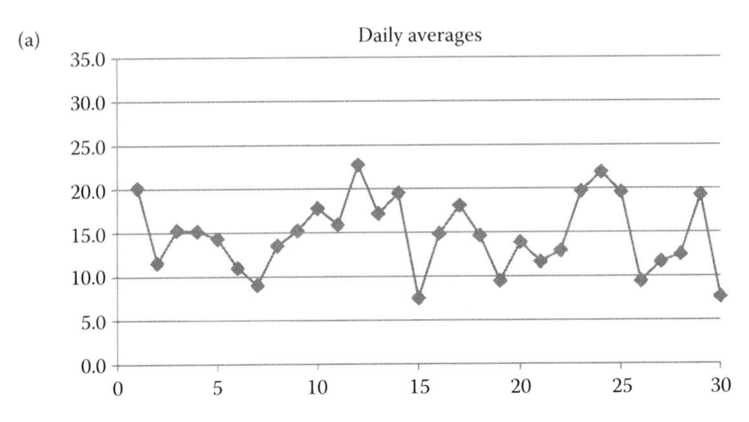

(b)

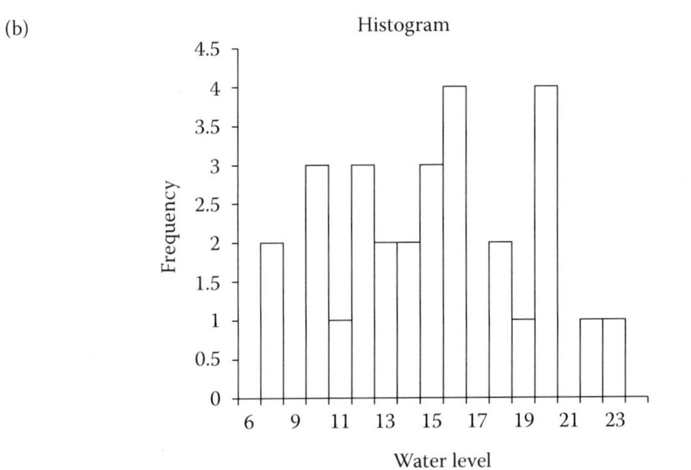

(c)

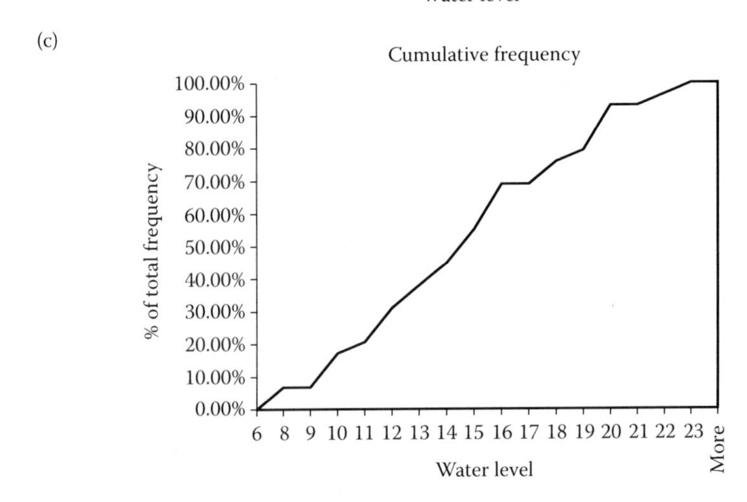

FIGURE 10.2
Analyses of data in Table 10.1a (a) Plot of daily averages with respect to their occurrence (data in Table 10.1a). (b) Histogram of daily averages (data in Table 10.1a). (c) Cumulative frequency of daily averages (data in Table 10.1a).

the tails of the distributions. The probability of exceedance is defined as the following:

Definition 10.2: Probability of Exceedance

Probability of exceedance of a random variable x is the probability that it will be exceeded.

From the axiomatic definition of probability, the exceedance probability can be calculated as

$$P(X > x) = 1 - P(X \leq x)$$

Note that in Chapter 3, the cumulative distribution function (CDF) was defined. Also note that the probability of exceedance is a complement of the CDF.

Exceedance probabilities may be used to answer questions regarding the likelihood of a variable being exceeded. Examples of such questions are the following:

- What is the likelihood of a flood exceeding 10 feet?
- What is the likelihood that a truck weighing more than 10 tons will pass over a bridge?
- What is the likelihood that the temperature will drop below freezing point?

All of these questions not only require information about the density functions but could also specifically pertain to the analysis of the extreme ends of the functions. As such, at the heart of these probabilistic analysis is the estimation of probability density and CDFs of random variables. This enables engineers to estimate how a particular system may perform under various situations, for example, the chance that a levee will break as flood level rises, or the chance a bridge will collapse as the magnitude of an earthquake increases.

However, when looking for extreme flood, earthquake, or other events that are in the tails of distributions, we often find that in real-life situations these tails are fatter (more heavy) than what classical distributions predict.

10.5 Extreme Event Probability Distributions

Significant advances have been made in analyzing extreme values, one of which is the work of Castillo (1988) on the tail equivalence of many continuous distributions. Johnson et al. (1994) presented the chronological development of the field of extreme value analysis. Extreme value theory (EVT) deals with the study of the asymptotic behavior of extreme observations of a random variable (maxima and minima).

In traditional data analysis, we specify probability distributions through knowledge of the phenomena (e.g., roll of a die, n sequential coin tosses) or based on empirical data and statistical methods (e.g., age of graduate students, rainfall for the last 50 years).

Unlike traditional data analysis, EVT focuses primarily on analyzing the extreme observations rather than the observations in the central region of the distribution. The fundamental result of EVT, known as the "extremal types theorem," identifies the possible classes of distributions of the extremes irrespective of the actual underlying distribution. EVT incorporates separate estimation of the upper and the lower tails due to possible existence of asymmetry.

Definition 10.3: Order Statistics

Order statistics is the arrangement of n random variables $X_1, X_2, X_3, \ldots, X_n$, in nondecreasing order, from 1 to n

$$X_{1:n}, X_{2:n}, X_{3:n}, \ldots, X_{n:n}$$

such that for sample size $n, X_{1:n}$ is the smallest, also called $X_{\min}$, and $X_{n:n}$ is the largest, called $X_{\max}$.

Several examples were provided in Section 10.2 of various extreme and rare events that may be of particular interest to systems engineers and risk analysts. These examples are revisited here in the context of order statistics.

1. *Quality of air in urban areas* is usually sampled during various times of the day, with special interest on the highest concentration of pollutants staying below a certain threshold. In this case, the interest may be on the actual highest measured value of pollutants, $X_{n:n}$. A city-wide health advisory may be issued when $X_{n:n}$ becomes higher than a mandated threshold to protect the health of the citizens with compromised or weak pulmonary health.
2. *Drinking water over a geographic area* may be sampled immediately after an earthquake to test for the presence of contaminants. Similarly, the interest may be on the actual highest measured value of contaminant $X_{n:n}$. Residents of that area may be advised to boil water before drinking when the highest measured value of contaminant $X_{n:n}$ is beyond the standard. This may also trigger the local city engineer to look closer into possible damage in the water purification or delivery pipes systems.
3. *Destructive testing of structural materials* such as concrete blocks and steel beams is used to ensure strong bridges and buildings. Keeping in mind that destructive testing is done both to describe the strength and the durability of a piece of material being tested and to predict that for related samples, a $X_{1:n}$ that is beyond the allowable statistical

limits may imply inferior quality of materials related to that one particularly tested. This may result in more testing of the related structural materials or even lead to rejecting and scrapping of whole batches of materials.

4. MTBF is a common measure of the reliability of equipment including those used in safety-critical applications in hospitals. Similarly, for an extremely short observed time between failures for a particular type of equipment, $X_{1:n}$ may be predictive of a systematic failure of an entire type of equipment. This may be a sufficient reason to issue a recall on all equipment currently in use, resulting in the immediate stoppage of its use that can disrupt operations in a hospital.

All these examples provide scenarios in which an analysis of the extreme values may lead to decisions that can be costly and disruptive. Nonetheless, the measurements $X_{1:n}, X_{2:n}, X_{3:n}, \ldots, X_{n:n}$ are random variables, and so are the extremes $X_{1:n}$ and $X_{n:n}$, which can be more appropriately described in a probabilistic manner.

For example, consider the water readings on day 1 from Table 10.1a.

	Reading 1	**Reading 2**	**Reading 3**	**Reading 4**
Day 1	29.8	16.7	24.3	9.6

If these water level readings are considered random variables with sample size $n = 4$, then the order statistics can be represented by arranging the water levels in increasing order such that

	$X_{1:4}$	$X_{2:4}$	$X_{3:4}$	$X_{4:4}$
Day 1	9.6	16.7	24.3	29.8

If this process of arranging n samples in increasing order is done for each of the 30 days, then we will have Table 10.2.

10.5.1 Independent Single-Order Statistic

However, first suppose the case where the random variables $X_{1:n}, X_{2:n}, X_{3:n}, \ldots, X_{n:n}$ are independent, each with identical CDF, $F(x)$. Furthermore, denote the CDF of the rth-order statistic $X_{r:n}$ to be $F_{(r)}(x), r = 1, 2, \ldots, n$. Then the CDF of the largest-order statistic $X_{n:n}$ is given by

$$\begin{aligned} F_{(n)}(x) &= P(X_{m:n} \leq x) = P(all\ X_i \leq x) \\ &= F(x) \cdot F(x) \cdot \cdots \cdot F(x),\ n \text{ times} \\ F_{(n)}(x) &= F^n(x) \end{aligned} \tag{10.1}$$

TABLE 10.2

Order Statistics of Four Daily Samples of Water Levels at Kiln Bend of the James River

	$X_{1:4}$	$X_{2:4}$	$X_{3:4}$	$X_{4:4}$
Day 1	9.6	16.7	24.3	29.8
Day 2	3.7	6.9	10.7	25
Day 3	4.1	15.7	16.3	25
Day 4	10.1	11.9	14.6	24.1
Day 5	3.2	5.1	19.9	29.1
Day 6	6.8	10.7	12.4	14.1
Day 7	3	3.1	8.2	21.8
Day 8	10	10.3	13.7	20.1
Day 9	4.3	15.2	17.1	24.3
Day 10	7.4	7.7	26.8	29.2
Day 11	2.3	9	23.7	28.6
Day 12	13	20.1	28	29.9
Day 13	9.9	13.8	18.1	26.8
Day 14	4.8	17	26.9	29.3
Day 15	5.8	6.5	6.7	11.1
Day 16	9.8	13.9	14.1	21.6
Day 17	9.5	12.5	24.4	25.9
Day 18	2.5	4.7	22.2	29.1
Day 19	4.6	6.4	7.7	19.2
Day 20	6.5	10.6	15.1	23.3
Day 21	1.8	12.9	13.9	18
Day 22	3.9	8.1	18.1	21.6
Day 23	7.9	18.8	24.8	27.3
Day 24	13.7	23.1	24.6	26.2
Day 25	6.7	22.2	23.4	26
Day 26	2.4	4.2	10.2	20.9
Day 27	3.6	7	8.2	27.8
Day 28	7.7	8.3	14.1	19.8
Day 29	5.2	21.2	23.1	27.6
Day 30	2	4.1	4.5	20

This equation essentially provides the CDF of the largest-order statistic in terms of the common CDF of all the order statistics.

Similarly for the smallest-order statistic,

$$\begin{aligned} F_{(1)}(x) &= P(X_{1:n} \leq x) = 1 - P(X_{1:n} > x) \\ &= 1 - P(all\ X_i > x) = 1 - [1 - F(x)]^n \end{aligned} \tag{10.2}$$

PROBLEM 10.1

Consider the example of air quality in an urban area measured in terms of parts per million (ppm). The standard threshold for pollen particulate is 300 ppm, which is known to have CDF of $P(pollutants \leq 300\ ppm)$ to be 0.07. What would be the probability that the largest-order statistic of 5 samples will be less than or equal to this threshold?

Using Equation 10.1,

$$F_{(5)}(300) = F^5(300) = (0.07)^5 = 1.68 \times 10^{-6}$$

PROBLEM 10.2

Consider the water level measurements at Kiln Bend of the James River. Similar to the measurement process resulting in Table 10.1a and b, four samples are taken each day. Suppose that it is known that the water level measurement can be described by a Weibull distribution, such that

$$F(x;k,\lambda) = \begin{cases} 1 - e^{-\left(\frac{x}{\lambda}\right)^k} & \text{for } x \geq 0 \\ 0 & \text{for } x < 0 \end{cases}$$

where $\lambda = 21$ is the scale parameter, and $k = 2$ is the shape parameter.

The engineer is contemplating on building a 30-foot dam and would like to know the probability that the largest-order statistic is less than or equal to 30 feet. From Equation 10.1, this can be written as $P(X_{4:4} \leq 30) = F_{(n)}(x) = F^n(x)$.

Calculating the CDF of a 30-foot water level,

$$\begin{aligned} F(30;2,21) &= 1 - e^{-\left(\frac{30}{21}\right)^2} \\ &= 1 - e^{-2.04} \\ &= 0.87 \end{aligned}$$

Therefore,

$$\begin{aligned} F_{(4)}(30) &= F^4(30) \\ &= 0.87^4 \\ &= 0.573 \end{aligned}$$

The engineer would also like to provide boaters and fishermen guidance on how shallow that point of the river can be and would like to know the probability that the smallest-order statistic will be less than 5 feet.

Calculating CDF of a 5-foot water level,

$$F(5;2,21) = 1 - e^{-\left(\frac{5}{21}\right)^2}$$
$$= 1 - e^{-5.67\times10^{-2}}$$
$$= 5.51\times10^{-2}$$

Therefore, from Equation 10.2

$$F_{(1)}(5) = P(X_{1:n} \le 5) = 1 - P(X_{1:n} > 5)$$
$$= 1 - P(all\ X_i > 5) = 1 - [1 - 5.51\times10^{-2}]^n$$
$$= 0.203$$

Looking at the columns in Table 10.2, it is evident that these random variables are not independent contrary to the underlying assumption of Equations 10.1 and 10.2. The level of water at any time of the day is very much related to other times of that same day and possible with those of the other days as well. Consider a scenario of a strong rainfall at the start of day 9. One can conclude that the rainfall event may result in higher water levels for several succeeding days—hence measurement for those days will be dependent on each other. Nonetheless, assuming the random variables $X_{1:n}, X_{2:n}, X_{3:n}, \ldots, X_{n:n}$ are independent each with identical CDF of $F(x)$ greatly simplifies the analysis and finds extensive valid and acceptable applications in engineering, albeit an approximation. For the more advanced cases of dependent-order statistics, see Castillo et al. (2005).

10.6 Limit Distributions

The previous section has shown that the CDF of largest- and smallest-order statistics of a sample size n from a population with known CDF $F(x)$ is expressed as Equations 10.1 and (10.2), respectively, as

$$F_{(n)}(x) = F^n(x)$$

$$F_{(1)}(x) = 1 - [1 - F(x)]^n$$

Consider the case when the sample size n is so large that it tends to approach infinity. This case gives

$$\lim_{n\to\infty} F_{(n)}(x) = \lim_{n\to\infty} F_{(n)}(x) = \begin{cases} 1, & \text{if } F(x) = 1 \\ 0, & \text{if } F(x) < 1 \end{cases}$$

and

$$\lim_{n\to\infty} F_{(1)}(x) = \lim_{n\to\infty} 1-[1-F(x)]^n = \begin{cases} 0, & \text{if } F(x)=0 \\ 1, & \text{if } F(x)>0 \end{cases}$$

This shows that the limit distributions only take values of 0 and 1 and thus are degenerate and may not be useful for most applications. To avoid degeneracy, a linear transformation is established using constants a_n, b_n, and c_n that are all dependent on sample size n. Castillo et al. (2005) consequently defined the domain of attraction of a given distribution.

Definition 10.4: Domain of Attraction

A distribution is defined to belong to the domain of attraction of $F_{\max}(x)$ or $F_{\min}(x)$ if

For at least one pair of sequences $\{a_n\}$ and $\{b_n > 0\}$, the following is true:

$$\lim_{n\to\infty} F_{(n)}(a_n + b_n x) = \lim_{n\to\infty} F^n(a_n + b_n x) = F_{\max(x)} \forall x$$

and similarly,

$$\lim_{n\to\infty} F_{(1)}(c_n + d_n x) = \lim_{n\to\infty} 1-[1-F(c_n + d_n X)]^n = F_{\min(x)} \forall x$$

Furthermore, Castillo et al. (2005) established that only the family of distributions that are nondegenerate are called the generalized extreme value distributions (GEVD).

The CDF of GEVD for the largest-order statistic is

$$GEVD_{\max;K}(x;\lambda,\delta) = \begin{cases} \exp\left[-\left[1-K\left(\dfrac{x-\lambda}{\delta}\right)\right]^{1/K}\right] & \text{if } 1-K\left(\dfrac{x-\lambda}{\delta}\right) \geq 0, K \neq 0 \\ \exp\left[-\exp\left(\dfrac{\lambda-x}{\delta}\right)\right] & -\infty < x < \infty, K=0 \end{cases}$$

The CDF of GEVD for the smallest-order statistic is

$$GEVD_{\max;K}(x;\lambda,\delta) = \begin{cases} 1-\exp\left[-\left[1+K\left(\dfrac{x-\lambda}{\delta}\right)\right]^{1/K}\right] & \text{if } 1+K\left(\dfrac{x-\lambda}{\delta}\right) \geq 0, K \neq 0 \\ 1-\exp\left[-\exp\left(\dfrac{\lambda-x}{\delta}\right)\right] & -\infty < x < \infty, K=0 \end{cases}$$

This family of distributions includes the Weibull, Gumbel, and Frechet. Tables 10.3 and 10.4 provide the CDF, mean, median, mode, and variance for the maxima and minima of these three distributions, respectively (adapted from Castillo et al., 2005). Overall, there are actually six unique distributions, namely maximal Weibull, maximal Gumbel, maximal Frechet, and minimal Weibull, minimal Gumbel, and minimal Frechet.

The primary and practical significant difference among these three distributions with respect to extreme event analysis is the shape of their tails. Figure 10.4 shows the density curves of the three distributions and provides a comparison of the tails. Shown are the truncated tail for the Weibull, the exponentially decaying tail for the Gumbel, and the polynomial decaying tail for the Frechet.

10.7 Determining Domain of Attraction Using Inverse Function

The previous section described the possible domain of attraction of maxima and minima for samples from population with a known CDF. Nonetheless, it still remains to be explored exactly which of these domains of attraction can be used in analyzing extreme events.

Castillo et al. (2005) showed that if the CDF of the population is known to be of the form $F(x)$, then the maximal domain of attraction $F_{\max}(x)$ can be identified by obtaining the inverse of $F(x)$ and obtaining the limit based on

$$\lim_{n\to 0}\frac{F^{-1}(1-\varepsilon)-F^{-1}(1-2\varepsilon)}{F^{-1}(1-2\varepsilon)-F^{-1}(1-4\varepsilon)}=2^{-z} \tag{10.9}$$

where $F^{-1}(x)$ is the inverse of $F(x)$, and z is the shape parameter of the associated limit distribution such that

If $z>0, F(x)$ belongs to the maximal Weibull domain of attraction
If $z=0, F(x)$ belongs to the maximal Gumbel domain of attraction
If $z<0, F(x)$ belongs to the maximal Frechet domain of attraction

Similarly, for CDF of the population known to be of the form $F(x)$, the minimal domain of attraction $F_{\min}(x)$ can be obtained by obtaining the inverse of $F(x)$ and obtaining the limit based on

$$\lim_{n\to 0}\frac{F^{-1}(\varepsilon)-F^{-1}(2\varepsilon)}{F^{-1}(2\varepsilon)-F^{-1}(4\varepsilon)}=2^{-z} \tag{10.10}$$

where z is the shape parameter of the associated limit distribution such that

If $z>0, F(x)$ belongs to the minimal Weibull domain of attraction
If $z=0, F(x)$ belongs to the minimal Gumbel domain of attraction
If $z<0, F(x)$ belongs to the minimal Frechet domain of attraction

TABLE 10.3

CDF, Mean, Median, Mode, and Variance for the Maxima of the Weibull, Gumbel, and Frechet Distributions

		Maxima	
Weibull			
	CDF	$F_{\max(x)} = \begin{cases} \exp\left[-\left(\frac{\lambda - x}{\delta}\right)^{\beta}\right] & \text{if } x \le \lambda, 0 < \beta \\ 1, & \text{otherwise} \end{cases}$	(10.3)
	Mean	$\lambda - \delta\Gamma\left(1 + \frac{1}{\beta}\right)$	
	Median	$\lambda - (\delta \times 0.693)^{\frac{1}{\beta}}$	
	Mode	$\lambda - \delta\left(\frac{\beta - 1}{\beta}\right)^{\frac{1}{\beta}} \quad \beta > 1$ $\lambda \quad \beta \le 1$	
	Variance	$\delta^2\left(\Gamma\left[1 + \frac{2}{\beta}\right] - \Gamma^2\left[1 + \frac{1}{\beta}\right]\right)$	
Gumbel			
	CDF	$F_{\max(x)} = \exp\left[-\exp\left(\frac{\lambda - x}{\delta}\right)\right] \quad -\infty < x < \infty$	(10.4)
	Mean	$\lambda + 0.57772\delta$	
	Median	$\lambda + 0.3665\delta$	
	Mode	λ	
	Variance	$\frac{\pi^2\delta^2}{6}$	
Frechet			
	CDF	$F_{\max(x)} = \begin{cases} \exp\left[1 - \left(\frac{\delta}{x - \lambda}\right)^{\beta}\right] & \text{if } x \ge \lambda, 0 < \beta \\ 0, & \text{if } x < \lambda \end{cases}$	(10.5)
	Mean	$\lambda + \delta\Gamma\left(1 - \frac{1}{\beta}\right) \quad \beta > 1$	
	Median	$\lambda + \delta\, 0.693^{-\frac{1}{\beta}}$	
	Mode	$\left(1 + \frac{1}{\beta}\right)^{-\frac{1}{\beta}}\left(\lambda\left[1 + \frac{1}{\beta}\right]^{\frac{1}{\beta}} + \delta\right)$	
	Variance	$\delta^2\left(\Gamma\left(1 - \frac{2}{\beta}\right) - \Gamma^2\left(1 - \frac{1}{\beta}\right)\right) \quad \beta > 2$	

Source: Adapted from Castillo et al. (2005), p. 193–212.

TABLE 10.4

CDF, Mean, Median, Mode, and Variance for the Minima of the Weibull, Gumbel, and Frechet Distributions

		Minima	
Weibull	CDF	$F_{\min(x)} = \begin{cases} 0, & \text{if } x < \lambda \\ 1 - \exp\left[-\left(\frac{x-\lambda}{\delta}\right)^{\beta}\right] & \text{otherwise} \end{cases}$	(10.6)
	Mean	$\lambda + \delta\Gamma\left(1 + \frac{1}{\beta}\right)$	
	Median	$\lambda + (\delta \times 0.693)^{\frac{1}{\beta}}$	
	Mode	$\lambda + \delta\left(\frac{\beta-1}{\beta}\right)^{\frac{1}{\beta}} \quad \beta > 1$ $\lambda \quad \beta \leq 1$	
	Variance	$\delta^2\left(\Gamma\left[1 + \frac{2}{\beta}\right] - \Gamma^2\left[1 + \frac{1}{\beta}\right]\right)$	
Gumbel	CDF	$F_{\min(x)} = 1 - \exp\left[-\exp\left(\frac{x-\lambda}{\delta}\right)\right] \quad -\infty < x < \infty$	(10.7)
	Mean	$\lambda - 0.57772\delta$	
	Median	$\lambda - 0.3665\delta$	
	Mode	λ	
	Variance	$\frac{\pi^2\delta^2}{6}$	
Frechet	CDF	$F_{\min(x)} = \begin{cases} 1 - \exp\left[-\left(\frac{\delta}{\lambda - x}\right)^{\beta}\right] & \text{if } x \leq \lambda \\ 1, & \text{otherwise} \end{cases}$	(10.8)
	Mean	$\lambda - \delta\Gamma\left(1 - \frac{1}{\beta}\right) \quad \beta > 1$	
	Median	$\lambda - \delta 0.693^{\frac{1}{\beta}}$	
	Mode	$\left(1 + \frac{1}{\beta}\right)^{-\frac{1}{\beta}}\left(\lambda\left[1 + \frac{1}{\beta}\right]^{\frac{1}{\beta}} - \delta\right)$	
	Variance	$\delta^2\left(\Gamma\left(1 - \frac{2}{\beta}\right) - \Gamma^2\left(1 - \frac{1}{\beta}\right)\right) \quad \beta > 2$	

Source: Adapted from Castillo et al. (2005), p. 193–212.

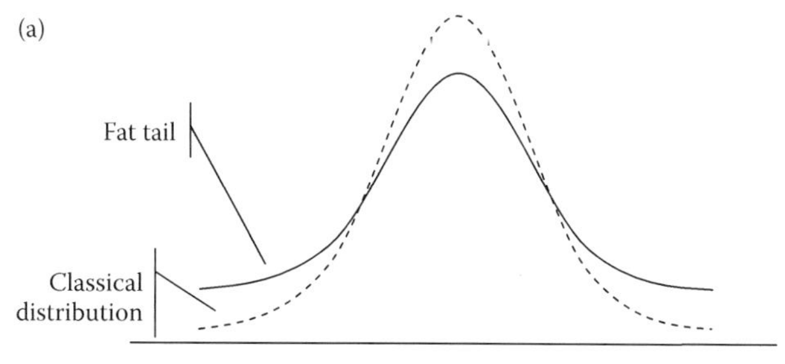

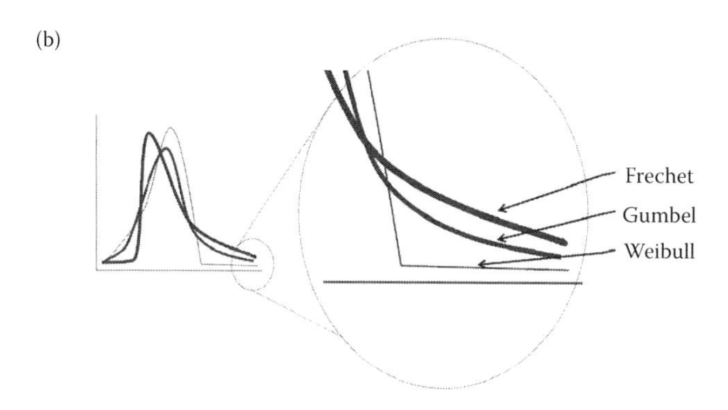

FIGURE 10.3
(a) Variation between classical parametric distributions and real-life empirical distributions. (b) Significant differences among the tails of the Weibull, Gumbel, and Frechet distributions.

PROBLEM 10.3

Consider the people lining up in front of a bank's automated teller machine (ATM). Suppose that this system is modeled as a M/M/1-FCFS queuing model with waiting times, w. If the CDF of waiting times is known to be

$$F(w) = P(W < w) = 1 - \rho e^{-\mu(1-\rho)w}$$

What is the form of the distribution of extreme waiting times w?

Solution

$$F(w) = 1 - \rho e^{-\mu(1-\rho)w}$$

$$F^{-1}(w) = \frac{\log(1 - w/\rho)}{\mu(-1+\rho)}$$

$$= \lim_{\varepsilon \to 0} \frac{\log(1-(1-\varepsilon)) - \log(1-(1-2\varepsilon))}{\log(1-(1-2\varepsilon)) - \log(1-(1-4\varepsilon))} = 1 = 2^0$$

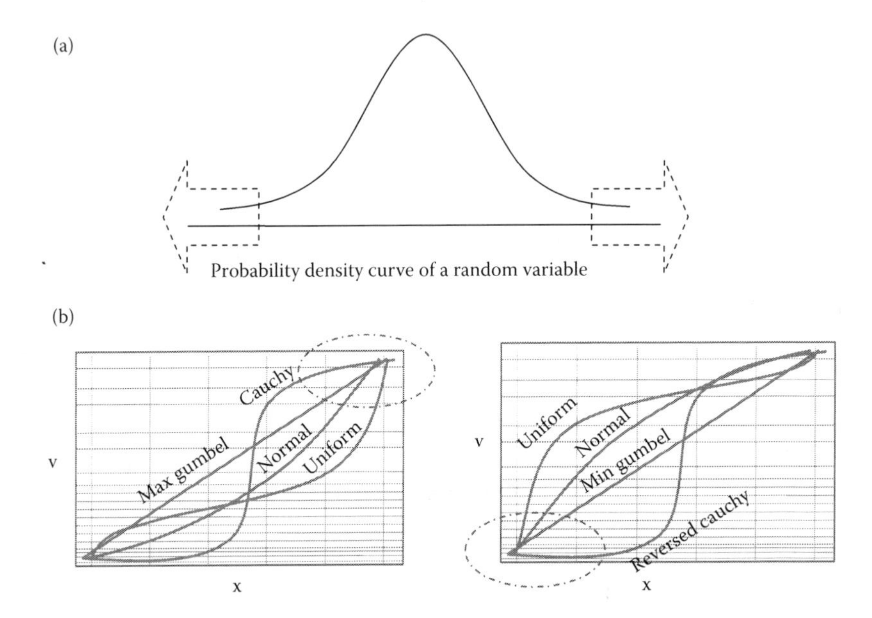

FIGURE 10.4
(a) Foci of EVT are the extreme observations. (b) Sample Gumbel plots for maxima (left) and minima (right) for some commonly known distributions.

TABLE 10.5a

Common Parametric Distributions and the Domain of Attraction of its Largest- and Smallest-Order Statistics

Common Parametric Distribution	Domain of Attraction of the Largest-(Smallest)-Order Statistics
Normal	Gumbel (Gumbel)
Exponential	Gumbel (Weibull)
Lognormal	Gumbel (Gumbel)
Gamma	Gumbel (Weibull)
Rayleigh	Gumbel (Weibull)
Uniform	Weibull (Weibull)
Cauchy	Frechet (Frechet)
Pareto	Frechet (Weibull)

Source: Adapted from Castillo et al. (2005), p. 207.

Since $z = 0, F(w)$ belongs to the domain of attraction of maximal Gumbel distribution.

Shown in Table 10.5a are some of the more commonly used parametric distributions in engineering and the domain of attraction of its largest- and smallest-order statistics.

10.8 Determining Domain of Attraction Using Graphical Method

The previous section provided a method for identifying the asymptotic distribution of maxima and minima using the inverse function of the underlying distribution. However, there may be cases where obtaining the inverse function is unwieldy or that the underlying distribution may be unknown or indeterminate. This may be particularly true for cases where the phenomenon being analyzed is not well understood or that there are not many data to analyze.

A less precise but nonetheless equally useful method for determining domain of attraction provided some data that is available is the graphical method. Similar to many graphical methods for analyzing data, this method relies very much on plotting data points and observing patterns or trend produced. The method described in this section combines the process of elimination and the technique of graphing using modified axes to identify a likely domain of attraction of the maxima or minima of the given set of empirical data points.

10.8.1 Steps in Visual Analysis of Empirical Data

1. *Eliminate:* As shown in Figure 10.3, the tails of the three asymptotic distributions may have two very discernable characteristics: either it is truncated or it is not. The analyst can use knowledge of phenomenon to conveniently eliminate possible domains of attractions; such may be the case for naturally truncated phenomenon like minimum rainfall where the empirical data will obviously be truncated at the zero rainfall level. This elimination process may conveniently limit the choices to possibly one or two: Weibull, if it is truncated, or between Gumbel and Frechet, if it is not truncated.
2. *Plot:* Plotting empirical data can be performed using a maximal or minimal Gumbel plot paper. Equivalently, the plot can be done using spreadsheets with modified plotting positions (skip this step if already using G-I plotting paper):
   a. If plotting maxima-order statistic, plot $X_{n:n}$ versus plotting position v, where $v = -\log[-\log(p_{i:m})]$ and i is the rank in $m\, X_{n:n}$, $p_{i:m} = i/(m+1)$.
   b. If plotting minima-order statistic, plot $X_{1:n}$ versus plotting position v, where $v = \log[-\log(1 - p_{i:m})]$.
3. *Diagonal/Linear?* If the tail of interest on the plot shows diagonal linear trend, then domain of attraction is Gumbel (G-I). Since the Gumbel plotting position was used, a straight plot essentially validates that the data do come from a Gumbel distribution.

4. *Vertical?* If the tail of interest shows a vertical asymptote, then the domain of attraction is Weibull (G-III).
5. *Horizontal?* If the tail of interest shows a horizontal asymptote, then the domain of attraction is Frechet (G-II).
6. *Be safe*: There may be cases where the plot does not clearly show a trend. This may be possible due to the actual data itself or the way the plot is visually presented, for example, disproportionately stretched or squeezed. If there is no obvious asymptote, use the distribution that will provide better margin of safety. Consider a case where further into the extreme is undesirable such as extreme rainfall. One may elect to choose as an asymptotic distribution such as the Frechet that will provide the heaviest tail among the three options. The expected effect of such a choice is that the risk due to the extreme rainfall will likely be overestimated rather than underestimated and thus reflect a "be safe" approach.

Recall that listed in Table 10.5a are some commonly known distributions and their asymptotic distributions for maxima and minima. If data sets known to be coming from these distributions are plotted in a Gumbel plot in the way described above, then their plots may look similar to those in Figure 10.4. Consider as an example the uniform distribution that by definition would be truncated on both tails of the distribution. As expected, data coming from the uniform distribution plotted in a Gumbel plot will show trends that are vertical, both for the maxima and the minima.

PROBLEM 10.4

Consider the order statistics of 4 daily samples of water levels at Kiln Bend of the James River in Table 10.2, which shows the maxima $X_{4:4}$ for all 30 days. Following each of the steps in determining domain of attraction provided some data, which is the graphical method.

Step 1: *Eliminate*. There seems to be no reason to conclude that extremely high water levels at the Kiln Bend of the James River is truncated. Therefore, analysis proceeds to the next step.

Step 2: *Plot*. To obtain the rank i of each in m readings, simply sort the maxima from Table 10.2 in an increasing order. This sorted maxima and their corresponding ranks are shown in the first two columns of Table 10.5b. Recognizing that there are distinct plotting positions for maxima and minima, choosing the appropriate one is important, and this is shown in the third and fourth columns of Table 10.5b.

TABLE 10.5b

Data for Plotting Maxima $X_{4:4}$ Using Gumbel Plotting Position v

Sorted Maxima	Rank i	$p_{i:m} = i/(m+1)$	$v = -\log[-\log(p_{i:m})]$
11.1	1	0.0323	–1.23372
14.1	2	0.0645	–1.00826
18.0	3	0.0968	–0.84817
19.2	4	0.1290	–0.71671
19.8	5	0.1613	–0.60133
20.0	6	0.1935	–0.49605
20.1	7	0.2258	–0.39748
20.9	8	0.2581	–0.30347
21.6	9	0.2903	–0.2125
21.6	10	0.3226	–0.12346
21.8	11	0.3548	–0.03546
23.3	12	0.3871	0.052262
24.1	13	0.4194	0.140369
24.3	14	0.4516	0.229501
25.0	15	0.4839	0.320292
25.0	16	0.5161	0.413399
25.9	17	0.5484	0.509537
26.0	18	0.5806	0.609513
26.2	19	0.6129	0.714272
26.8	20	0.6452	0.824955
27.3	21	0.6774	0.942982
27.6	22	0.7097	1.070186
27.8	23	0.7419	1.209009
28.6	24	0.7742	1.362838
29.1	25	0.8065	1.536599
29.1	26	0.8387	1.737893
29.2	27	0.8710	1.979413
29.3	28	0.9032	2.284915
29.8	29	0.9355	2.70768
29.9	30	0.9677	3.417637

Using the sorted $X_{4:4}$ values and plotting them with their corresponding plotting position v produces the plot shown in Figure 10.5. Keeping in mind that the focal part of the plot is the higher water level value as highlighted in Figure 10.5, there seems to be a vertical trend of the plot. This suggests that the asymptotic distribution is the maxima Weibull, for example, empirical data shows that the maxima water level has a truncated distribution.

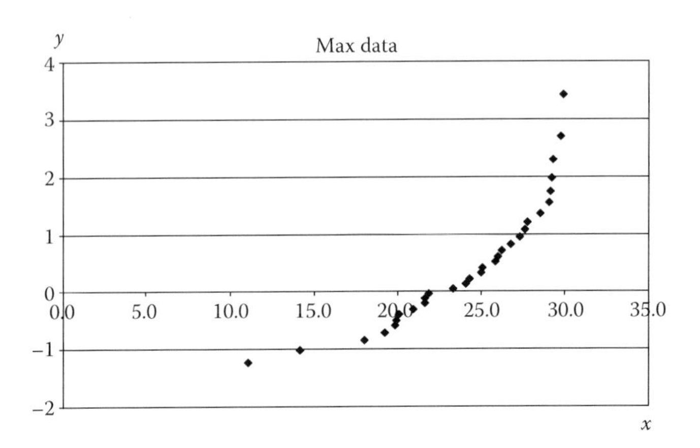

FIGURE 10.5
Gumbel plot of $X_{4:4}$ from Table 10.5b showing an asymptotic distribution of maximal Weibull.

PROBLEM 10.5

Consider the analysis where the interest may be the minima $X_{1:4}$ of the same data set described in Table 10.2. Application of the steps discussed above, particularly the *eliminate* step, may lead to conclusion that the asymptotic distribution is minima Weibull. That is, knowledge of the phenomenon of water level at a river will suggest that empirical data will show an asymptotic distribution of the minima to be truncated Weibull.

Nonetheless, this conclusion can be further reinforced by actually plotting the $X_{1:4}$ data using the plotting position v for minima. These data are shown in Table 10.6.

Similar to Problem 10.4, and using the sorted $X_{1:4}$ values and plotting them with their corresponding plotting position v produces the plot shown in Figure 10.6. Keeping in mind that the focal part of the plot is the lowest water level value as highlighted in Figure 10.6, there seems to be a vertical trend of the plot. This suggests that the asymptotic distribution is the minima Weibull, for example, empirical data shows that the minima water level has a truncated distribution.

Recall that in Table 10.5a are listed a number of fairly dissimilar known distributions having similar asymptotic distributions. This is for the reason why asymptotic distributions are mainly used for the tails of the distributions and not on the central regions. It is notable that there are many dissimilar distributions. Consider two distributions that may be very dissimilar in their central regions as shown in Figure 10.5. However, since the Gumbel plot is meant to highlight only the tails of the distributions, Figure 10.7 shows the same asymptotic distribution of the minima.

TABLE 10.6

Data for Plotting Minima $X_{1:4}$ Using Gumbel Plotting Position v

Sorted Minima	Rank i	$p_{i:m} = i/(m+1)$	v
1.8	1	0.0323	–3.418
2.0	2	0.0645	–2.708
2.3	3	0.0968	–2.285
2.4	4	0.1290	–1.979
2.5	5	0.1613	–1.738
3.0	6	0.1935	–1.537
3.2	7	0.2258	–1.363
3.6	8	0.2581	–1.209
3.7	9	0.2903	–1.070
3.9	10	0.3226	–0.943
4.1	11	0.3548	–0.825
4.3	12	0.3871	–0.714
4.6	13	0.4194	–0.610
4.8	14	0.4516	–0.510
5.2	15	0.4839	–0.413
5.8	16	0.5161	–0.320
6.5	17	0.5484	–0.230
6.7	18	0.5806	–0.140
6.8	19	0.6129	–0.052
7.4	20	0.6452	0.035
7.7	21	0.6774	0.123
7.9	22	0.7097	0.212
9.5	23	0.7419	0.303
9.6	24	0.7742	0.397
9.8	25	0.8065	0.496
9.9	26	0.8387	0.601
10.0	27	0.8710	0.717
10.1	28	0.9032	0.848
13.0	29	0.9355	1.008
13.7	30	0.9677	1.234

10.8.2 Estimating Parameters of GEVD

The previous section described that knowing the CDF of the underlying population, the domain of attraction of its largest- and smallest-order statistics of samples from that population can be identified. It was shown that even without knowledge of the underlying distribution, as long as there are empirical data, it is possible to identify the family of the asymptotic distribution of the

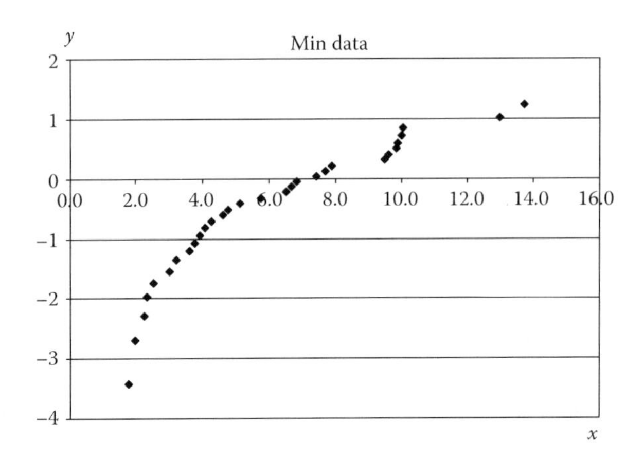

FIGURE 10.6
Gumbel plot of $X_{1:4}$ from Table 10.6 showing an asymptotic distribution of minimal Weibull.

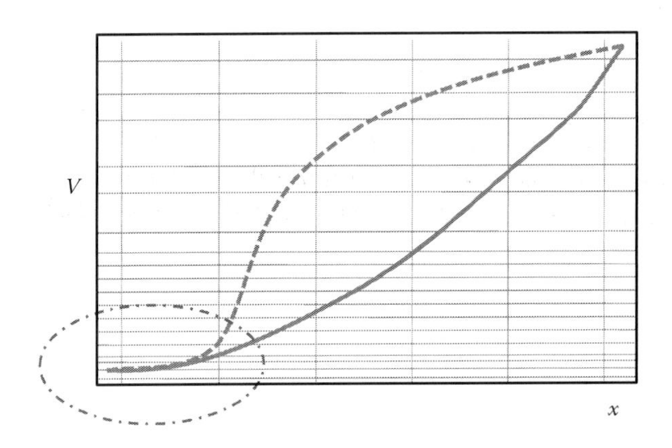

FIGURE 10.7
Two dissimilar distributions showing the same asymptotic distribution of the minima.

maxima and minima. The reasonable succeeding step is to estimate parameters of the domain of attraction distribution. Nonetheless, there are cases wherein the population is infinite or very large. Consider the water level represented by samples in Table 10.1a. There are many more water levels that are not measured and those that are on record will keep increasing in numbers as time goes on. For this particular case and other similar cases, the real population parameters are not observable, simply because analysis will always be based merely on samples taken from the entire population. Essentially, population parameters are approximated by using estimates based on samples rather than the entire population.

There are many statistical techniques or methods that can be employed in this estimation process. The more commonly used ones are the following:

- Least square method
- Method of moments
- Maximum likelihood method
- Kalman filtering technique

For the most part, the choice of which tool or technique to use may be determined by one or all of the following factors:

- Ease of use (including analytical complexity and computing need)
- Efficiency (i.e., lower variance)
- Consequences of getting the estimates wrong

The first two factors are common to many estimation processes in which techniques or methods may each present their own advantages and disadvantages over the other. The third factor is specific from the perspective of risk analysis where the parameters being estimated may be construed to have a direct effect on how risk may be later managed. This factor definitely relies heavily on the context of the phenomenon being analyzed. A context of estimating parameters to eventually predict flood levels in a highly developed geographic region will differ if it was in an inhabited and undeveloped wetland. This is simply because the consequence of getting the wrong estimates may be different for the two contexts.

10.9 Complex Systems and Extreme and Rare Events

A complex system is described by Keating et al. (2005, p. 1) as "a bounded set of richly interrelated elements for which the system performance emerges over time and through interaction between the elements and with the environment ... and the appearance of new and unforeseen system properties which cannot be known before the system operates, regardless of how thoroughly they are designed."

A complex system has the following discernable properties (adapted from Keating et al., 2005):

- Large number of richly interrelated elements
- Dynamically emerging behavior and structure that can dramatically change over time
- Uncertainty in outcomes

- Incomplete understanding of the system
- Multiple, and possibly divergent, system stakeholder perspectives
- Constrained resources and shifting requirements/expectations
- Urgency for immediate responses with dire, potentially catastrophic consequences for "getting it wrong"

The best analogy to describe risks in complex systems is the pebble-in-the-pond analogy, modified from Hofstetter et al. (2002).

At the most basic risk management scenario—even in the absence of any structured risk management process and notion of complexity—there will always exists a situation where a decision has to be made as a response to a perceived risk. This particular perceived risk can be more accurately described as original risk, compared to other related risks that may later arise or perceived. The response to this original risk can then be described as a pebble thrown into a pond that creates ripples in the water. These ripples can symbolize the various effects of the response to the original risk. These ripple effects can be described with respect to a particular objective as either favorable or unfavorable, the latter being synonymous to risk effects. In a complex system, made up of multiple, and possibly divergent, system stakeholder perspectives, these ripple effects may be favorable to some stakeholders and at the same time unfavorable to others.

There is also the temporal domain of the analogy—as time goes on, the ripples in the pond resulting from the response to the original risk spreads throughout the pond at varying magnitudes. In essence, certain characteristics of complex systems—such as dynamically emerging behavior and structure that can dramatically change over time, uncertainty in outcomes, incomplete understanding of the system—to name a few, result in new benefits and risks as well.

10.9.1 Extreme and Rare Events in a Complex System

Extreme and rare events have been defined earlier as those having relatively very low frequency of occurrence and at the same time have relatively extreme high or extreme low degree of magnitude. Through several examples, extreme and rare events have been exemplified in both natural and engineering systems. In data measurement, some extreme data may not be recorded for various reasons such as limitation of measuring equipment or too few samples to capture the truly rare events. All these result in what analysts may refer to as data censoring. For the few possible instances that rare events do get recorded, classical data analysis may be performed resulting in extreme and rare measurements being labeled as outliers. This is often the result of hasty data filtering and forcing the assumption of stationarity. What may not be very apparent is the relationship among these notions: extreme and rare events and the challenges in their analysis, system complexity, probabilistic causation (Chapter 9), and evidence-based analysis (i.e., Bayes' rule, Chapter 3). These notions can converge and pose themselves as two familiar problems of causalities and correlations.

10.9.2 Complexity and Causality

Establishing causalities among events occurring in the context of complex systems is never a trivial exercise. Recall from Chapter 9 that the notion of sufficient condition, for example, event A is said to be sufficient to cause event B if and only if

$$A \rightarrow B$$

Using the concept of conditional probability to express such causality,

$$P(B \mid A) = 1$$

However, inherent in complex systems is the incomplete understanding of the coupling and interactions among elements. Even more difficult would be the case when the event of interest, B, or the evidence A, or both A and B happen to be rare in occurrence. That is, the unconditional probability of B occurring is very low,

$$P(B) \cong 0$$

Since $P(B|A) = 1$, then it must be true that similar to B, the unconditional probability of A occurring is very low,

$$P(A) \leq P(B) \cong 0$$

This further shed light on the importance of not only the sufficient evidences but the necessary conditions as well in identifying risk scenarios.

10.9.3 Complexity and Correlation

Where there is causation, there is most likely correlation. Observations of the more-common events dominate the estimation process and since extreme observations consist of only a small part of the data, their contribution to the estimation is relatively smaller. Therefore in such an approach, the tail regions are not accurately estimated and causation is more difficult to establish, or may have low reliability, because of rarity of extreme data. Nonetheless, the analyst will necessarily have to recognize that evidences to support occurrence of a rare event of interest will show, at best, probabilistic causation. In essence, the observation of apparent correlation is merely being extended to imply causation. And therein lies a dilemma. Consider these statements.

Statement 1: A occurs in correlation with B.

Statement 2: Therefore, A causes B.

Statement 1 is essentially what is really being recorded when an analyst is observing a phenomenon in a complex system. However, this apparent correlation between A and B may be due to a number of reasons:

a. A indeed causes of B, in which case Statement 2 is true.
b. B may actually be the cause of A, in which case Statement 2 is false.
c. A causes B in the same way B causes A, without precluding a possible third cause, say C, in which case Statement 2 is true but incomplete.
d. Another event, say C, is actually causing both A and B, in which case Statement 2 is false.
e. The occurrence of A and B is a mere coincidence, in which case Statement 2 is false.

Reasons c and d introduce the notion of a possible common-cause variable C among the preestablished events A and B. Consider the example of correlation established between the use of a certain brand Y of automotive engine oil and the rapid accumulation of sludge inside the engine. After an extensive study, this apparent correlation may become so convincing that it is extended to be taken as causality, that is,

Brand Y automotive engine oil causes rapid accumulation of sludge.

Nonetheless, a common-cause variable C may actually be later discovered. As an example, a closer look at the geographic locations of the automobiles sampled in the study showed that brand Y motor oil was actually given out free of charge to those who bought air filter of the same brand. It is known that inefficient air filter contributes to buildup of engine sludge. As such, what may have been construed to be a clear causation between two events may actually be less clear after discovery of new information.

10.9.4 Final Words on Causation

One of the eternal challenges for systems engineers and risk analysts is on establishing causalities. The importance of establishing causalities in effectively managing risks was discussed in Chapter 9. Nonetheless, it was also shown and exemplified by the notion of probabilistic causation that there exists a causal relationship in complex systems that may never be ultimately established.

Here are several criteria that can help in establishing causation (adapted from Hill's Criteria of Causation, see Goodman and Phillips, 2005, for more details).

1. Precedence relationship: Cause always precede the effect
2. Strength: Association as measured by appropriate statistical tests
3. Robustness: Same relationship exists in various operating environment
4. Plausibility: In agreement with current understanding of underlying processes.

Nonetheless, some have argued that since proving causation is based on experience, and experience is only valid if the causation between past and the future experiences is assumed—that proving causation is a circular argument (and hence invalid).

10.10 Summary

This chapter presented the notion of extreme and rare events and the difficulties in analyzing such events for the purpose of risk analysis. This difficulty is based on two reasons. The first difficulty is on the inappropriateness of traditional data analysis, which highlights the more common events and masks insights into the rare and extreme data. Nonetheless, the use of extreme value analysis is well established, but not commonly known extreme value distributions can overcome such difficulty. The second difficulty is inherent in the rarity of some events coupled with the dilemma of needing to establish causalities using the latest and most convincing information at hand. Complex systems, being characterized by emergence and incomplete knowledge of the actual interaction among elements, are always subject to the discovery of new information that can either reinforce or weaken causalities and correlations currently held as true.

Questions and Exercises

1. When are EVT tools applied? When is data analysis based on central tendency applied?
2. An engineer is trying to determine the suitability of two metal alloys, alloy A and alloy B, in fabricating electrical fuse elements,

TABLE 10.7a

Melting Temperatures of Alloy A

Batch	Sample 1	Sample 2	Sample 3	Sample 4	Sample 5
1	258.8	144.6	264.2	151.7	164.3
2	158.6	168.6	280.0	154.7	169.3
3	217.9	216.8	203.6	163.1	177.8
4	241.0	227.3	249.6	246.8	144.4
5	167.3	243.7	141.6	108.3	114.6
6	215.9	168.6	277.4	269.5	138.4
7	189.3	195.0	179.5	254.4	202.5
8	205.4	276.6	167.4	169.6	199.9
9	191.0	233.4	224.5	127.0	190.0
10	172.5	140.9	177.1	184.0	285.2
11	262.0	206.5	215.9	176.9	166.1
12	294.6	131.8	230.6	250.1	226.3
13	154.9	229.7	301.9	156.5	165.2
14	191.8	200.8	202.7	199.2	166.6
15	154.9	155.4	214.3	191.7	169.6
16	177.7	172.7	182.5	185.4	126.8
17	232.6	180.5	198.0	136.7	170.7
18	183.3	166.0	257.9	110.3	226.0
19	161.7	195.3	260.3	195.5	235.6
20	175.2	301.2	205.6	201.0	144.7
21	188.8	150.5	158.5	143.3	232.3
22	181.5	214.0	268.4	185.4	265.3
23	181.6	131.8	144.5	220.1	245.2
24	198.0	238.2	187.5	133.6	373.1
25	267.7	155.3	258.3	105.2	150.6
26	222.9	225.4	179.4	168.2	251.7
27	166.9	183.2	186.3	193.1	233.8
28	143.0	187.0	153.5	156.3	191.9
29	225.2	309.9	188.5	198.8	131.9
30	179.1	186.0	153.8	125.0	244.7
31	286.9	270.7	181.0	153.3	158.5
32	153.4	134.5	130.1	235.4	183.7
33	175.3	194.8	208.8	294.9	248.9
34	269.4	199.7	253.0	223.4	224.5
35	209.3	209.0	175.8	182.5	229.5
36	250.6	164.2	205.1	172.2	220.1
37	164.3	258.2	212.2	135.2	145.6
38	205.5	180.0	185.2	246.5	194.6
39	168.0	153.2	156.5	181.7	218.5

TABLE 10.7a (Continued)

Melting Temperatures of Alloy A

Batch	Sample 1	Sample 2	Sample 3	Sample 4	Sample 5
40	232.6	226.3	226.3	169.8	154.4
41	155.8	200.0	209.6	214.3	135.2
42	168.1	271.7	154.8	149.5	224.6
43	317.9	140.7	181.3	224.1	158.1
44	258.0	216.0	179.0	100.6	156.1
45	184.8	164.1	294.9	141.0	225.8
46	151.9	156.0	248.6	220.1	220.8
47	156.7	312.2	193.3	204.2	192.0
48	264.9	169.9	248.9	275.9	150.9
49	258.4	174.9	225.0	194.1	203.6
50	173.1	171.6	237.9	223.2	227.9
51	216.0	180.1	214.6	209.9	163.4
52	163.7	194.9	253.5	163.8	187.6
53	170.7	250.1	192.7	161.9	170.4
54	217.3	196.2	128.9	194.7	171.0
55	203.3	249.1	262.1	240.0	202.1
56	218.7	168.7	151.3	130.5	181.1
57	135.3	149.5	234.2	115.7	173.1
58	157.7	188.1	229.7	104.2	253.1
59	214.7	204.4	176.3	193.2	217.9
60	247.1	186.7	273.2	248.2	160.9

the part of the electrical fuse that melts after reaching a certain temperature or current. Samples are taken from batches of alloys and are subjected to laboratory test where the melting temperatures are recorded. The melting temperatures for both alloys are provided in Table 10.7a and b. The engineer had asked your help in describing the extremely high and extremely low melting temperatures of the alloys by using G-I plot for minima and maxima.

(A) For alloy A, what is the plotting position v for the largest of the minimal values?

(B) What is the plotting position v for the largest of the maximal values?

(C) What is the asymptotic distribution of the minimal melting temperature of alloy A?

(D) What is the asymptotic distribution of the maximal melting temperature of alloy A?

(E) For alloy B, what is the plotting position v for the largest of the minimal values?

TABLE 10.7b

Melting Temperatures of Alloy B

Batch	Sample 1	Sample 2	Sample 3	Sample 4	Sample 5
1	252.3	243.9	174.3	210.9	192.4
2	266.7	196.0	56.0	187.8	236.7
3	278.9	59.3	219.5	187.0	198.5
4	206.9	196.6	97.5	129.4	181.0
5	160.8	212.2	287.5	125.9	308.7
6	196.0	201.1	198.3	189.4	204.1
7	154.3	341.8	115.8	196.5	159.8
8	153.3	215.0	153.1	290.6	168.9
9	132.1	253.4	252.9	247.6	112.8
10	203.1	286.5	123.1	223.2	284.6
11	113.3	241.6	230.0	184.8	235.7
12	213.9	143.5	240.7	133.9	275.2
13	152.8	250.4	244.5	179.7	268.4
14	221.8	167.7	199.6	246.5	243.3
15	172.3	168.0	302.1	154.4	302.3
16	366.3	209.2	208.2	235.1	210.2
17	246.8	242.9	233.3	273.3	207.9
18	112.0	251.7	173.4	260.1	207.7
19	274.8	256.5	159.2	174.9	280.2
20	202.7	275.1	115.1	233.7	150.6
21	177.3	203.3	288.8	256.6	176.5
22	223.1	157.0	146.1	182.3	266.7
23	206.2	211.9	198.9	239.0	184.9
24	197.2	203.9	230.4	250.1	234.0
25	160.4	185.0	235.0	203.6	172.1
26	242.1	185.8	218.6	262.9	181.0
27	149.0	201.3	244.4	135.9	196.9
28	208.1	191.3	208.2	260.4	162.4
29	122.1	184.7	222.2	166.9	259.9
30	159.4	138.0	199.5	134.1	152.8
31	252.8	207.1	222.2	165.5	191.3
32	193.7	315.1	133.0	276.0	187.3
33	182.3	224.3	136.2	202.0	233.0
34	313.9	264.1	310.7	177.5	219.7
35	182.3	290.4	127.7	213.3	175.9
36	179.7	216.3	159.5	207.9	154.7
37	192.7	174.8	210.9	169.6	201.9
38	171.2	311.0	205.7	263.3	251.5
39	127.8	185.4	148.0	239.0	214.9
40	227.8	195.4	214.6	202.0	222.8

TABLE 10.7b (Continued)

Melting Temperatures of Alloy B

Batch	Sample 1	Sample 2	Sample 3	Sample 4	Sample 5
41	248.9	122.1	189.2	260.4	228.2
42	240.6	224.4	157.7	149.0	206.3
43	244.0	256.8	180.7	256.3	171.2
44	251.2	276.6	299.5	285.5	147.2
45	217.1	207.1	278.1	248.8	246.3
46	170.0	194.6	231.8	243.4	160.3
47	292.0	185.7	237.4	148.0	272.2
48	238.5	245.9	199.3	129.9	271.0
49	175.8	192.5	224.0	208.1	176.0
50	259.6	226.3	228.7	228.4	151.5
51	223.9	240.8	273.0	191.2	226.8
52	215.5	170.6	276.9	175.9	213.6
53	181.6	275.4	231.3	134.9	324.9
54	161.9	224.5	251.3	162.1	174.0
55	232.0	197.5	209.5	286.9	237.3
56	236.3	273.7	247.4	198.5	171.1
57	159.9	211.7	298.2	258.3	130.3
58	279.7	137.8	191.8	250.2	189.2
59	225.2	92.2	158.2	242.3	211.2
60	119.2	111.9	246.8	221.2	290.0
61	220.2	150.1	228.6	194.2	235.8
62	173.5	263.1	229.1	158.1	219.2
63	196.8	120.0	219.9	220.8	177.3
64	277.8	248.2	143.6	235.4	175.0
65	198.3	262.3	211.6	238.0	125.9
66	162.5	230.0	75.7	127.3	277.2
67	201.8	194.7	270.2	155.3	208.7
68	186.5	187.4	188.4	256.3	199.3
69	230.3	120.7	204.0	166.7	131.5
70	197.7	196.2	164.2	172.9	231.5
71	106.2	251.8	210.6	293.8	289.6
72	277.4	250.3	147.7	199.1	277.3
73	203.6	268.1	238.0	281.7	247.3
74	146.0	154.1	157.6	229.8	263.2
75	144.5	222.9	188.1	236.4	250.3
76	201.7	206.1	155.9	269.6	190.3
77	171.4	168.9	97.6	228.0	85.3
78	194.1	308.6	206.5	233.2	176.2
79	201.7	207.5	69.3	234.3	261.8
80	222.7	196.5	259.6	192.5	170.9

(F) What is the plotting position v for the largest of the maximal values?

(G) What is the asymptotic distribution of the minimal melting temperature of alloy B?

(H) What is the asymptotic distribution of the maximal melting temperature of alloy B?

3. Describe the importance of causality and correlation in complex systems.
4. Provide real cases where apparent correlation between events is being extended to imply causality.
5. Describe the significance of probabilistic causation and emergence in complex systems.
6. How can the four criteria for establishing causation complement the analysis of extreme and rare events in complex systems?

11

Prioritization Systems in Highly Networked Environments

11.1 Introduction

In Chapter 9, the popularity of the queuing system to model many engineering systems was discussed. In particular, it was emphasized how real entities such as production machines, human operators, computer processors, and other resources can be modeled as a server and the entities such as raw materials, operator tasks, and computer tasks can be modeled as entities that need to be served. It was also emphasized how fairly simple queuing systems can be the building blocks to model complex systems made up of combinations of servers and entities termed as network of queues.

This chapter describes in more detail how real systems can be modeled as network of queuing systems and how such model can be used to manage risks in terms of entities waiting too long to be served.

11.2 Priority Systems

Priority systems (PS), also called priority disciplines in the field of queuing, are composed of entities and one or more servers. As entities arrive into the system, they are labeled according to some attribute or system state at the time of entry into the system. The labels are then used to distinguish the entities into classes. Each of the classes is assigned a priority level that serves as the basis for choosing from which class the next entity to be served will be coming from (Jaiswal, 1968). Within a class, the typical entities are chosen on a first come first serve (FCFS) basis. The grouping can be based on some measure related to (Kleinrock, 1976):

- Arrival times of entities
- Service time required or already received by the entities
- Functions of group membership of entities

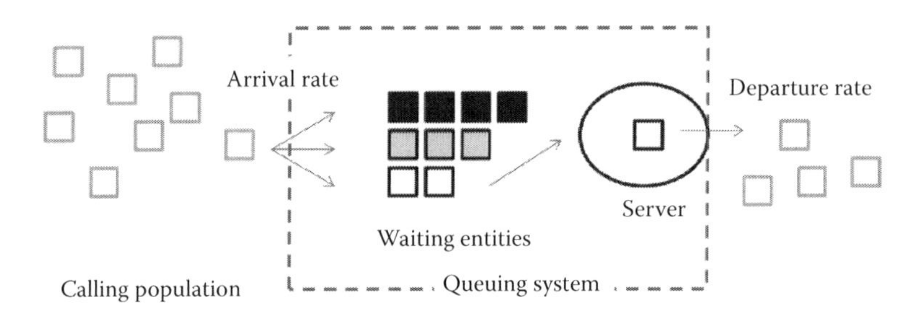

FIGURE 11.1
Schematic diagram of priority system showing arriving entities grouped into three classes with a single server processing an entity from one of the classes.

Examples of decisions based on arrival times are FCFS and last come first serve (LCFS). Examples of preferences based on service times only are shortest job first (SJF) and longest job first (LJF). The decision can also be based on a mixture of these measures, as will be discussed in the following sections.

Figure 11.1 is a schematic diagram of a PS showing (from left to right) the stream of arriving entities from the calling population, the grouping of entities into classes, and a server with an entity currently being served. Comparing Figure 11.1 with Figure 9.2 shows that PS is essentially built from the simple queuing model as the foundation.

11.2.1 PS Notation

This section introduces a notation for describing variations of PS that enables the use of both analytical and simulation models. To facilitate managing of latency times through configuration, the notation is based on configuration parameters. The parameters include the following:

N: the number of classes of entities
I_0: the set of initial priorities
$I_f(*)$: the priority increment/decrement function

The notation is flexible enough to enable the modeling of PS of increasing complexity. In addition, the notation facilitates a discrete event simulation model of a given PS. The following sections define the parameters of a PS and are exemplified in a FCFS system. It can be argued that a FCFS system is a PS and can be modeled with infinitely many classes of entities, with a set of initial priorities and a static increment function.

Number of Classes

Arriving entities are classified into one of several classes as shown in Figure 11.1. The number of classes is dependent on the application of the system. For relatively simple applications, a two-class system may be possible, whereas for complex applications, the number of classes may be more than 100. There can also be as many classes as there are entities in a PS. However, the common practice is the classification of entities into numbers of classes convenient for analysis.

As an example in an e-business application, the number of classes may be dependent on which part of the e-business organization the PS is being implemented. On a Web server, a meaningful classification can be according to web server requests (*Enter, Browse, Search,* etc.). Another example is on a database server where the classification can be made based on the type of information or part of the database being accessed (*customer profile, product information, account information,* etc.). Classifications are made such that all entities belonging to a class are homogeneous with respect to some attributes. It is assumed that a class is fixed for an entity. The grouping of entities into classes affords differentiated service among the classes, the first class having the priority over the lower classes. Such grouping of entities into N number of classes is a configuration parameter of a PS defined as follows:

Definition 11.1: Number of classes

A PS parameter N is the number of classes of entity requiring service.

It is evident from the definition that $N \in I^{+}$, where I^{+} is the set of positive integer numbers. The definition of N does not require that there are actually entities belonging to a class for a class to be defined, and thus an empty class is a possibility. Such is the case for types of entity that seldom occurs in a system. If such type of entity is the only one that can belong to a certain class, then the class may be empty for most of the time.

Initial Priority Levels

Each of the classes has a priority level, which is the basis for the decision from which class the next entity will be taken. The range of possible values of the priority level is the field of real numbers. The priority level of a class can be defined as follows:

Definition 11.2: Priority Level of a Class

A priority level $p_{i,t}$ is the priority level of class i indexed on t, such that $p_{i,t} \in \Re$, $0 < i \leq N, t \geq 0$ where the index t is time.

We now introduce the set based on this definition that describes the priority levels of the classes at any time t.

Definition 11.3: Set of Priority Levels

The set I_t is the set of priority levels for all the classes in a PS at time t, that is,

$$I_t = \{p_{1,t}, p_{2,t}, p_{3,t}, \ldots, p_{N,t}\}$$

An important set of priority levels is the *initial priority levels* at $t = 0, I_0$. Succeeding sections will focus on the behavior of the PS if I_0 plays a role in the choice of which class is served next.

Priority Increment Function

The rules or logic for implementing the decision on which class to serve next can be summarized into a parameter of the PS, termed as the increment function. If there is more than one entity of the same class waiting for service, then the discipline within the class is FCFS. The function describes how the priority levels of the classes will change and is indexed with t. The parameter can be defined as follows:

Definition 11.4: Increment Function

The increment function $I_f(*)$ is the set of rules that describes the change in the priority levels such that $I_{t+\Delta t} = I_f(I_t; w, x, ..., z)$, where Δt is some increment of index t, and $w, x, ..., z$ are system states, entity attributes, or some other derived parameters.

The increment function will also be expressed as $I_f(I_t)$ in cases when the other arguments (e.g., system states, entity attributes) are not significant. The increment function can range in complexity from the simple static case to more complex rules, as will be discussed in the succeeding sections. The form of the increment function can be linear or nonlinear. $I_f(x)$ is a linear function of x if it has a form $I_f(x) = ax + b$ where a and b are constants. When the increment function cannot be written in this form, then it is a nonlinear function with respect to the argument.

Related to the increment function is the operator used to choose which among the priority levels are considered high and low priority. A typical rule is to choose the class with the highest priority level value:

$$\text{Class to be served at time } t + 1 = \text{idx max/min}\{I_t\} \qquad (11.1)$$

where the operator idx max/min{} is the search for the index of the maximum or the minimum elements(s) in the set in { }, that is, $z = \text{idx max/min}\{x_1, x_2, ...\}$ if $x_z = \text{max/min}\{x_1, x_2, ...\}$. Since $\max\{I_t\}$ is the same as $\min\{-I_t\}$, then any maximization rule can be translated into an equal minimization rule. For the remainder of the book, the rule for choosing among classes is of the maximization form.

Length of Waiting Line

The lengths of waiting lines in a PS are the number of entities waiting for service for each class. Each class can have a unique waiting line and can be defined for each instance of the index as follows:

Definition 11.5: Lengths of Waiting Lines

The length of waiting line $Q_n(t)$ is the number of entities of Class n waiting for service at time t.

As introduced in Chapter 9, there are two basic types of PS with reference to the lengths of waiting lines: infinite and finite.

$Q_n(t)$ can be allowed to increase without bound or assumed not to reach a bound even if such bound exists. Analytical models such as the birth-and-death process in queuing models generally have assumed infinite waiting line and can be defined as follows:

Definition 11.6: Infinite Waiting Line

The number of entities of Class n waiting for service at time t is infinite if $0 \leq Q_n(t) \leq \infty$ for all t.

Some practical reasons for assuming infinite waiting line are large memory buffer and high-speed servers common in high-efficiency communication systems. The assumption of infinite waiting line is also typical of analysis of systems with low server utilization wherein actual limits on the length of waiting lines are never reached.

Nonetheless, there can be conditions that may warrant assuming a finite waiting line. The physical constraints in space, time, and other resources may put an upper limit on the possible or allowable number of waiting entities. A combination of factors, some of which are the frequency of arrival of entities, the length of service, and the frequency the server is able to perform service, can contribute to reaching the maximum allowable length of waiting lines. Such an upper limit is defined as follows:

Definition 11.7: Finite Waiting Line

The maximum allowable number of entities of Class n waiting for service is $Q\max_n$, such that $0 \leq Q_n(t) \leq Q\max_n$ for all t.

All arriving entities of Class n that find there is already $Q\max_n$ in the system will be turned away. Such entities can be dropped and forever lost or can be processed in a variety of ways. Entities turned away can be redirected to other servers, consigned to wait in another system used for storage or rerouted to another class of entities whose limit has not yet been reached.

Since there can be limited or unlimited waiting lines for each class $n \in N$ of a PS, the possible number of waiting entities from all classes in the entire PS is the sum of the $Q\max_n$ for all classes and can be defined as follows:

Definition 11.8: Waiting Entities in a PS

The maximum allowable number of entities in a PS is $Q\max = \sum_{n=1}^{N} Q\max_n$ if Qmax$_n$ exists. It follows that Qmax is finite only if $Q\max_n < \infty$ for classes $n = 1, 2, \ldots, N$, and $N < \infty$.

PROBLEM 11.1

Consider a FCFS system. Suppose that time is measured in discrete time length Δt. Furthermore, suppose that every entity that arrived within a particular time within the period $(t, t+\Delta t)$ belongs to the same class. Since time can be measured infinitely far in the past, then there can be infinitely many Δt, and equivalently infinite number of classes of entities. That is,

$$\text{for FCFS, } \dot{N} \to \infty$$

Often times in the analysis of FCFS system, only one entity is allowed to arrive within a time length Δt, either by assumption or by making Δt infinitesimally small. The result is that there can only be at most one entity in each class.

The FCFS discipline is equivalent to having the time of arrival as a measure of priority of entities; the earlier the arrival, the higher the priority. The set of priority levels can then be modeled as follows:

$$I_t = \{p_{1,t}, p_{2,t}, p_{3,t}, \ldots, p_{n,t} \ldots\} \quad \text{where } p_{n,t} = n \times \Delta t, \quad \text{for all } n \times \Delta t \leq t$$

This is true assuming that time is measured towards the direction of positive infinity. Note that the number of classes increases as the time of observation lengthens. The increment function can now be formulated as follows

$$I_{t+\Delta t} = I_f(I_t) = I_t$$

This is a static priority system since there are no evident changes on the relative priorities of the classes. The rule for choosing the next class to be served is formulated based on Equation 11.1 and is used to select the entities that arrived earliest among those in the system:

$$\text{Class to be served at time } t+1 = \text{idx} \max\{-I_t\}$$

Since there can only be one entity for each class, then the waiting line for each class is finite and have a maximum of one, that is,

$$Q\max_n = 1 \quad \text{for } n = 1, 2, \ldots, N$$

Thus, a FCFS system can be modeled as a PS with parameters N, I_t, and I_f.

Entity Arrival Process and Service Process

In this section, characteristics of entity arrivals and departures are defined and described. The utilization of a PS server is also defined in terms of the arrival and departure processes. Consider a PS with N classes and define the arrival of entities of class $n, n \in N$ during the time interval $[0,t)$ to be $A_n(t)$ and can be described by a probabilistic function (modified from Liebeherr et al. 2001).

A natural extension is the departure function of class n, and the number of entities waiting for service assumes that each class has its own waiting line. The departure of entities of class $n, n \in N$ during the time interval $[0,t)$ is described by the function $D_n(t)$ (modified from Liebeherr et al. 2001). A related and often used concept in queuing modeling is the rate at which a resource can serve entities or mean service rate μ. The service rate describes the rate at which a resource can perform the service required by the entities. The unit is in entities per unit time.

Definition 11.9: Saturated PS

The PS is saturated if there is always at least one entity in each class waiting for service, that is, $Q_n(t) \geq 1$ for all n and t.

Definition 11.10: Supersaturated PS

The PS is supersaturated if the lengths of waiting lines for all classes are increasing without bounds, that is,

$$\underset{t \to \infty}{Q_n(t) \to \infty} \quad \text{for all } n \text{ and } t.$$

PS described in Definitions 11.9 and 11.10 can be clearly compared and contrasted to the usual assumption of a stable queue. A stable queue assumes that the queue length is always finite.

In the area of product-form queuing models, a common assumption on the arrival and service process is that both possess the Markovian property. However, cases can be such that the distributions of arrivals are bounded, that is, Weibull distributed arrival. An example of such cases is when upstream server in a network has early failure.

11.3 Types of Priority Systems

The remainder of this chapter discusses the various types of priority systems using the succeeding definitions as the foundation. Figure 11.2 shows the various types of priority systems and how each are related.

11.3.1 Static Priority Systems

Static PS (SPS), also known as *strict priority* or *head-of-line* discipline, is the simplest type of PS, where class priorities do not change. In such cases, the

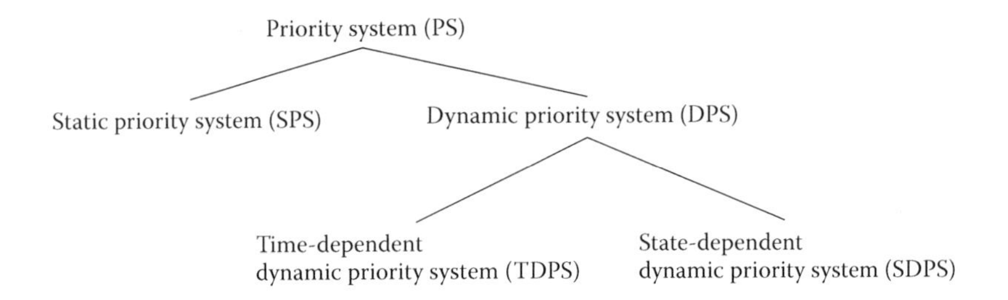

FIGURE 11.2
Types of priority systems.

lower class entities will be accommodated only after all the higher class entities have been exhausted. The result is the short average latency for the higher class entities in comparison to the lower class entities. There are special cases where the arrivals of entities and the service provided by the server have Markovian property*: Poisson arrival rate for each class λ_i, $i = 1, 2, ..., N$, and one exponential service rate for all classes.

Chapter 9 shows that the distribution of the latency times of entities in an SPS may require the Laplace transforms of the probability functions of the service times of the entity. However, the transforms of such function may not always be readily obtainable (Kleinrock, 1976).

Definition 11.11: Static Priority System (SPS)

An SPS is a type of PS wherein one class of entity is always preferred over another. That is, for each pair of classes i and $j, i \neq j \in N$, one and only one of the following conditions is true:

(i) $(p_{i,t} - p_{j,t}) > 0$ for all $t > 0$,

(ii) $(p_{i,t} - p_{j,t}) < 0$ for all $t > 0$, or

(iii) $(p_{i,t} - p_{j,t}) = 0$ for all $t > 0$.

Consider a telephone switch, an equipment used to route phone calls in systems where there can be a large number of phone users such as hotels, hospitals, and office buildings. It is recognized that not all phone calls may be of the same importance. At a common office building, most phone calls may be treated the same, except for incoming calls into the sales or customer service departments that are deemed business critical and always have higher priority than calls coming into other departments.

Another example is airline passengers lining up at the airport gate on their way to board an airplane. First to board are first-class passengers, those with special needs, those with infants and children, and others with preferred services. After all preferred passengers have boarded, the general classes of

* Series of events have a Markov property if the past states can be completely summarized by the current state.

passengers start boarding. However, when a preferred passenger approaches the gate, this passenger will quickly move into the head of the line.

11.3.2 Dynamic Priority Systems

Dynamic PS (DPS) are such that the priority of a class of entities changes. In effect, a class can have the priority over the other classes only for some time, after which another class will have the priority. Hahn and Shapiro (1967) discussed the flexibility offered by DPS in cases when there is a need for system administrators to have more control over system performance. Effectively, the lower class entities do not have to wait for the higher class entities to be exhausted. Other studies that dealt with DPS in the field of communications are by Vazquez-Abad and Jacobson (1994), Greiner et al. (1998), Berger and Whitt (2000), Ho and Sharma (2000), and Bolch et al. (1998). A common observation among these studies is the difficulty of determining the distribution of the performances of DPS. The difficulty is further complicated by the numerous variants of such systems, as can be observed from the succeeding two sections. DPS can be further classified according to the basis of changes in the priorities of the classes. In modeling of queues, the typical classification is time-dependent and state-dependent DPS. The classification came about due to the prevalence of analysis of time-dependent queues in the early development of queue analysis. State-dependent DPS came about later in modeling of more complicated queuing disciplines.

Definition 11.12: Dynamic Priority System

A DPS is a type of PS wherein no one class of entity is always preferred over another. That is, for each pair of classes i and j, $i \neq j \in N$, none of the following conditions are true:

(i) $(p_{i,t} - p_{j,t}) > 0$ for all $t > 0$,

(ii) $(p_{i,t} - p_{j,t}) < 0$ for all $t > 0$, or

(iii) $(p_{i,t} - p_{j,t}) = 0$ for all $t > 0$.

11.3.3 State-Dependent DPS

This section introduces a state-dependent DPS (SDPS), which is modeled according to the notation presented in Section 11.2. The first section describes in detail the dynamic nature of the priority levels and how they translate into the increment function notation. The second section presents propositions on the class latencies of a two-class SDPS. The propositions provide knowledge on the distribution of the class waiting time that can be used to describe the latencies of the entities.

SDPS are such that the priorities of the classes change based on some combinations of system states and entity attributes *and not* directly based on the arrival time of the entities. Jaiswal (1968) called such implementation as *endogenous models.*

Some examples of states that can serve as basis for changes in priority are as follows:

- *Origin of entity.* The origin of entity can provide knowledge on the urgency of providing service. Since origin of entities does not generally change prior to service, this is used in combination with other more dynamic system states or entity attributes.
- *Nature or length of required service.* A strategy to shorten the overall latency times of entities for all classes is to serve those with shortest required service times. For systems that partially serve entities without necessarily completing the service (e.g., round-robin service discipline), the priority can be based on remaining service times.
- *Expected result of service or lack of service.* The expected result of service or lack of service pertains to the benefits of providing service or the cost or penalty of delaying service. There can be a relationship between the penalty and the origin of entities such as in the case of airline passenger classes.
- *Length of waiting line.* The length of waiting line or the number of entities waiting for service is important for cases wherein there is a limit on space or memory buffer that is being occupied by the waiting entities. Length of waiting line is affected by the rate of arrival of entities and the rate of service being performed by the resource.

SDPS Increment Function

A specific implementation of the state-dependent type is wherein the priority level of each class at time $t+1$ is dependent on the class from which the last entity served belongs and on an increment variable a_n.

Definition 11.13: State-dependent Dynamic Priority System

SDPS is a state-dependent DPS with an increment function described by

$$p_{i,t+1} = I_f(p_t; n_t^*, a_i, \text{ class queue}) = \begin{cases} p_{i,0} & \text{if } i = n_t^* \text{ or if } Q_i(t+1) = 0 \\ p_{i,t} + a_i & \text{if } i \neq n_t^* \text{ and } Q_i(t+1) > 0 \end{cases}$$

where $Q_i(t+1)$ is the length of waiting line for Class i at time $(t+1)$, and a_i is the increment variable, $i = 1, 2, \ldots, N$. Within a class, FCFS discipline is used.

Service differentiation is implemented by choosing the entity to be served next from the class with the highest, $p_{n^*,t}$, where

$$n_t^* = \arg_{n=1,2,3,...,N} \max[p_{n,t}]$$

n_t^* is the class from where the next entity to be served will be taken. The operator $\max[p_{n,t}]$ chooses the highest value among the priority levels $p_{n,t}$ and returns as an argument the class number n. Inside a class, the discipline is usually FCFS. Ties among class priority levels are broken based on the initial priority level $p_{n,0}$.

SDPS Class Latency

This section develops a proposition on the distribution of the waiting time of entities in a two-class SDPS. Consider a SDPS described by two classes of entities ($N = 2$), with sets of initial priority levels $I_0 = \{p_{1,0}, p_{2,0}\}$ such that $p_{1,0} - p_{2,0} = C$, and the increment variable is unity ($a_1 = a_2 = 1$). Consider a period of analysis such that the SDPS is saturated, and the time between arrivals and time for service of entities of class n are random variables x_n with density functions f_n.

Proposition 11.1: Waiting Time for Class 1 of a Two-class Saturated SDPS

In a two-class SDPS with $I_0 = (p+C, p)$ and $a_1 = a_2 = 1$, Class 1 entity has waiting time distribution of

$$f(w_1) = \begin{cases} \dfrac{C+1}{C+2}, w_1 = 0, \\ \dfrac{1}{C+2} \times f_2(w_1), w_1 > 0 \end{cases}$$

and

Proposition 11.2: Waiting Time for Class 2 of a Two-class Saturated SDPS

Class 2 entities have a waiting time distribution of

$$f(w_2) = \int_0^{W_2} f_1^C(w_2 - u) f(u)\, du$$

Proof of propositions 11.1 and 11.2:

Define $O(t) \in \{1,2,3,...,N\}$ to denote the class from which the next entity to be served is taken at time t. Consider w_n as the waiting time of class n as the time elapsed between similar $O(*) = n$, that is,

$$w_n = t' - t$$

such that $O(t') = O(t) = n$ and there exists a $t'', t < t'' < t'$ such that $O(t'') \neq n$.

TABLE 11.1

Waiting Times for a Two-class Saturated SDPS with $N = 2$, $I_0 = \{p_{1,0}, p_{2,0}\}$ Such That $p_{1,0} - p_{2,0} = C, a_1 = a_2 = 1$

Time	Current Priority of Class 1 $[p_{1,t}]$	Current Priority of Class 2 $[p_{2,t}]$	Class to Be Served $[O(t)]$	Waiting Time of Class to be Served $[w_{0(t)}]$
0	$p+C$	p	1	—
x_1	$p+C$	$p+1$	1	0
$x_1 + x_2$	$p+C$	$p+2$	1	0
$\sum_{i=0}^{C} x_i$	$p+C$	$p+C$	1	0
$\sum_{i=0}^{C+1} x_i$	$p+C$	$p+C+1$	2	—
$\sum_{i=0}^{C+2} x_i$	$p+C+1$	p	1	x_{C+2}
$\sum_{i=0}^{C+3} x_i$	$p+C$	$p+1$	1	0
$\sum_{i=0}^{C+4} x_i$	$p+C$	$p+2$	1	0
...	...	...	...	...
$\sum_{i=0}^{2C+2} x_i$	$p+C$	$p+C$	1	0
$\sum_{i=0}^{2C+3} x_i$	$p+C$	$p+C+1$	2	$\sum_{i=C+2}^{2C+3} x_i$

Table 11.1 shows values of $O(t)$ from $t = 0$ to $t = \sum_{i=0}^{2C+3} x_i$ where $x_1, x_2, \ldots$ represents service times of entities and the priority levels of the classes, $O(t)$.

Consider the first row of Table 11.1. At time $t = 0$, Class 1 has a higher priority level than Class 2, and an entity of Class 1 is served for a duration of x_1. Consider the second row. At time x_1, the service of the first entity ended. Since the SDPS is saturated, there is always at least one entity of each class waiting for service, and the priority levels of the classes are incremented based on Definition 11.13 to

$$p_{1,x_1} = p + C \quad \text{and}$$

$$p_{2,x_1} = p + 1$$

Class 1 still has a higher priority level that Class 2, an entity from Class 1, is served until completion at time $t = (x_1 + x_2)$. An entity from Class 2 is served only after C-numbers of entity from Class 1 have been served, as shown in the fourth row coinciding with time $t = \sum_{i=0}^{C} x_i$.

From Table 11.1, it can be generalized that $w_1 \in \{0, y\}$, where y is the service time for entities from Class 2, and with corresponding frequencies $\{C+1/C+2, 1/C+2\}$. A $w_1 = 0$ corresponds to events when Class 1 entities are served successively, and a $w_1 = y$ corresponds to events when the immediately preceding entity served is from Class 2. It is notable that the waiting time of one class corresponds to the service times of the other class. Thus, the waiting times for Class 2 is $w_2 = \sum^{C+1} x_i$ every time, where x is the service time of entities from Class 1. Therefore, the density function of w_1 is expressed as

$$f(w_1) = \begin{cases} \dfrac{C+1}{C+2}, w_1 = 0 \\ \dfrac{1}{C+2} \times f_2(w_1), w_1 > 0 \end{cases}$$

where $f_2(w_1)$ is the density function associated with Class 2 service times. For Class 2, the density function of w_2 is the $(C+1)$-fold convolution of f_1 (Bierlant et al., 1996) expressed as:

$$f(w_2) = \int_0^{W_2} f_1^C(w_2 - u) f(u)\, du$$

For a Markovian service process, (i.e., exponentially distributed with mean $1/\mu$), the $(C+1)$-fold convolution is a gamma distribution with a mean $C+1/\mu$ (Ang and Tang, 1975).

Propositions 11.1 and 11.2 describe the distribution of waiting times of classes in a two-class saturated SDPS. By treating the waiting times of the classes as a type of server failure, the waiting times of the entities can be obtained by modeling a FCFS-SPS with server failure.

Analysis of SDPS using a simulation model showed that during time spans when system is saturated (there is always at least one element in each class waiting for service), the order of servicing of the classes is cyclic. The cycle of servicing is affected by relative priorities and not by the absolute priorities. Examples of the cycle of servicing are shown in Table 11.1 for three classes $n = 1,2,3$ with initial priorities $p_{1,0}, p_{2,0}$, and $p_{3,0}$, respectively, denoted as (p_1, p_2, p_3). The presence of service cycles in a saturated system suggests that for a specific set of priority levels, the ratio of elements served from each class is constant. From Table 11.1, the ratio of entities served from each class for (10,5,2) and (15,10,7) is 14:3:2, that is, for every 19 elements served, 14 elements are from Class 1, three elements are from Class 2, and two elements are from Class 3 (Table 11.2).

The ratio for $I_0 = (10,5,3)$ is 13:3:2. Therefore, it can be deduced that in a saturated SDPS, the probability that an element in the lowest priority class will be served is 1 as long as the service times are finite. This can be compared to a nonpreemptive SPS where only the elements in the highest class are assured of service.

Another approach in analyzing SDPS is to consider the change in the priority levels of the classes. Figure 11.3 shows the plot of the priority levels for each class. For simplicity, the service times were assumed equal and constant for all three classes. The plot shows that there is a warm-up period after which the cycle of priority levels stabilizes (see arrow on plot).

TABLE 11.2

Classes Served in an SDPS. Cycle of Servicing for a Saturated SDPS with Different Initial Priority Levels $I_0 = (10,5,2), (15,10,70)$ and $(10,5,3)$

(10,5,2)	(15,10,7)	(10,5,3)
1	1	1
1	1	3
1	1	1
1	1	1
3	3	1
1	1	2
2	2	1
1	1	1
1	1	1
1	1	3
1	1	1
1	1	2
2	2	1
1	1	1
3	3	1
1	1	1
1	1	1
1	1	2
2	2	

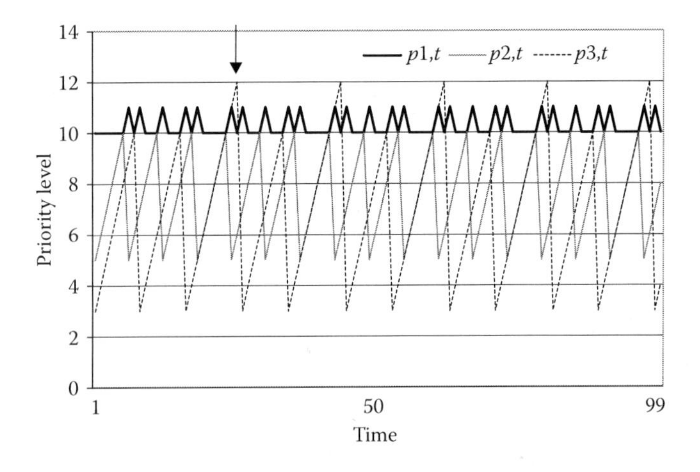

FIGURE 11.3
Priority levels for three-class SDPS.

The plot shows a warm-up time for the priority levels of a saturated SDPS. The initial priority levels are (10,5,3); the arrow indicates the end of the warm-up period.

The peaks on the level of the class priorities correspond to an entity being served. As expressed in Definition 11.1, each service of entity is followed by the return of the priority to the initial level. Similar to Table 11.1, Figure 11.3 suggests that all classes will be served as long as the service times of entities are finite.

11.3.4 Time-Dependent DPS

Time-dependent DPS (TDPS) is such that priority levels are assigned based on the time of arrival of entities. The earliest analyses of DPS are made as time-dependent queuing models due to their affinity with existing queuing models for SPS. As entities enter the system, the times of arrival are recorded and each entity is labeled for class identification. When an entity needs to be chosen among those waiting for service, the time-dependent priority functions are examined. The entity with the highest priority is then served.

Definition 11.14: Time-dependent Dynamic Priority System

The time-dependent PS described by Bolch et al. (1998) can be equivalently described as TDPS with the following configurations:

$$N < \infty$$
$$I_0 = \text{time of arrival}$$

$$I_f(*) = I_{t+\Delta t} = I_f(I_t, \Delta t; b_n) = I_t + \Delta t \times b_n,\ I_t = (t - t_0) \times b_n \text{ for Equation 11.4}$$

$$= I_{t+\Delta t} = I_f(I_t, \Delta t; r_n) = I_t + \Delta t,\ I_t = r_n + t - t_0 \text{ for Equation 11.5}$$

where t_0 is the time the entities arrived in the system.

The rule for choosing the next entity for service is $\max\{I_t\}$. For cases of tied priority levels I_t (when two or more entities of the same class arrived at exactly the same time), the choice is made randomly.

TDPS is different from SDPS in two aspects:

(1) Priority levels for SDPS are assigned on a class-wide basis. All entities of the same class have the same priority level. For TDPS, priority levels are assigned for each entity and are dependent on the time of arrival.

(2) SDPS priority levels revert to the initial levels once entities from that class is served.

TDPS Compared to FCFS

Recall the example in Section 11.3 that reformulated a FCFS system into the DPS notation such that

$$I_{t+\Delta t} = I_f(I_t) = I_t$$

and

$$I_t = n \times \Delta t$$

where $n \times \Delta t$ is the time when the entity arrived. From Definition 11.14, substituting $t_0 = n \times \Delta t$:

$$I_t = b_n(n \times \Delta t) \quad \text{and} \quad I_t = r_n + (n \times \Delta t)$$

The equations are translations of the priority levels for classes in the FCFS model. The difference is the presence of b_n, which transforms the rate of change of the priority level, and r_n, which shifts the intercept of the rate of change (see Figure 11.3).

Bolch et al. (1998) describe two systems with time-dependent priorities where the priorities of the classes of entities are expressed as follows:

$$q1_r(t) = (t - t_0) \times b_r$$

$$q2_r(t) = r_r + t - t_0,$$

where $q1_r(t)$ and $q2_r(t)$ express the two systems of time-dependent priorities, t_0 is the time when the entity entered the system, $0 \le b_1 \le b_2 \le \ldots \le b_r$, and $0 \le r_1 \le r_2 \le \ldots \le r_r$ are system parameters.

It is observed that for both functions of the time-dependent priority levels, the system behaves like a FCFS system when parameters b_r and r_r are very close together. On the other hand, values of b_r and r_r far from each other will behave like a static PS. Figure 11.4 shows the plot of the priority level with time.

TDPS Class Latency

The mean latency time for an r-class entity is given by

$$E(x_r) = L_r = \frac{E(L_{FIFO}) - \sum_{i=1}^{r-1} \rho_i E(L_i)\left(1 - \frac{b_i}{b_r}\right)}{1 - \sum_{i=r+1}^{R} \rho_i \left(1 - \frac{b_r}{b_i}\right)} \tag{11.2}$$

where L_i denotes the latency times for entities of Class i, L_{FCFS} denotes the latency times of entities in a similar FCFS discipline, and ρ_i denotes the

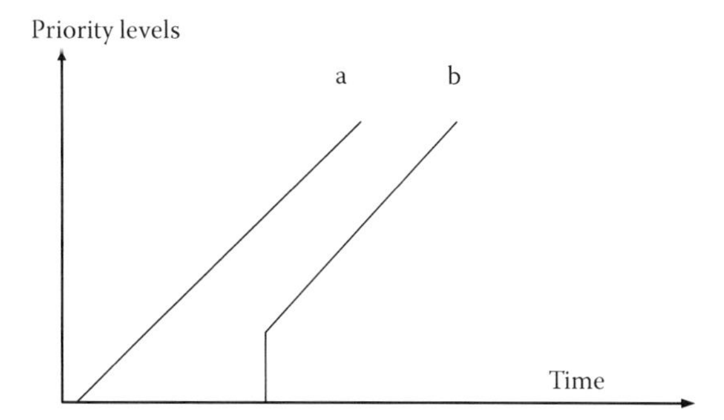

FIGURE 11.4
Values of priority versus time for (a) $q1_r(t)$ and (b) $q2_r(t)$ showing the differentiation typical of priority systems.

utilization in class $i, 1 \le i \le R$. Equation 11.2 shows that to determine the mean latency time of one class of entity requires a system of equations of the mean latency times of the other classes of entities.

TDPS Class Differentiation

Suppose that the performance of a system is measured by the average latency times $L_n, n = 1,2,\ldots,N$. Then the relative differentiation model imposes for all pairs of classes $(a,b), a,b = 1,2,\ldots,N, a \ne b$:

$$\frac{L_a}{L_b} = \frac{c_a}{c_b} \tag{11.3}$$

where $c_1 < c_2 < \ldots < c_N$ are the quality differentiation parameters for each of the classes (Essafi et al., 2001).

A time-dependent PS proposed by Bolch et al. (1998) and Essafi et al. (2001) that implements the relative differentiation of Equation 11.3 is such that priority levels are assigned based on the time of arrival of entities. Two systems with time-dependent priorities where the priority of a class of entities and the time-dependent priority functions are expressed as follows:

$p1_{n,t}$ and $p2_{n,t}$ are the priority levels of class n at time t for systems 1 and 2:

$$p1_{n,t} = (t - t_0) \times b_n \tag{11.4}$$

$$p2_{n,t} = r_n + t - t_0 \tag{11.5}$$

where t_0 is the time when the entity entered the system, $0 \le b_1 \le b_2 \le \ldots \le b_N$, and $0 \le r_1 \le r_2 \le \ldots \le r_N$ are system parameters. For Equations 11.4 and 11.5, the system behaves like a FCFS system when parameters b_r or r_r are very close together. On the other hand, values of b_n or r_n far from each other will behave like a head-of-line discipline.

Figure 11.5 shows the comparison of the priority levels of a typical entity in a TDPS with increment functions given by Equations 11.4 and 11.5 with

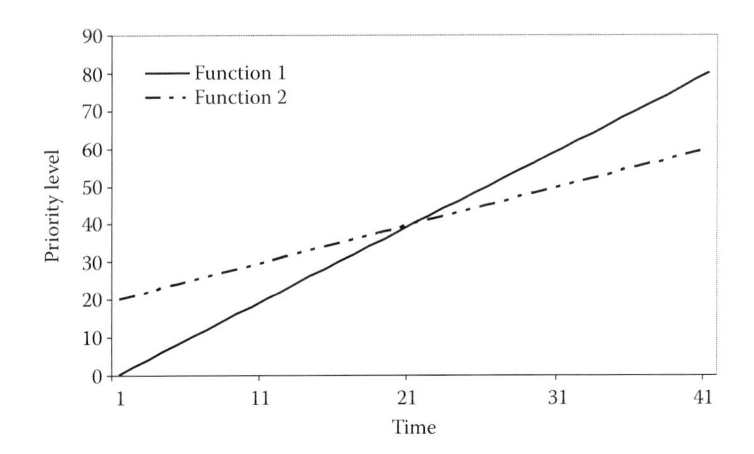

FIGURE 11.5
TDPS priority levels: Comparison of TDPS priority levels for functions 1 and function 2.

$b_n = 2$ and $r_n = 20$, respectively. From the figure and the form of the functions, it is notable that b_r affects the rate of increase of the priority level with respect to an increase in the waiting time of entities—the higher the value of b_n, the more rapid the increase of the priority level. On the other hand, r_n affects the minimum value of the priority level and not the rate of increase. Furthermore, Figure 11.5 implies that this PS falls under the classification of a DPS since none of the two functions dominate the other over the whole period of time (see definition of a DPS).

The mean latency time is expressed as (Bolch et al. 1998):

$$L_r = \frac{L_{FIFO} - \sum_{i=1}^{r-1} \rho_i L_i \left(1 - \frac{b_i}{b_r}\right)}{1 - \sum_{i=r+1}^{R} \rho_i \left(1 - \frac{b_r}{b_i}\right)} \tag{11.6}$$

Using the notation presented in Section 11.2, the increment functions equivalent to Equations 11.4 and 11.5 are as follows:

For Equation 11.4,

$$\begin{aligned} I_{t+\Delta t} &= I_f(I_t, \Delta t; b_n) \\ &= I_t + \Delta t \times b_n \quad \text{for } I_t = (t - t_0) \times b_n \end{aligned}$$

For Equation 11.5,

$$\begin{aligned} I_{t+\Delta t} &= I_f(I_t, \Delta t; r_n) \\ &= I_t + \Delta t, \qquad \text{for } I_t = r_n + t - t_0 \end{aligned}$$

The new expression is obtained by first changing the notation such that $q_r(t) \Leftrightarrow I_t$. Then for Equation 11.4,

$$\begin{aligned} I_t &= q_n(t) = (t - t_0) \times b_n, \quad \text{moreover} \\ I_{t+\Delta t} &= ([t + \Delta t] - t_0) \times b_n \\ &= (t + \Delta t - t_0) \times b_n \\ &= (t - t_0 + \Delta t) \times b_n \\ &= ([t - t_0] + \Delta t) \times b_n \\ &= (t - t_0) \times b_n + \Delta t \times b_n \\ &= I_t + \Delta t \times b_n \end{aligned}$$

For Equation 11.5,

$$\begin{aligned} I_t &= q_n(t) = (r_n + t - t_0), \quad \text{moreover} \\ I_{t+\Delta t} &= (r_n + [t + \Delta t] - t_0) \\ &= (r_n + t - t_0 + \Delta t) \\ &= (r_n + t - t_0) + \Delta t \\ &= I_t + \Delta t \end{aligned}$$

11.4 Summary

This chapter developed a state-dependent DPS uniquely identified by its increment function and consequently defined as SDPS later used in the demonstration of the framework forwarded in the book. Propositions were presented that show the effect of increment function to the frequency that an entity from one class is served in relation to another class. In the process, the distribution of the waiting times of the classes was derived. It was shown that in a SDPS, the proportions of entities served from the various classes of entity can be controlled by the initial priority level parameter, and by the increment variable a_i. The result is a configuration of a DPS that can implement relative differentiation among classes that has a fairly simple distribution of class latency times.

Questions and Exercises

True or False

1. Priority systems (PS) are distinctively different from priority disciplines in the field of queuing.

2. Arriving entities into a PS are always classified into a single class.
3. A PS parameter N is the number of classes of entity requiring service.
4. The definition of N requires that there are actually entities belonging in a class for a class to be defined, thus an empty class is not permissible.
5. Each of the classes has an arrival rate, which is the basis for the decision from which class the next entity will be served.
6. The rules for implementing the decision on which class to serve next can be summarized into a parameter of the PS, the increment function.
7. Since $\max\{I_t\}$ is the same as $\min\{-I_t\}$, then any maximization rule can be translated into an equal minimization rule.
8. There are two basic types of PS with reference to the lengths of waiting lines for the classes of entities.
9. Some practical reasons for assuming infinite waiting line are large memory buffer and high-speed servers common in communication systems.
10. In a PS with finite waiting line, entities turned away can be redirected to other servers, consigned to wait in another system used for storage, or rerouted to another class of entities whose limit has not been yet reached.
11. A traditional FCFS system can be modeled as a PS with parameters $N, I_t, \text{and } I_f$.
12. State-dependent DPS are such that the priorities of the classes change based on some combinations of system states and entity attributes and not directly on the arrival time of the entities.
13. SDPS is a particular type of DPS wherein the priority levels of the classes of entities increments by a_i every time waiting entities of the class is not served.

Questions

1. Define the following terms in a narrative manner, that is, not using equations.
   (A) Number of classes
   (B) Priority level of a class
   (C) Set of priority levels
   (D) Increment function
   (E) Infinite waiting line

(F) Finite waiting line

(G) Waiting entities in a PS

(H) Saturated PS

(I) Supersaturated PS

2. Provide three examples of real systems that can be modeled as PS having infinite waiting line. Justify this assumption.
3. Provide three examples of real systems that can be modeled as PS having finite waiting line. Justify this assumption.
4. Provide three examples of real systems that can be modeled as PS having finite waiting line and describe how entities turned away are managed.

12

Risks of Extreme Events in Complex Queuing Systems

12.1 Introduction

This chapter formulates the measurement of the risk of extreme latencies for various classes of entities in priority systems (PS). The first section investigates the conditions for the occurrence of extreme latencies, namely, the arrival and service rates of entities. The second section describes a basis for the measurement of risk in PS by defining nonexceedance probabilities of latency times. Latency thresholds are defined for each of the classes in a PS to identify the level of latency times wherein unnecessary cost is incurred, and the reliability of the PS is affected. The last section summarizes the contributions of this chapter to the book.

12.2 Risk of Extreme Latency

Chapter 11 presented two types of dynamic priority systems (DPS) that implements relative differentiation among classes of entities. It was also shown that configuration parameters such as initial priority levels and increment variables can be used to affect the latency times of entities in a DPS. Particularly important is the risk of entities from extremely long latency times due to its ability to drive the cost of operation unnecessarily high. Section 11.3 discussed the deficiency of the mean of latency times in describing the risk of extremes of latencies. There is a need to describe the extremely long albeit rare latency times without being overwhelmed by the more frequent shorter latency times. Furthermore, measurements of risk of extremes of latency times need to be made for each of the classes in a PS for the following reasons:

- The heterogeneity across classes often warrants different treatment and cost considerations for each class.
- The flexibility and the potential to improve system performance of the PS can be maximized if configuration is based on classes.

In systems like digital equipment and software-intensive systems, the probabilistic behavior of the system is not the only cause of failure or extreme events (Garrett and Apostolakis, 1999). System failure and extreme events also occur as results of encountering combinations of inputs, particular operating environment, and some particular configuration. Any randomness in the performance of PS is brought about by the randomness of the input and the operating environment. Extremes of latencies are known to cause system failures (see Klungle, 1999; Miller and Bapat, 1999; Tanir and Booth, 1999). Therefore, it is worthwhile to investigate the contributing factors in the occurrence of extremes of latencies in PS.

Consider an emergency room in a hospital. Suppose there is a sudden arrival of several patients in critical conditions. The medical personnel on duty can be overwhelmed, and patients waiting in the emergency room with less severe conditions can experience delay on their treatment because patients in critical conditions are served first. The situation represents an extreme increase in the arrival of high-priority entities (critical patients), which results to longer latency times for low-priority entities (noncritical patients). But after a while, other medical personnel may be deployed in the emergency room to aid those on duty, effectively increasing the overall service capacity of the emergency room. The latency times of low-class patients can then revert to the normal lengths. Suppose that all the commotions in the emergency room had unduly stressed the medical personnel on duty. The service to each patient given by the medical personnel now takes a longer time than before due to fatigue. Such degraded service rate can also result into longer latency times for patients.

The above is a case in point of conditions under which extreme latency times of entities in PS can occur, namely, under the following conditions:

- The traffic rates of entities are extremely high
- Service rates are extremely low
- Presence of both these conditions

The first condition is more of concern for entities of low-priority classes than to high-priority classes. As described in the above example, extremely high traffic rates of high-priority entities will definitely affect the latency times of low-priority entities. However, extremely high traffic rates for low-class entities do not necessarily affect the latency times of high-class entities. State-dependant dynamic priority system (SDPS) is an example of a PS in which high arrival rates of low-class entities do not affect the latency times of high-class entities.

The result of the conditions is the *saturation* of the PS when more entities arrive than what can be served by the resource (see Definition 11.9 in Chapter 11). During saturation, there can be lengthening of the waiting lines and increase in latency times of entities. Figure 12.1 shows a possible relationship of the rate of arrival of entities, the service rate, and the latency times in a PS. From time period $(0,e)$, the PS is *stable* and has finite average latency times.

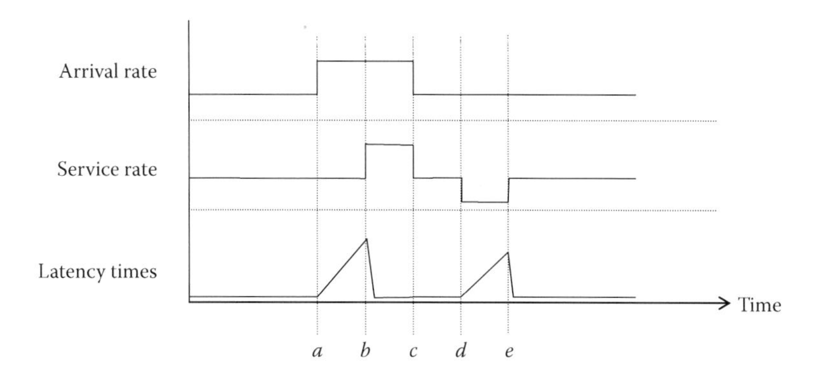

FIGURE 12.1
Arrival, service, and latency times. Variations in latency times brought about by variations in arrival and service rates.

However, within the long-term period lies shorter periods, like time period (a,b) when arrival rates of entities increases drastically. Extremes of latencies can occur if period (a,b) of increasing latency times persists long enough. The PS becomes stable once the service rate increases correspondingly during period (b,c) or when the arrival rate returns to previous level, like during period (c,d). Another possible period of increasing latency times is when the service rate drops considerably, like in period (d,e).

The challenge to PS designers and administrators is to measure the risk of extremes of latency times during periods (a,b) and (d,e) as illustrated in Figure 12.1. Furthermore, the measurement of risk must lend to the configuration of PS for each of the classes.

12.2.1 Methodology for Measurement of Risk

Chapter 11 described several approaches to gathering and analyzing information on the latency times of a PS, as well as possible effects of the configuration of PS to the latency times of entities of various classes. The measurement of risk is performed to each of the N classes in a PS with the objective of analyzing the configuration of a PS. Measurement of risk of extremes of latencies is based on the cumulative probability function $\Pr\{X_n \leq x_n\} = F(x_n)$ of the latency times x_n of entities from Class n.

First, for a given PS configuration f, define the nonexceedance probability of the latency time x_n of entity of Class n.

Definition 12.1: Nonexceedance Probability

The nonexceedance probability of latency time x_n for entities of Class $n, n = 1,2,\ldots,N$ is defined as

$$NE_n^f(x_n) = P\{X_n \leq x_n\}$$

Definition 12.1 describes a concept closely related to the cumulative probability function for latency time x_n. Most applications of PS impose thresholds on the latency times for various classes, such as in the case of the e-business example in Section 1.2. The threshold signifies the limit on latency times, which is intended for a class of entities. If *reliability* is defined as the ability to perform the intended function in a given environment at a specified period, then the thresholds can determine the reliability of a PS.

A definition of a threshold in terms of reliability and failure is given below.

Definition 12.2a: Latency Threshold Based on Reliability C_n

Latency threshold based on reliability C_n is the threshold for latency times of Class n such that

$$\{X_n > C_n\} \Rightarrow \text{PS failure in Class } n$$

$$\{X_n \le C_n\} = 1 - \{X_n > C_n\} \Rightarrow \text{PS non-failure in Class } n$$

It follows from Definition 12.2a that the probability of failure in Class n is

$$P(\text{failure in class } n) = \mathrm{P}\{X_n > C_n\}$$

By Definitions 12.1 and 12.2a, the reliability of the PS due to Class n can be expressed as

$$\begin{aligned} \text{Realiability in class } n &= 1 - P\{X_n > C_n\} \\ &= P\{X_n < C_n\} \\ &= NE_n^f(C_n) \end{aligned}$$

The reliability in Class n of a PS is equal to the nonexceedance of the latency time equal to C_n. An alternative to Definition 12.2a is a definition of the latency threshold based on cost functions. The thresholds on latency times can be determined by the cost incurred due to extremes of latencies, examples of which are penalties for late deliveries of entities. The threshold on the latency times of a class of entities can be defined in terms of a cost function.

Definition 12.2b: Latency Threshold Based on Cost Function C_n

Latency threshold based on cost function C_n is the threshold for latency times of Class n such that for a cost function,

$$\text{Cost} = g(X_n; C_n) = 0 \quad \text{for } X_n \le C_n, \text{ and}$$

$$\text{Cost} = g(X_n; C_n) > 0 \quad \text{for } X_n > C_n$$

Since risk can be defined as the complement of reliability (Ang and Tang, 1984), then the risk associated with threshold C_n for Class n in a PS with configuration f is defined below.

Definition 12.3: Risk of a Class

Risk associated with the threshold C_n is

$$R_n^f(C_n) = 1 - NE_n^f(C_n)$$

A risk function of $R_n^f(C_n)$ for the range of C_n equal to the range of the latency time is

$$R_n^f(x_n) = 1 - NE_n^f(x_n)$$

Figure 12.2 illustrates the latency time threshold and the nonexceedance curve and the risk of extreme latencies. Figure 12.2a shows the two definitions of the latency time threshold C_n as determinant of the failed or operational state of the PS and also as determinant of the value of cost incurred due to extremes of latencies. Figure 12.2b shows the nonexceedance curve and its complementary and the risk curve for the range. It is important to note that the risk can be determined for an extreme value of latency only if the probability density function of the latency is known, especially at the extremes.

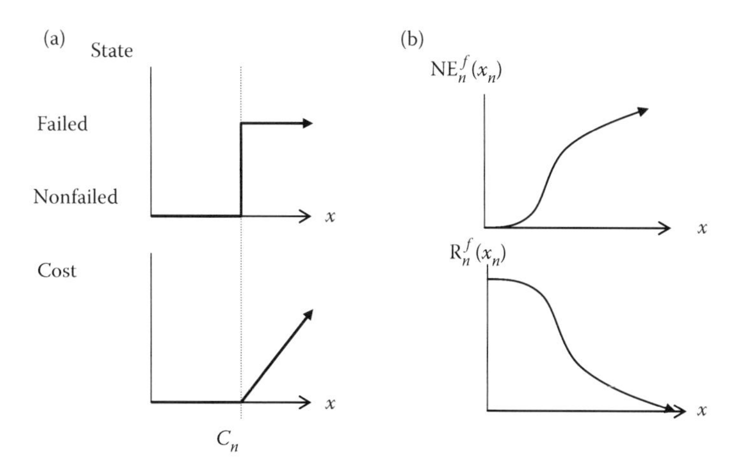

FIGURE 12.2
Definitions of risk of extreme latencies. The comparison between (a) Definitions 12.2a and 12.2b, and (b) the relationship between the nonexceedance curve $NE_n^f(x)$ and the corresponding risk $R_n^f(x)$.

PROBLEM 12.1 (Application of $R_n^f((C_n))$)

Consider a PS with several classes of entities each with known performance for a specific configuration. The classes of entities may each has different requirement in the level of performance. An example can be three classes of entities on an e-business server: *Browse, Add-to-Cart,* and *Buy*. The three classes of entities may require different levels of performance; the *Buy* entities have the strictest requirement, followed by the *Add-to-Cart*, and then by the *Browse* entities. Suppose a PS with a certain configuration f has three classes with each class having a nonexceedance function:

$$NE_n^f(x_n) = \exp\left(-\exp\left(\frac{-(x_n - \lambda_n)}{\delta_n}\right)\right)$$

with

$$\begin{aligned}\lambda_1 &= 38.6, \delta_1 = 63.5,\\ \lambda_2 &= 85, \delta_2 = 165,\\ \lambda_3 &= 1280, \delta_3 = 922\end{aligned}$$

For threshold $C_1 = 400, C_2 = 1500$, and $C_3 = 8000$, the risks of extremes of latencies are obtained by Definition 12.3:

$$\begin{aligned}R_1^f(400) &= 1 - NE_1^f(400) = 0.0034\\ R_2^f(1500) &= 1 - NE_2^f(1500) = 0.00018\\ R_3^f(8000) &= 1 - NE_3^f(8000) = 0.00063\end{aligned}$$

The risk for individual classes in a PS is expressed by Definition 12.3. More precisely, Definition 12.3 describes the probability that an entity of Class n have a latency time longer than C_n. Another way of describing the risk of extreme latencies is for the entire PS rather than for a particular class. Analogous to Definition 12.2, the probability of failure of the PS is the probability that an entity, regardless of the class to which it belongs, has a latency time longer than the threshold of that class.

Proposition 12.1: (Risk of a PS)

A PS with configuration f and N classes, each class having $R_n^f(C_n), n \in N$, has a probability of failure due to extremes of latency:

$$R_{PS}^f = \sum_{n=1}^{N}\left[R_n^f(C_n) \times \phi_n\right]$$

where $\phi_n > 0$ are the relative frequencies of entity from each class, and $\sum_{n=1}^{N} \phi_n = 1$.

Proof:

The risk of extremes of latency for the entire PS is the risk for each class taken together, that is,

$$R_{PS}^{f} = P(X_k > C_k \mid k \in 1,2,\ldots,N) \tag{12.1}$$

$R_n^f(C_n)$ defined in Definition 12.3 is the probability of an entity latency time to be longer than the threshold C_n, provided that the entity is of Class n. That is, consider a particular Class n,

$$R_n^f(C_n) = P(X_n > C_n) \tag{12.2}$$

Equation 12.2 can be equally expressed in terms of any class $k \in 1,2,\ldots,N$ as

$$R_n^f(C_n) = P([(X_k > C_k) \mid k \in 1,2,\ldots,N] \mid k = n) \tag{12.3}$$

Let

$$[(X_k > C_k) \mid k \in 1,2,\ldots,N] = A$$

and

$$\{k = n\} = B$$

Then Equation 12.3 becomes

$$R_n^f(C_n) = P(A \mid B)$$

The *Total Probability Theorem* states that $\Pr(A) = \sum \Pr(A \mid B_i) \times \Pr(B_i)$ for all i. Then, by the same reasoning,

$$\begin{aligned} R_{PS}^{f} &= P\{X_k > C_k \mid k \in 1,2,\ldots,N\} \\ &= P(\mathrm{A}) \\ &= \sum P(A \mid B_i) \times \Pr(B_i) \\ &= \sum [P\{[(X_k > C_k) \mid k \in 1,2,\ldots,N] \mid k = n\} \times P\{k = n\}] \end{aligned} \tag{12.4}$$

for any class $k \in 1,2,\ldots,N$. By substitution of Equation 12.3 into Equation 12.4,

$$R_{PS}^{f} = \sum_{n=1}^{N} [R_n^f(C_n) \times P(k = n)] \tag{12.5}$$

Suppose that the probability that an entity k belongs to Class n is expressed as ϕ_n, then Equation 12.5 becomes

$$R_{PS}^{f} = \sum_{n=1}^{N} [R_n^f(C_n) \times \phi_n] \tag{12.6}$$

with $\sum_{n=1}^{N} \phi_n = 1$.

The proportion of all entities that belong to Class n is equal to ϕ_n and can be deduced from several sources. One possible source of this information is the known departure rates of entities. That is, if entities of Class n are known to depart at a rate of Y_n (entities per minute), then the probability that a departing entity belongs to Class n is

$$\phi_n = \frac{Y_n}{\sum_{i=1}^{N} Y_i} \tag{12.7}$$

PROBLEM 12.2 (Application of R_{PS}^f)

Consider the PS described in Problem 12.1. Suppose that the probability that an entity belongs to one of the three classes are

$$\phi_1 = 0.367$$
$$\phi_2 = 0.232$$
$$\phi_3 = 0.401$$

The risk of the PS as expressed in Proposition 12.1 is

$$\begin{aligned} R_{PS}^f &= \sum_{n=1}^{3} [R_n^f(C_n) \times \phi_n] \\ &= [0.0034 \times 0.367 + 0.00018 \times 0.232 + 0.00063 \times 0.401] \\ &= 0.001542 \end{aligned}$$

The result suggests that about 15 out of 10,000 entities will have latency times beyond the threshold, regardless of the class to which the entities belong.

12.3 Conditions for Unbounded Latency

This section develops the conditions on the configuration of PS such that the latencies for a class of entities are unbounded. This section provides a background on the unbounded latencies and the sufficient conditions for unbounded latency times for a saturated PS.

Chapter 10 describes the possible characteristics of the tails of a distribution of latency times of entities: exponentially decaying, polynomially decaying, and bounded (also referred to having the domain of attraction of Gumbel types I, II, and III, respectively). A method presented by Castillo (1988) and subsequently adapted in Chapter 10 can determine the domain of attraction of continuous cumulative probability functions, effectively describing the characteristic of the tails of the distributions. The method states the following:

A necessary and sufficient condition for the tail of a continuous cumulative probability function, $F(x)$, to belong to one of the Gumbel type distributions is

$$\lim_{\varepsilon\to 0}\frac{F^{-1}(1-\varepsilon)-F^{-1}(1-2\varepsilon)}{F^{-1}(1-2\varepsilon)-F^{-1}(1-4\varepsilon)}=2^{d} \tag{12.8}$$

More specifically, $F(x)$ belongs to the domain of attraction for maxima of the Gumbel type I if $d=0$, Gumbel type II if $d>0$, and Gumbel type III if $d<0$.

PROBLEM 12.3 (Domain of Attraction for M/M/1-FCFS)

Consider an M/M/1-FCFS. The distribution of waiting times and latency times is (Hillier and Lieberman, 1986) as follows:

$$P(W<w)=1-\rho e^{-\mu(1-\rho)w}\quad \text{for } w>0 \tag{12.9}$$

$$P(L<\ell)=1-e^{-\mu(1-\rho)\ell}\quad \text{for } \ell>0 \tag{12.10}$$

By applying the necessary and sufficient conditions of Equation 12.8, consider Equation 12.9, with an inverse function of

$$F^{-1}(w)=\frac{\log\left(\dfrac{1-w}{\rho}\right)}{\mu(-1+\rho)}$$

Equation 12.8 becomes

$$\begin{aligned}&\lim_{\varepsilon\to 0}\frac{F^{-1}(1-\varepsilon)-F^{-1}(1-2\varepsilon)}{F^{-1}(1-2\varepsilon)-F^{-1}(1-4\varepsilon)}\\&=\lim_{\varepsilon\to 0}\frac{\log(1-(1-\varepsilon))-\log(1-(1-2\varepsilon))}{\log(1-(1-2\varepsilon))-\log(1-(1-4\varepsilon))}=1=2^{0}\end{aligned}$$

Since $d=0$, the distribution of waiting times for a SPS (M/M/1) is unbounded and has a domain of attraction for maxima of Gumbel type I. The result can be generalized to all cumulative density function with exponential form like the latency function of Equation 12.10.

PROBLEM 12.4 (Domain of Attraction for (G/G/1))

Consider a generally distributed arrival and service processes, equivalent to an SPS (G/G/1). Furthermore, consider the case of heavy entity traffic such that utilization* $\rho\cong 1$. Section 12.2 showed the effects of heavy entity traffic

* Utilization is defined as $\rho=\mu/\lambda$, where μ and λ are the steady state service and arrival rate of entities, respectively.

to the extremes of latency times. The latency times can be approximated by (Kleinrock, 1976):

$$P(W < w) \cong 1 - e^{-\frac{2(1-\rho)\lambda^{-1}}{\sigma_a^2 + \sigma_b^2} w} \tag{12.11}$$

where σ_a^2 is the variance of interarrival times, and σ_b^2 is the variance of service times.

Equation 12.11 also has an exponential form; therefore, applying Equation 12.8 results in $d = 0$, implying that the distribution of the latency times is unbounded and belongs to Gumbel type I.

12.3.1 Saturated PS

There can be difficulties in assessing the true characteristics of the tails of the distribution of the latency times for variations in PS, as discussed in Chapter 11. The difficulties arise from lack of knowledge on the cumulative probability function of the latency times due to incomplete information on the arrival and service of entities and on the possible variations in PS. This section presents a set of conditions that will assure that the tails of the distribution of latency times of classes in a PS are unbounded.

Proposition 12.2 (Sufficiency Conditions for Unbounded Latency Times in PS)

Sufficient conditions for unbounded entity latency times in a saturated PS are as follows:

- The entity service times are mutually independent.
- The entity service times have common distribution.

Proof:

Recall definition of saturated PS in Definition 11.9 and consider the *m*th entity to be served in the PS. The latency time of the entity is composed of the following:

- Remaining service time of the entity being served when m arrived, assuming there is no preemption
- Service times of all entity that will be served ahead of m
- Service time of entity m.

Define the length of the remaining service time of the entities being served when m arrived as y_0 and $y_i, i = 1, 2, \ldots, m-1$, are the service times of entities that are served ahead of m. The latency of entity m is therefore the sum of these service times, that is,

$$\begin{aligned} x_m &= y_0 + y_1 + y_2 + \cdots + y_{m-1} + y_m \\ &= \sum_{i=0}^{m} y_i \end{aligned}$$

TABLE 12.1

Domains of Attraction for SPS[a]

System Description	Domain of Attraction of Arrival and Service Distribution Are Both:		
	G-I	G-II	G-III
	Domain of Attraction of the Waiting Time Distribution		
Unsaturated	?	?	Not G-II
Saturated	G-I	G-I	G-I

[a] Domains of attraction for SPS: single-server FCFS model (see Problems 12.3 and 12.4).

If the service times and the number of entities $(m-1)$ already in waiting for service are assumed to be random variables, then x_m is a sum of a series of random variables i and is by itself also a random variable. That is,

If y_i is a random variable, $i = 0,1,2,\ldots,m-1$, and m is a random variable, then $x_m = \Sigma_{i=0}^{m} y_i$ is a random variable.

By the central limit theorem,* x_m have a normal distribution function for large values of m. Since supersaturated PS is characterized by increasing the number of waiting entities (see Definition 11.10), that is, $m \to \infty$ as $t \to \infty$, the central limit theorem applies. The domain of attraction of a normal distribution is Gumbel type I (Castillo, 1988).

Note that Proposition 12.1 is applicable, regardless of the priority class of the mth entity or the underlying distribution of the service times. For SPS, the domain of attraction of the distribution of latency times can be summarized in Table 12.1. For a saturated case, regardless of the underlying distributions of the arrival and service of entities, the latency times are unbounded and belong to the domain of attraction of the Gumbel type I. However, for the unsaturated case, the domain of attraction has been determined only for the arrival and service process both belong to the domain of attraction of the Gumbel type III (bounded). For other types of arrival and service process, the domain of attraction has not been determined.

12.4 Conditions for Bounded Latency

This section establishes the conditions for a general SPS to have bounded entity latency times, the sufficient conditions for entities of a particular class in an SPS to have bounded latency times, as well as for a particular class in a SDPS.

12.4.1 Bounded Latency Times in Saturated Static PS

Section 12.4 established sufficient conditions for entities in a PS to have unbounded latency times. However, there are situations wherein there is need

* Central Limit Theorem states that if x_i are mutually independent with common distribution, none of which is dominant, then for the sequence $\{x_i\}$ and for every fixed β, $\Pr(\underline{L}_m < \beta) \to N(\beta)$ as $m \to \infty$, where $N(\beta)$ is the normal distribution function $\frac{1}{\sqrt{2\pi}} \int_{-\infty}^{\beta} e^{-\frac{1}{2}y^2} dy$; $\mu = E(x_i)$ and $\sigma^2 = \mathrm{Var}(x_i)$.

for a guaranteed maximum latency times, such as in health services. Patients in an emergency room cannot wait for an infinitely long time. Thus, to model such types of PS requires configurations that guarantee bounded latency times. As shown in Table 12.1 for an SPS, bounded latency times can be obtained only if the system is not saturated and both the service and interarrival times of entities are bounded. However, a saturated SPS will have unbounded latency times even if both the interarrival time and service times are bounded. Therefore, it is of interest to PS designers and administrators to determine configurations of SPS to effectively implement bounded latency times.

Proposition 12.3 (Sufficient Conditions for Bounded Latency Times in SPS)

For an entity of a particular Class n in an SPS, the sufficient conditions for bounded latency time are given as follows:

- The service times of entities for all classes are bounded.
- The waiting line for Class n is bounded ($Q\max_n < \infty$).
- The number of entities with higher priority than Class n is finite.

Proof:

Consider an SPS with N classes of entities and with particular interest to Class n. The latency time of the m*th* entity of Class n is composed of the following:

- The remaining service time of the entity being served when m arrived, assuming there is no preemption
- Service times of all entity that will be served ahead of m
- Service time of entity m

Consider two states of the server as busy and none-busy. A *busy* state is when the server is serving entities belonging to class other than n, whereas a nonbusy state is when the server is serving an entity from Class n. Figure 12.3 illustrates the states where y_n represents the service time of entities from Class n, and z represents the service times of entities served in-between the service of entities from the Class n.

Therefore, the latency time of the mth entity of Class n is

$$\begin{aligned} x_m &= z_1 + y_1 + \cdots + z_{m-1} + y_{m-1} + z_m + y_m \\ &= \sum_{i=1}^{m} y_i + \sum_{j=1}^{m} z_j \end{aligned}$$

Consider the possible properties of the service times y_i, z_i, and m (the number of terms in the sequence). The random variables can be either bounded or unbounded, and the number of terms m in the sequence can also be bounded

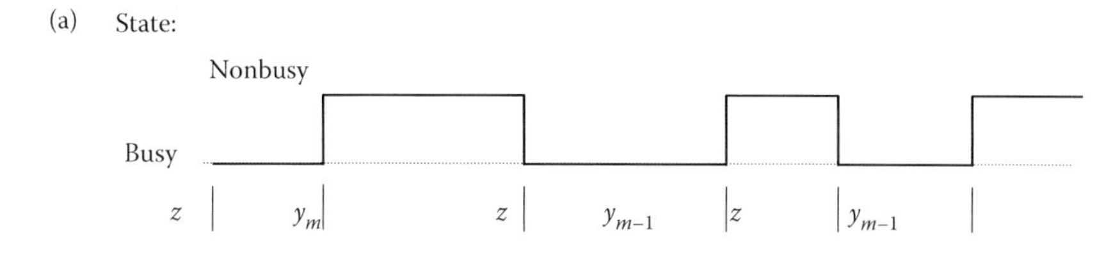

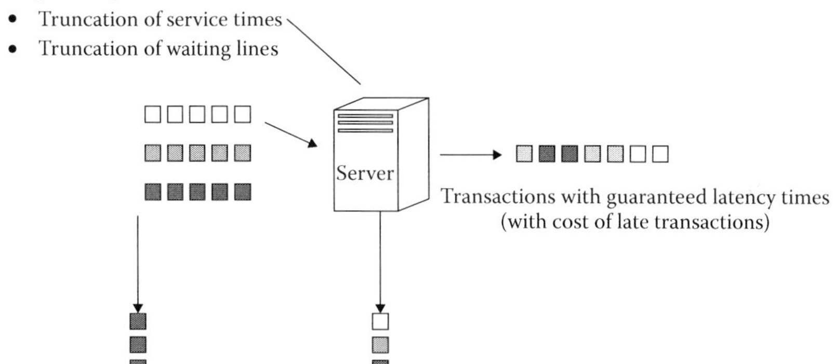

FIGURE 12.3
(a) Two-state server, and (b) modified PS with bounded latency times and three streams for outgoing entities: served, refused, and entities with truncated service.

TABLE 12.2

Bounded and Unbounded Latency Times[a]

y_i, z_i	m	x_m
Bounded	Bounded	Bounded
Unbounded	Bounded	Unbounded
Unbounded	Bounded	Unbounded
Unbounded	Unbounded	Unbounded
Bounded	Unbounded	Unbounded
Bounded	Unbounded	Unbounded
Unbounded	Unbounded	Unbounded
Unbounded	Bounded	Unbounded

[a] Combinations of bounded and unbounded service times and number of entities ahead of m.

or unbounded. Table 12.2 shows that only for the case of bounded service times y_i, z_i, and bounded m, the latency time x_m can be bounded.*

* x_m, y_i, and z_i are bounded if there exists an X, Y, Z such that $x_m < X$, $y_i < Y$, and $z_i < Z$ for all possible values of m, and i.

Consider the case when x_m is bounded. Such is the case if y_i, z_i, and m are bounded. A bounded m implies two conditions: the waiting line of Class n is bounded, and the number of entities that can possibly be served ahead of the entity of interest (i.e., has a higher priority than Class n) has an upper limit. A bounded y_i and z_i implies that the service times for all classes are bounded.

The first condition of Proposition 12.2 is satisfied only if service times are known to be bounded or are artificially bounded by ejecting entities that have been served for a long period of time. The second condition can be implemented by configuring the SPS with a limit on the number of waiting elements for Class n such that all entities arriving after the limit is reached are dropped or rerouted. The third condition is more difficult to satisfy since there is very little control over the arrival of entities, and most SPS applications have an infinitely many potential entities.

Systems such as banks and offices that have a defined beginning and end of availability times can satisfy the third condition. However, for systems such as e-business servers with an expected availability of almost 100% of the time, the number of potential entities is unlimited.

12.4.2 Bounded Latency Times in a Saturated SDPS

The conditions of bounded service times and waiting lines can be reasonably implemented by configuring the SPS accordingly. However, the condition of finite number of entities with higher priorities is a challenge to implement. Recall the SDPS (see Definition 11.13) where the priority increment function is defined as

$$p_{i,t+1} = I_f(p_t; n_t^*, a_i, \text{class queue}) = \begin{cases} p_{i,0} & \text{if } i = n_t^* \text{ or class } i \text{ is empty} \\ p_{i,t} + a_i & \text{if } i \neq n_t^* \text{ and class } i \text{ is not empty} \end{cases}$$

A unique characteristic of the SDPS is the increment of priority levels of classes based on the number of high-priority entity served.

Proposition 12.4: (Sufficiency Conditions for Bounded Latency Times in a Saturated SDPS)

For an entity of a particular Class n in an SDPS, the sufficient conditions for bounded entity latency times are as follows:

- The service times of entities for all classes are bounded.
- The class of interest has a bounded waiting line ($Q\max_n < \infty$).
- The differences among the initial priorities are finite, $-\infty < (p_{i,0} - p_{j,0}) < +\infty$, for all $i, j = 1, 2, \ldots, N$.

Proof:

Consider parts of the proof for Proposition 12.1 where the latency time of the mth entity of Class n is

$$x_m = z_1 + y_1 + \cdots + z_{m-1} + y_{m-1} + z_m + y_m$$
$$= \sum_{i=1}^{m} y_i + \sum_{j=1}^{m} z_j$$

where y represents the service time of entities from Class n, and z represents the service times of entities served in between the service of entities from the Class n. It was established that for x_m to be bounded, y_i, z_i, and m must be bounded. Bounded y_i and z_i implies that the service times of entities of all classes must be bounded. Recall that m is the sum of the number of entities of higher classes served before the entities of interest and the number of entities of Class n, which arrived ahead of the entity of interest. For m to be bounded, the number of Class n entities ahead of m must be finite. Recall that the initial priority level $I_0 = \{p_{1,0}, p_{2,0}, p_{3,0}, \ldots, p_{N,0}\}$ and the increment variable a_i determine how many entities of the higher class are served before an entity of a lower class is served. Therefore, a bounded m implies a finite difference among the initial priority levels $\{p_{1,0}, p_{2,0}, p_{3,0}, \ldots, p_{N,0}\}$ and a positive-valued increment variable a_i. A bounded m implies that the waiting line for Class n must be bounded.

It is important to note that under specific conditions, the limit imposed on the waiting line and the relative initial priorities of the classes may determine whether the distribution function of the latency is bounded. Since the central limit theorem do not distinguish among distribution of the random variables but only the number of terms in the sequence, a very large number m and very disparate initial priority levels of the classes may result in an approximate normal distribution for the latency. Thus, m must be defined such that a bounded distribution is assured.

Let the minimum upper bound of latency times associated with m exist and is equal to $x_{\max}(m)$. The modified distribution of the latency times F' with the imposed limit m as a function of the distribution F without the restriction on the waiting line is

$$F'(\omega) = P(X < \omega \mid X < x_{\max}(m))$$
$$= \frac{P(X < \omega)}{P(X < x_{\max}(m))}$$
$$= \frac{F(\omega)}{F(x_{\max}(m))}$$

where F is the cdf of x.

$x_{\max}(m)$ can be deduced from $x_m = z_1 + y_1 + \cdots + z_{m-1} + y_{m-1} + z_m + y_m$. Therefore,

$$\begin{aligned} x_{\max}(m) &= z_{1(\max)} + y_{1(\max)} + \cdots + z_{m-1(\max)} + y_{m-1(\max)} + z_{m(\max)} + y_{m(\max)} \\ &= \sum_{i=1}^{m} y_{i(\max)} + \sum_{j=1}^{m} z_{j(\max)} \end{aligned}$$

The value for $y_{i(\max)}$ is the minimum upper bound on the service times for entities of the class of interest. The value for $y_{i(\max)}$ is a function of the minimum upper bound on the service times for entities of the classes and the number of terms in the sequence. The number of terms in this sequence is known to be 1 for Class 1 and (C + 1) for Class 2 in a PS with $N = 2$, and $Ip = (p + C, p)$.

Propositions 12.2 and 12.3 imply modification to the PS illustrated in Figure 12.3. The streams of entities that were refused entry into the system and those whose services are truncated can be rerouted to another server giving rise to a multiserver PS.

12.4.3 Combinations of Gumbel Types

Conditions for PS configurations such that bounded or unbounded latency times for classes of entities are assured were developed in Section 12.3. The conditions establish guidelines for the engineering of PS and its configuration. Furthermore, the conditions can eventually affect the domain of attraction of the latency times as shown in Tables 12.1 and 12.2. However, it will be very useful for PS designers if the domain of attraction of the resulting latency times can be drawn from the domains of attraction of the distributions of service times.

Consider a latency time of the mth entity as $x_m = y + z$, where y and z are random variables with probability density functions f_y and f_z and cumulative probability functions F_y and F_z, respectively. Suppose that the domain of attractions of F_y and F_z are known, then it would be interesting for PS administrators to know the domain of attraction of the cumulative probability function of x_m, F_x. The knowledge on the unbounded and the bounded properties y and z can be sufficient to determine whether x is bounded or unbounded. Consider an example if both F_y and F_z have bounded distributions. Then, F_x also has a bounded distribution. Table 12.3 shows the resulting property of F_x for known tails of F_y and F_z.

For cases when the functions f_y and f_z are known, the domain of attraction of f_x can be analytically determined by making use of convolution. The cumulative probability function of x_m is

$$F(x_m) = \iint_{y+z \le x_m} f_{Y,Z}(y,z)\,dy\,dz$$

For unknown functions of f_y and f_z, possible sources of information are the limit theorems of Mucci (1976), also presented by Galambos (1987).

TABLE 12.3

Bounded or Unbounded Tail of the Distribution of Sum of Random Variables[a]

y	z	x_m
Bounded	Bounded	Bounded
Bounded	Unbounded	Unbounded
Unbounded	Unbounded	Unbounded

[a] Resulting property of the tail of the distribution of the sum of random variables with known underlying properties.

12.5 Derived Performance Measures

This section develops various performance measures relevant to the trade-off of configuration in SDPS. Section 12.5.1 develops tolerance levels for the risk of extremes of latencies of classes in PS. Section 12.5.2 discusses the degree of deficit of satisfying the maximum level of risk of extremes of latencies. Section 12.5.3 defines the relative risks between pairs of classes in a PS. Section 12.5.4 develops a measure for the differentiation between pairs of classes of entity. Section 12.5.5 develops the cost functions for delayed and refused entities in a PS.

12.5.1 Tolerance Level for Risk

Section 12.2 established the risk of extremes of latencies for a class in PS with a particular configuration *f*. The risk $R_n^f(C_n)$ is a function of the threshold C_n, which determines incurring unnecessary cost or system failure. The risk $R_n^f(C_n)$ is the probability that an entity of Class n will have a latency beyond the threshold C_n. The risk measure can then be used to gauge the appropriateness of the particular PS configuration *f*. Suppose that there is a prescribed level of risk that a PS administrator is willing to tolerate. That is, there exists a maximum tolerable level on the risk $R_n^f(C_n)$ such that a risk beyond the tolerable level is *unacceptable*. Unacceptable risk warrants reconfiguring the PS.

Consider the example of e-business entities. The *Log-on* entities and other entities of the same class has a $C_n = 8\,\text{seconds}$. The tolerable level for latency times of this class of entity is 1 in 10,000 entities or Pr = 0.0001. A configuration for the e-business server is being proposed for better utilization of the server hardware. However, the proposed configuration will result in longer latency times for the Log-on entities, with $R_n^{\text{new}}(8\,\text{seconds}) = 0.0005$. The new configuration is called an *unacceptable* configuration, since $R_n^{\text{new}}(8\,\text{seconds}) > 0.0001$.

Definition 12.4: Tolerance Level for Risk

For a PS with N classes, the tolerance levels β_n for Class $n = 1,2,\ldots,N$ is a real number between 0 and 1, inclusive, such that an acceptable configuration f is characterized by

$$R_n^f(C_n) \leq \beta \quad \text{for all } n \in N$$

From Definition 12.4, it follows that if $R_n^f(C_n) > \beta_n$ for any $n \in N$, the configuration f is unacceptable.

PROBLEM 12.5 (Application of β_n)

Suppose an SDPS with three classes whose risks $R_n^f(C_n)$ are measured for two configurations, namely $I_0^2 = (2,5,10)$ and $I_0^2 = (4,5,10)$. The risks of extremes of latencies are

$$\begin{array}{ll} R_1^1(C_1) = 3\times10^{-3} & R_1^2(C_1) = 1\times10^{-3} \\ R_2^1(C_2) = 2\times10^{-4} & R_2^2(C_2) = 3\times10^{-4} \\ R_3^1(C_3) = 6\times10^{-6} & R_3^2(C_3) = 9\times10^{-6} \end{array}$$

However, the tolerance levels for the risks of extremes of latencies are

$$\beta_1 = 2\times10^{-3} \quad \beta_2 = 2\times10^{-4} \quad \beta_3 = 2\times10^{-5}$$

Comparing $R_n^1(C_n)$ and $R_n^2(C_n)$ with the tolerance levels β_n:

$$\begin{array}{ll} R_1^1(C_1) > \beta_1 & R_1^2(C_1) < \beta_1 \\ R_2^1(C_2) < \beta_2 & R_2^2(C_2) > \beta_2 \\ R_3^1(C_3) < \beta_3 & R_3^2(C_3) < \beta_3 \end{array}$$

Configuration $f = 1$, SDPS with initial priority levels $I_0^1 = (2,5,10)$, results in a risk of extremes of latencies for Class 1 entities, which is beyond the tolerance level ($R_1^1(C_1) > \beta_1$). By definition, an SDPS with initial priority levels $I_0^1 = (2,5,10)$ is an unacceptable configuration. Reconfiguring the SDPS specifically to affect the infeasibility of the configuration $f = 1$, the initial priority levels are adjusted to $I_0^2 = (4,5,10)$. With the new configuration, $R_1^2(C_1) < \beta_1$. However, $R_2^2(C_2) > \beta_2$, which still results in an infeasible configuration.

Problem 12.5 illustrates the case when a change in PS configuration produces an improvement in the risk of one class but deterioration in the risk of one or more other classes. The result is typical of PS with high utilization and is known as the *law of conservation* (see Kleinrock (1976)).

12.5.2 Degree of Deficit

To facilitate comparison between configurations, the incremental changes in the risks of the classes with respect to the tolerance level can be measured. Particularly important is the case when the risk of extremes of latencies for a class of entities is greater than the tolerance level. Such a case is described as unacceptable and is important in evaluating a PS configuration. The degree of deficit between the measured risk and the tolerance level can be used to further gauge among the effectiveness of PS configurations in addressing the risk of extremes of latencies.

Definition 12.5: Degree of Deficit on Tolerance Level for Risk

For a configuration f of a PS, the degree of deficit of the tolerance levels for risk β_n of Class $n, n = 1, 2, \ldots, N$ is

$$\begin{aligned} G_n^f &= R_n^f(C_n) - \beta_n \quad \text{if } R_n^f(C_n) > \beta_n, \\ &= 0 \quad \text{otherwise} \end{aligned}$$

The degree of deficit G_n^f is a measure of how much more entities of Class n can be late than the tolerable number. Thus, it is desirable to have the lowest degrees of deficit for all classes of entities.

PROBLEM 12.6 (Application of G_n^f)

Consider the resulting risks $R_n^f(C_n)$ in Problem 12.5. The degree of deficits G_n^1 and G_n^2 for the three classes of the SDPS are

$$\begin{aligned} G_1^1 &= 3\times10^{-3} - 2\times10^{-3} = 1\times10^{-3} \\ G_2^1 &= 0 \\ G_3^1 &= 0 \\ G_1^2 &= 0 \\ G_2^2 &= 3\times10^{-4} - 2\times10^{-4} = 1\times10^{-4} \\ G_3^2 &= 0 \end{aligned}$$

A comparison of the two configurations can then be made based on the trade-off between the degree of deficits of Class 1 and Class 2 as shown in Figure 12.4.

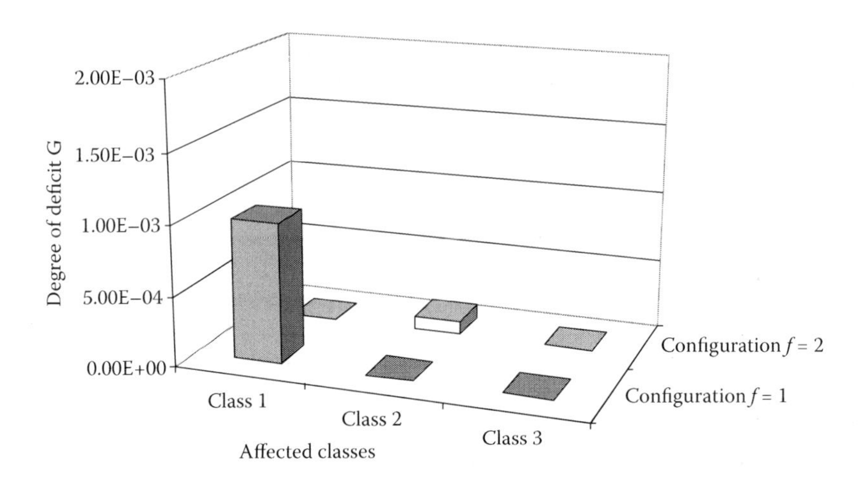

FIGURE 12.4
Deficits on tolerance levels. Trade-off between degree of deficits of Classes 1 and 2 associated with two infeasible configurations, $f = 1$ and $f = 2$.

12.5.3 Relative Risks

Problem 12.6 shows the changes in the risk of extremes of latencies that can result from changes in the configuration of a PS. Another measure of the acceptability of a configuration is the comparison of the risk among classes. There are applications that require some degrees of preference of one class over another to be maintained, as in the case where such preference is being paid for. Consider the application of PS to Internet service providers (ISP). ISP often provides various levels of service to suits of clients (i.e., corporate, small businesses, households). The differentiation among levels of service can be based on guarantees on relative latency times of entities for each client. Clients who want to have shorter latency times often pay a premium for the service. Such differentiation can be measured through *relative risk between classes*. However, the comparison for the risk of extremes of latencies among classes is meaningful only if all the risks are measured based on the same threshold value. That is, a condition has to be imposed on the values of the latency threshold C_n for the risk measure $R_n^f(C_n)$ to be comparable.

Definition 12.6: Relative Risk between Classes

The relative risk of extremes of latencies between Class a and Class b for a threshold value C is

$$H^f(a,b) = R_a^f(C) - R_b^f(C)$$

such that $a,b = 1,2,\ldots,N, a \neq b$.

It is evident that unless $H^f(a,b) = 0$, $H^f(a,b) \neq H^f(b,a)$ and $H^f(a,b) = -H^f(b,a)$. Therefore, for a PS with N number of classes, there are $\binom{N}{2} = N!/(N-2)!\,2!$ meaningful combinations of a and b for $H^f(a,b)$.

PROBLEM 12.7 (Application of $H^f(a,b)$)

Consider Problem 12.5. Suppose that the risk of extremes of latencies $R_n^f(C_n)$ are measured for a single value $C_n = C, n = 1,2,...,N$. The relative risks for configuration $f = 1$ are

$$H^1(1,2) = R_1^1(C_1) - R_2^1(C_2) = 30\times10^{-4} - 2\times10^{-4} = 2.8\times10^{-3}$$
$$H^1(1,3) = R_1^1(C_1) - R_3^1(C_3) = 3000\times10^{-6} - 2\times10^{-6} = 2.994\times10^{-3}$$
$$H^1(2,3) = R_2^1(C_2) - R_3^1(C_3) = 200\times10^{-6} - 6\times10^{-6} = 0.194\times10^{-3}$$

For configuration $f = 2$, the relative risks are

$$H^2(1,2) = R_1^2(C_1) - R_2^2(C_2) = 10\times10^{-4} - 3\times10^{-4} = 1.7\times10^{-3}$$
$$H^2(1,3) = R_1^2(C_1) - R_3^2(C_3) = 3000\times10^{-6} - 9\times10^{-6} = 2.991\times10^{-3}$$
$$H^2(2,3) = R_2^2(C_2) - R_3^2(C_3) = 300\times10^{-6} - 9\times10^{-6} = 0.291\times10^{-3}$$

Figure 12.5 shows the resulting $H^f(a,b)$ values. With the change in the initial priority level of Class 1 from 2 to 4, the differentiation in risk of extremes of latency for Classes 1 and 2 narrowed from 2.8×10^{-3} down to 1.7×10^{-3} while the differentiation between Classes 2 and 3 widened from 0.194×10^{-3}

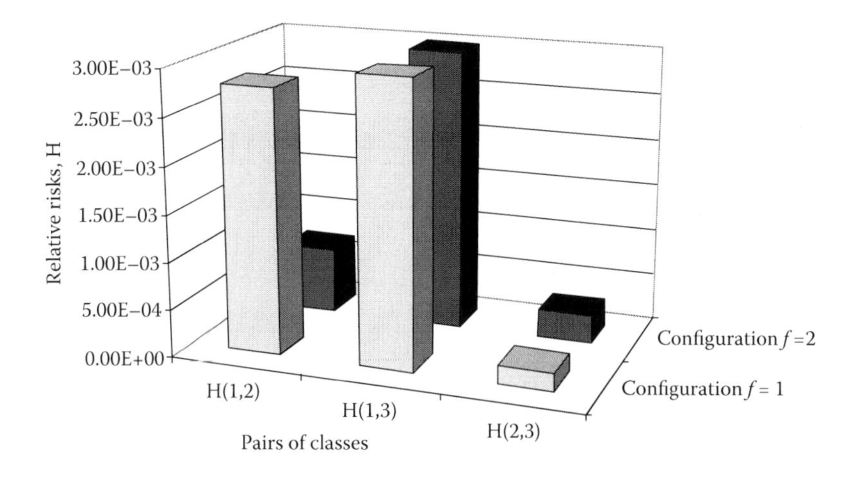

FIGURE 12.5
Relative risks, $H^f(a,b)$, between classes showing the most significant change between Classes 1 and 2.

up to 0.291×10^{-3}. Differentiation between Classes 1 and 3 does not vary much for both configurations.

12.5.4 Differentiation Tolerance Level

This section establishes a risk tolerance level that can describe how a PS configuration satisfies a maximum risk of extremes of latencies $R_n^f(C_n)$. An analogous measure can be developed for the relative risk between two classes $H^f(a,b)$. The acceptability of a configuration f of a PS can be judged by whether the minimum differentiation between two classes is satisfied.

Definition 12.7: Differentiation Tolerance Level

For a PS with N classes, the differentiation tolerance levels $\phi(a,b)$ for Class $n = 1,2,\ldots,N$ is a real number between 0 and 1, inclusive, such that an acceptable configuration f is characterized by

$$H^f(a,b) \geq \phi(a,b), \quad \text{for all } a,\, b \in N$$

Analogous to degree of deficit of the tolerance level for risk established is the degree of deficit on the tolerance level for class differentiation.

Definition 12.8: Degree of Deficit on the Tolerance Level for Class Differentiation

For a configuration f of a PS, the degree of deficit of the tolerance levels for differentiation $\phi(a,b)$ of Class a and Class b is

$$\begin{aligned} J^f(a,b) &= \phi(a,b) - H^f(a,b) && \text{if } \phi(a,b) > H^f(a,b) \\ &= 0 && \text{otherwise} \end{aligned}$$

$J^f(a,b)$ can be used to gauge if configuration f is acceptable. By Definition 12.6, a $J^f(a,b) > 0$ implies that configuration f is unacceptable since the minimum differentiation between Class a and Class b is not satisfied. Thus, it is desirable for a configuration to result in the lowest degrees of deficit for all classes of entities.

PROBLEM 12.8 (Application of $J^f(a,b)$)

Consider Problem 12.7 in which the relative risks are obtained for two configurations. For configuration $f = 1$,

$$H^1(1,2) = 2.8 \times 10^{-3} \quad H^1(1,3) = 2.994 \times 10^{-3} \quad H^1(2,3) = 0.194 \times 10^{-3}$$

For configuration $f = 2$, the relative risks are

$$H^2(1,2) = 1.7 \times 10^{-3} \quad H^2(1,3) = 2.991 \times 10^{-3} \quad H^2(2,3) = 0.291 \times 10^{-3}$$

Suppose that the tolerance levels for differentiation $\phi(a,b)$ are

$$\phi(1,2) = 2\times10^{-3} \quad \phi(1,3) = 2\times10^{-3} \quad \phi(2,3) = 0.2\times10^{-3}$$

The degrees of deficit on tolerance level for class differentiation are

$$J^1(1,2) = 0$$
$$J^1(1,3) = 0$$
$$J^1(2,3) = 6\times10^{-6}$$
$$J^2(1,2) = 1.3\times10^{-3}$$
$$J^2(1,3) = 0$$
$$J^2(2,3) = 0$$

Shifting from configuration $f = 1$ to configuration $f = 2$, the deficit between Classes 1 and 2 is eliminated. However, a deficit between Classes 2 and 3 is created. Figure 12.6 shows magnitude of the deficit values $J^f(a,b)$ for the three pairs of classes.

12.5.5 Cost Functions

There may be application in which a bounded latency is preferred over the unbounded case, such as when there is heavy penalty for exceeding a specified latency threshold. Consider the condition of Propositions 12.1 and 12.2 when some entities are refused entry into the PS, by either dropping or

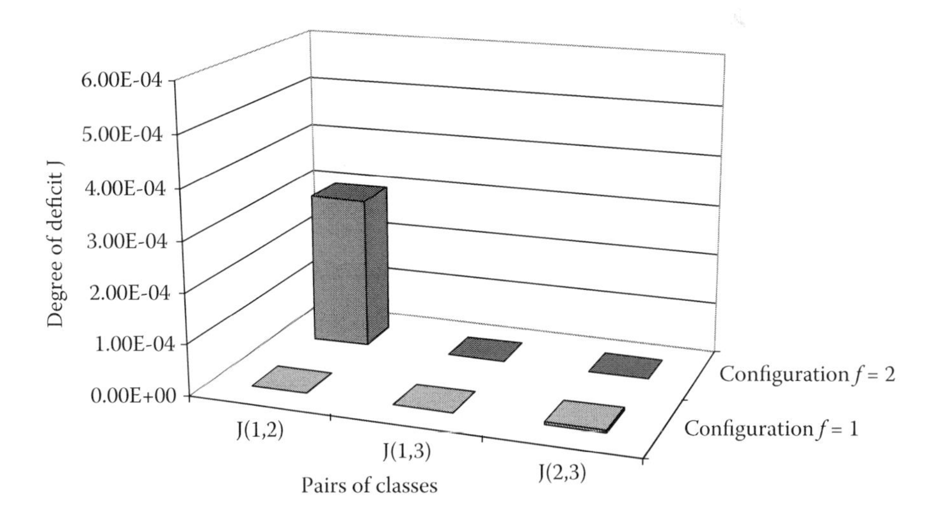

FIGURE 12.6
Deficits on class differentiation: degrees of deficit $J^f(a,b)$ between classes showing the most significant change between pairs of classes (1,2) and (2,3). Gumbel type I plots for distribution of latencies for the two configurations.

rerouting of entities to put a bound on the waiting lines. For a waiting line length with a cumulative probability function F_w and a limit on the waiting line m, the probability that an arriving entity will be refused entry into the PS is $1-F_w(m)$. In most applications, low chances of dropping or rerouting entities and short latency times are both desired. However, consider $L_{\max}(m)$ to be the minimum upper bound of latency times associated with the waiting line limit m. Then, as $m \to \infty$, $(1-F_w(m)) \to 0$ and $L_{\max}(m) \to \infty$. That is, as the limit on the waiting line increases, the frequency of dropped entities decreases, but there is a corresponding increase on the extremes of latency times of entities. The trade-off between the probability entities refused entry into the PS and the upper bound on the latency times is $\partial(1-F(m))/\partial(L_{\max}(m))$, which suggests a trade-off between the cost incurred due to late entities and cost incurred due to refused entities. The cost due to late entity can be attributed to a PS configuration and is a function for the probability that an entity is late.

Definition 12.9: Cost Due to Late Entities

The cost due to a late entity of Class $n, n = 1,2,\ldots,N$, is

$$S_L = \theta_n(x_n),$$

where θ is a function of the latency time of the entity. Since a late entity has latency beyond the threshold,

$$\theta_n(X_n) = 0 \quad \text{if } X_n < C_n, \text{ and}$$

$$\theta_n(X_n) > 0 \quad \text{otherwise}$$

Definition 12.10: Cost Due to Refused Entities

The cost incurred due to a refused entity of Class $n, n = 1,2,\ldots,N$ is

$$S_R = \psi_n V$$

where ψ_n is a cost multiplier, and V is an indicator variable such that
$V = 1$ if the entity is refused, and
$V = 0$ if entity is not refused.

PROBLEM 12.9 (Application of S_L and S_R)

Consider Problem 12.7, an SDPS with three classes of entity, configurations $f = 1$ and $f = 2$. Furthermore, consider bounded waiting line for entities of Class 1. The risks of extremes of latencies are

$$R_1^1(C_1) = 3\times10^{-3} \quad R_1^2(C_1) = 1\times10^{-3}$$
$$R_2^1(C_2) = 2\times10^{-4} \quad R_2^2(C_2) = 3\times10^{-4}$$
$$R_3^1(C_3) = 6\times10^{-6} \quad R_3^2(C_3) = 9\times10^{-6}$$

Suppose that 1,000,000 latency times are measured with the following breakdown among classes: 350,000 entities of Class 1, 350,000 entities of Class 2, and 300,000 entities of Class 3. Furthermore, a number of entities of Class 1 are refused for the two configurations: 1000 entities for $f = 1$ and 5000 entities for $f = 2$. Suppose that the cost functions are given as

$$\theta_1 = \$0.03,\ \theta_2 = \$0.02,\ \theta_3 = \$0.01, \text{ and } \psi_1 = \$0.5$$

Therefore, the costs for each of the two configurations are as follows:

For $f = 1$,

$$\begin{aligned}
(3\times 10^{-3})\times(3.5\times 10^{5})\times \$0.03 &= \$31.50\\
(2\times 10^{-4})\times(3.5\times 10^{5})\times \$0.02 &= \$1.40\\
(6\times 10^{-6})\times(3\times 10^{5})\times \$0.01 &= \underline{\$0.02}\\
S_L &= \$32.92\\
S_R = 1000\times \$0.5 &= \underline{\$500.00}\\
\text{Total Cost} &= \$532.92
\end{aligned}$$

For $f = 2$,

$$\begin{aligned}
(1\times 10^{-3})\times(3.5\times 10^{5})\times \$0.03 &= \$10.50\\
(3\times 10^{-4})\times(3.5\times 10^{5})\times \$0.02 &= \$\ 2.10\\
(9\times 10^{-6})\times(3\times 10^{5})\times \$0.01 &= \underline{\$\ 0.03}\\
S_L &= \$12.63\\
S_R = 5000\times \$0.5 &= \underline{\$2500.00}\\
\text{Total Cost} &= \$2512.63
\end{aligned}$$

Configuration $f = 1$ has a higher S_L than configuration $f = 2$, but the S_R for configuration $f = 2$ is higher than that for configuration $f = 1$, resulting in a higher total cost.

The cost incurred on refused entities is typically higher than for late entities of the same class, that is, $\psi_1 > \theta_1$. A refused entity can represent those that are forever lost resulting in lost revenue or rerouted to other servers requiring additional processing or server cost.

12.6 Optimization of PS

This section develops and formulates the minimization of costs due to late and refused entities subject to restriction on degrees of deficits. This section also shows the evaluation of the impact of the choice of Gumbel type distribution in representing the tails of the distributions of the latency times of the classes of entity in a PS. Furthermore, this section investigates both optimistic and pessimistic approaches to configuring a PS.

12.6.1 Cost Function Minimization

The basic optimization framework for a PS is to minimize costs subject to various constraints, where the decision variables are the configuration parameters of the PS. The choice among several configurations can be based on a chosen primary performance measure such as latency or length of waiting line, degrees of deficit G_n^f and $J^f(a,b)$, or derived measures such as penalties and cost S_L and S_R. An optimization framework for a PS with N classes of entity can be as follows:

$$\begin{aligned} &\text{Minimize} \quad Z(S_L, S_R; f) \\ &\text{subject to} \\ &\qquad G_n^f \le 0 \\ &\qquad J^f(a,b) \ge 0 \\ &\text{for all } n,\ a,\ b = 1,2,\ldots,N \end{aligned} \tag{12.12}$$

The total cost Z in Equation 12.12 is a function of the cost due to late entities S_L and the cost due to refused entity S_R, both of which are functions of the PS configuration variable f. The constraints are the required tolerance level for risk measured by G_n^f and the required level for class differentiation measured by $J^f(a,b)$ for all the classes of entities.

12.6.2 Bounds on Waiting Line

In general, a derived measure Y(i.e.,G_n^f and $J^f(a,b)$) can be obtained from a primary performance measure x (i.e., latency times) through the relation $Y = g(x)$ and for a known distribution function of the latency time x:

$$P(Y < y) = P[X < g^{-1}(y)] \text{ and}$$

$$F_Y(y) = F_X[g^{-1}(y)] \tag{12.13}$$

In Sections 12.2 through 12.5, it has been shown that the effectiveness of a configuration to control latencies below a threshold and the associated cost can be used to determine which among several configurations is adapted. Furthermore, it was also shown that to bound the latencies for classes of entities, the length of waiting lines have to be bounded. Putting bounds on waiting lines may result in costs associated with rerouting of entities to other servers or dropping of entities resulting in lost entities, both cases referred to as *refused entities*. As such, there is a trade-off between the cost of entity latencies exceeding thresholds and entities being rerouted or lost. Aside from the primary performance measure x, latency times previously discussed, consider another primary measure Q, number of entities waiting for service at a given time, and their respective derived cost measure $Y = g(X)$, and $Z = h(Q)$. Specifically,

$$Z = \begin{cases} h(Q) > 0 & Q > Q' \\ 0 & \text{otherwise} \end{cases}$$

Q' is the bound imposed on entity waiting line.

Then for given distributions F_X and $F_{Q'}$, Equation 12.13 becomes

$$P(Y < y) = P[X < g^{-1}(y)] \text{ and } F_Y(y) = F_X[g^{-1}(y)] \tag{12.14}$$

$$\begin{aligned} P(Z < z) &= P[Q < h^{-1}(z) \mid Q > Q'] \\ &= \frac{P[Q' < Q < h^{-1}(z)]}{P(Q > Q')} \\ &= \frac{P[Q < h^{-1}(z)] - P(Q < Q')}{1 - P(Q < Q')} \end{aligned}$$

$$F_Z(z;Q') = \frac{F_Q[h^{-1}(z)] - F_Q(Q')}{1 - F_Q(Q')} \tag{12.15}$$

Since the analysis covers all the classes of the SDPS, Equations 12.14 and 12.15 translate to the following equations with consideration to the classes, the configurations, and the hypothesized domain of attraction:

$$F_Y^{f,i,n}(y) = F_X^{f,i,n}[g^{-1}(y)] \tag{12.16}$$

$$F_Z^{f,i,n}(z;Q') = \frac{F_Q^{f,i,n}[h^{-1}(z)] - F_Q^{f,i,n}(Q')}{1 - F_Q^{f,i,n}(Q')} \tag{12.17}$$

where $f = \{1,2,\ldots\}$ are the configurations under consideration, $i \in \{\mathrm{I}, \mathrm{II}, \mathrm{III}\}$ is the hypothesized domain of attraction of F_X and F_Z, and $n \in \{1,2,\ldots,N\}$ are the classes of entities in the SDPS.

Assuming the density functions associated with Equations 12.16 and 12.17 exist, the problem of minimizing the total cost of refused and delayed entities of Equation 12.12 can be reformulated as

$$\text{Minimize total cost} = E[Y] + E[Z \mid Q > Q'] = \int y f(y) dy + \int z f(z;Q') dz$$

Subject to:

Derived measure constraints: $P(y > \Upsilon') \cong 0, \quad P(z > Z') \cong 0$

Primary measure constraints: $P(x > X') \cong 0,$

The decision variables are the bounds Q' and configuration f.

12.6.2.1 Impact of the Gumbel Type

From the previous sections, it was shown that the tail of the distribution function F_X can be approximated by an equivalent tail and can be described as belonging to the Gumbel type I, II, or III domain of attraction. The hypothesized domain of attraction of the distribution F_X may also affect the value obtained in Equation 12.12. The further one goes out to the tail ($F_Y^i(y) \to 1$ or $F_Y^i(y) \to 0$), the more significant the effect of the choice of domain of attraction. Therefore, Equation 12.12 is actually

$$F_Y^{f,i}(y) = F_L^{f,i}[g^{-1}(y)]$$

where $f = \{1, 2, \ldots\}$ are the configurations under consideration and $i \in \{\mathrm{I}, \mathrm{II}, \mathrm{III}\}$ is the hypothesized domain of attraction of F_X. One way to choose among the configuration is to consider the optimum Y for a given probability of nonexceedance NE, that is,

$$\underset{f}{\text{Min or Max}}\{Y_{f,i}\}$$

$$\text{such that } F_Y^{f,i-1}(\text{NE}) = Y_{f,i} \tag{12.17}$$

$Y_{f,i}$ is the value with a nonexceedance probability NE derived from X resulting from configuration f obtained by hypothesizing that the domain of attraction of F_X is Gumbel type i.

12.6.3 Pessimistic and Optimistic Decisions in Extremes

To minimize or to maximize in Equation 12.17 depends on the nature of the derived measure $Y_{i,f}$, whether it is a cost or penalty to be minimized or a revenue to be maximized. One can take a pessimistic approach and minimize the maximum penalties and maximize the minimum revenues, or an optimistic approach and minimize the minimum penalties and maximize the maximum revenues. The problem of choosing among configuration f can be formulated as follows:

For penalty $Y_{i,f}$,

$$\text{Pessimistic: } f = idx \min_{f}\{\max_{i}\{Y_{i,j}\}\} \tag{12.18}$$

$$\text{Optimistic: } f = idx \min_{f}\{\min_{i}\{\Upsilon_{i,j}\}\} \tag{12.19}$$

For revenue $Y_{i\ f}$,

$$\text{Pessimistic: } f = idx \max_f \{\min_i \{Y_{i,j}\}\} \tag{12.20}$$

$$\text{Optimistic: } f = idx \max_f \{\max_i \{Y_{i,j}\}\} \tag{12.21}$$

$i \in \{\text{I}, \text{II}, \text{III}\}$, $f \in \{1, 2, \ldots\}$.

The operator *idx* max/min{} is the search for the index of the maximum or the minimum elements(s) in the set in {}, that is, $z = idx \max/\min\{x_1, x_2, \ldots\}$ if $x_2 = \max/\min\{x_1, x_2, \ldots\}$.

PROBLEM 12.10 (Impact of Gumbel Type)

Consider the Class 2 entity latencies of the e-business example for two SDPS configurations $I_p^1 = (30,15,10)$ and $I_p^2 = (30,25,10)$. The Gumbel type-I distribution parameters are estimated using the method of moments, and the Gumbel types II and III parameters are estimated using the following relations for the two-parameter Gumbel types II and III (adapted from Castillo, 1988):

$$x_I = \log(x_{II} - \lambda_{II}), \quad \lambda_I = \log(\delta_{II}), \quad \delta_I = \frac{1}{\beta_{II}}$$

$$x_I = -\log(\lambda_{III} - x_{III}), \quad \lambda_I = -\log(\delta_{III}), \quad \delta_I = \frac{1}{\beta_{III}}$$

The subscripts II and III pertain to the Gumbel types. The parameters are shown in Table 12.5. The corresponding Gumbel type I plots are shown in Figure 12.7.

Consider a measure of damage given by $Y = 100L$ in \$ units of cost and a chosen probability of nonexceedance NE = 0.999. Then, $F_y(Y) = F_X[g^{-1}(Y)]$ for the two configurations are shown in Figure 12.8.

TABLE 12.4

Domains of Attraction of Latency Times

f_x	f_y	$f_{I,m}$
G-I	G-I	G-I, G-II, or G-III
G-I	G-II	G-I or G-II
G-I	G-III	G-I or G-II
G-II	G-II	G-I or G-II
G-II	G-III	G-I or G-II
G-III	G-III	G-III

TABLE 12.5

Estimated Parameters for Problem 12.9[a]

Configuration (f)	Gumbel Type I		Gumbel Type II			Gumbel Type III		
	λ_{I}	$_{I}$	λ_{II}	δ_{II}	β_{II}	λ_{III}	δ_{III}	β_{III}
1	563	165	0	552	4.42	5000	4436	25.5
2	463	117	0	457	5.05	5000	4536	37.2

[a] Estimated parameters for the Gumbel types I, II, and III domains of attraction for the maxima of latencies for two configurations of SDPS applied to e-commerce entities.

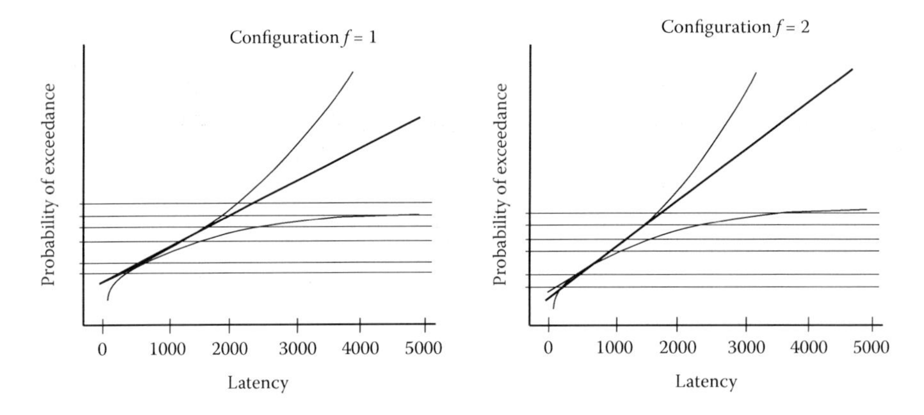

FIGURE 12.7
Gumbel type I plots for distribution of latencies for the two configurations.

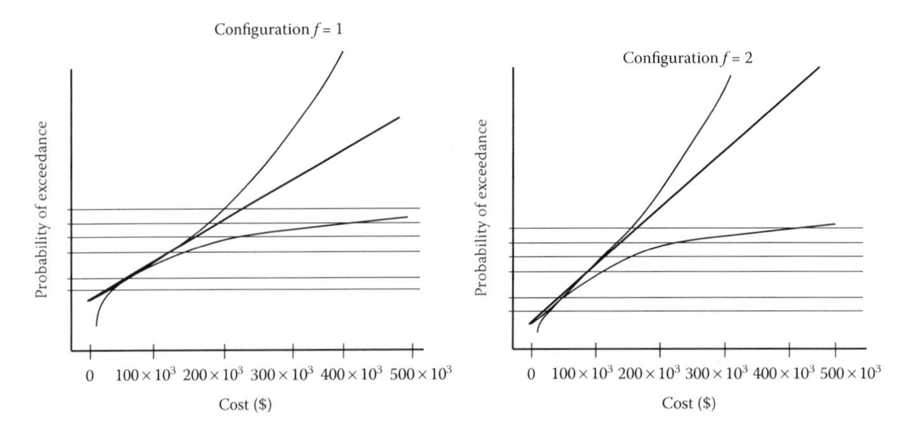

FIGURE 12.8
Gumbel type I plots for distribution of $ costs associated with the latencies of the two configurations.

The resulting values for NE = 0.999 are

$$Y_{1,\mathrm{I}} = \$172,000$$

$$Y_{1,\mathrm{II}} = \$264,000$$

$$Y_{1,\mathrm{III}} = \$162,000$$

$$Y_{2,\mathrm{I}} = \$128,000$$

$$Y_{2,\mathrm{II}} = \$180,000$$

$$Y_{2,\mathrm{III}} = \$124,000$$

Applying Equation 12.18 results in choice of $f = 2$ as illustrated in Table 12.6. First, the maximum cost for configurations 1 and 2 are identified (i.e., \$264 for $f = 1$ and \$180 for $f = 2$). Second, the minimum among the maximum cost is selected (i.e., \$180 for $f = 2$). Therefore, configuration 2 and the choice of the Gumbel type II are the optimal decisions based on a pessimistic cost analysis.

Another way to choose the appropriate configuration is to compare the probability of nonexceedance for a given value of $Y_{f,i}$. Then Equation 12.17 can be modified to

$$\underset{f}{\mathrm{Min}} / \mathrm{Max}\{P^{f,i}\}$$

$$\text{such that } F_Y^{f,i}(Y) = P^{f,i}.$$

$P^{f,i}$ is the nonexceedance probability of Y derived from L resulting from configuration f obtained by hypothesizing that the domain of attraction of F_L is Gumbel type i.

PROBLEM 12.11 (Comparison Based on Cost $Y_{f,i}$)

As an example, consider a Y = \$ 150,000. The corresponding $P^{f,i}$ for the two configurations with different hypothesized domain of attraction for the maxima of the latency x are

TABLE 12.6

Total Cost for Problem 12.9[a]

	Gumbel Type of F_X for Maxima		
Configuration	**I**	**II**	**III**
1	172,000	264,000	162,000
2	128,000	180,000	124,000

[a] Cost in \$ associated with NE = 0.999 and the result of applying $f = idx\ \min_f\{\max_i\{Y_{i,j}\}\}$ with $f = 2$ and G-II as the optimal solution (bold).

TABLE 12.7

Nonexceedance Probabilities of Costs[a]

	Gumbel Type of F_X for Maxima		
Configuration	**I**	**II**	**III**
1	9.966×10^{1}	9.880×10^{-1}	9.977×10^{-1}
2	9.999×10^{-1}	9.975×10^{-1}	9.999×10^{-1}

[a] Nonexceedance probabilities associated with cost = \$ 150,000 and the result of applying $f = idx \min_f\{\max_i\{Y_{i,j}\}\}$.

$$NE^1(G\text{–}I) = 9.966 \times 10^{-1}$$
$$NE^1(G\text{–}II) = 9.880 \times 10^{-1}$$
$$NE^1(G\text{–}III) = 9.977 \times 10^{-1}$$
$$NE^2(G\text{–}I) = 9.999 \times 10^{-1}$$
$$NE^2(G\text{–}II) = 9.975 \times 10^{-1}$$
$$NE^3(G\text{–}III) = 9.999 \times 10^{-1}$$

Since higher probability of nonexceedance for a cost measure is better, the problem is similar to choosing among measures of benefits, thus Equation 12.20, $f = idx \min_f\{\max_i\{Y_{i,j}\}\}$ can be used in choosing the configuration. Table 12.7 shows the results and highlights the chosen configuration, $f = 1$.

12.7 Summary

This chapter established sufficient conditions for unbounded latency times of entities in a saturated PS. It was established that regardless of the underlying distributions of the service times and arrival rates of entities, the latency times are unbounded for a saturated PS. A PS that is not saturated, the combinations of unbounded and bounded interarrival and service times, the domain of attraction of the waiting times has not been determined. These conditions provide insights in configuring PS wherein certain classes of entity do not have guaranteed latency times and form the foundation for the establishment of conditions for bounded latencies.

This chapter also established sufficient conditions for bounded latency times. Proposition 12.1 states that for a class of entities in an SPS to have bounded latency times, the waiting lines and service times of all entities have to be bounded. Difficulties on implementing the conditions arise due to infinitely many potential entities. However, an advantage presented by configuring a SDPS can result in bounded waiting lines even if there are infinitely many potential entities (Proposition 12.2). The establishment of the conditions provides a foundation in configuring SDPS wherein certain classes of entity need to have guaranteed latency times.

Furthermore, this chapter developed various performance measures relevant to the trade-off of configuration in SDPS for various field of applications. The performance measures are based on the risk of extremes of latencies for the classes of entities in PS, $R_n^f(C_n)$. These measures are tolerance levels for the risk of extremes of latencies of classes β_n, degree of deficit of satisfying the tolerance level of risk of extremes of latencies $G_n^f = R_n^f(C_n) - \beta_n$, relative risks between pairs of classes $H^f(a,b)$, differentiation $R_a^f(C) - R_b^f(C)$, limit on the differentiation between pairs of classes $J^f(a,b)$, cost functions for late entity S_L, and cost for refused entities S_R. Examples are provided to illustrate the use of the performance measures.

Finally, this chapter has presented the optimization of PS through configuration. The objective of choosing among configuration of the PS is formulated as minimization of total cost due to both late and refused entities. The constraints of the formulation are the restrictions on degrees of deficits developed in this Chapter. Moreover, the impact of the Gumbel type distribution as the hypothesized domain of attraction can also be examined in terms of the optimistic and pessimistic approaches. Since the tail of most continuous distributions can be approximated by one of the three Gumbel type distributions, an envelope of approximates can be created. The envelope then describes the *best* and *worst* estimates.

Questions and Exercises

1. Define the following terms in a narrative manner, that is, not using equations, and describe their relevance in the context of PS.
   (A) Nonexceedance probability
   (B) Latency threshold based on reliability C_n
   (C) Latency threshold based on cost function C_n
   (D) Risk of a Class
   (E) Tolerance level for risk
   (F) Degree of deficit on tolerance level for risk
   (G) Relative risk between classes
   (H) Differentiation tolerance level
   (I) Degree of deficit on tolerance level for class differentiation
   (J) Cost due to late entities
   (K) Cost due to refused entities

2. What can be the reasons for measuring risk of extremes of latency times for each of the classes in a PS?
3. Consider a dentist's office where patients wait for service of a dental doctor and a dental hygienist. Describe several scenarios and conditions under which patients may have extremely long waiting times.
4. A data-processing system used by a credit card company is considered to be reliable if it can process any request for a data within 5 seconds after a request has been placed (e.g., a request to verify a card holder's credit limit). Describe how this system reliability can be related to the latency threshold based on reliability as described in Definition 12.2a.
5. Costumers of fast-food restaurants lining up at the drive through servers are known to leave the line if made to wait for more than a certain time based on the time of the day and presence of other nearby fast-food restaurants. Describe how this waiting time threshold for costumers can be related to the latency threshold based on cost function as described in Definition 12.2b.
6. Consider the PS described in Problem 12.1. However, closer analysis of incoming entities suggests that any incoming entity has equal chances of being any of the three classes. What are the risks of extreme latencies for each of the three classes?
7. Consider a data-processing system used by a credit card company where length of times needed to accomplish a request for data are mutually independent and have common distribution.
   (A) Based on this information, are the times requests have to wait bounded or unbounded? Please support your answer.
   (B) What restrictions have to be placed on this data-processing system to change the times requests have to wait from bounded (or unbounded) to unbounded (or bounded)? Please support your answer.
8. Consider a dentist's office where patients wait for service of a dental hygienist. To improve the overall satisfaction of patients, the office manager decided to make sure no patient waits longer than a specified length of time even during the busiest periods of the day. Currently, there are two categories of patients, those that have emergency situations, for example, in pain, and nonemergency situations, for example, periodic check-up. Describe how the following sufficiency conditions for bounded latency times in a saturated SDPS (Proposition 12.3) can be implemented in this situation (the first of these conditions has already been described as an aid).
   (A) Sufficient condition: The service times of entities for all classes are bounded. Implementation: regardless of a patient's reasons

for being in the dental office or the procedures that needs to be performed, patients cannot be attended longer than a certain length of time.

(B) Sufficient condition: The class of interest has a bounded waiting line ($Q\max_n < \infty$)

(C) Sufficient condition: The differences among the initial priorities are finite, $-\infty < (pi,o - pj,o) < +\infty$, for all $i, j = 1, 2, \ldots, N$.

9. Consider Problem 12.5 (Application of β_n). Instead of what was initially provided, suppose that the tolerance levels for the risks of extremes of latencies are

$$\beta_1 = 7 \times 10^{-3} \quad \beta_2 = 2 \times 10^{-5} \quad \beta_3 = 5 \times 10^{-7}$$

Compare the two configurations based on the risks of each of the Classes.

Appendix

Bernoulli Utility and the St. Petersburg Paradox

As discussed in Chapter 2, the literature on risk and decision theory is deeply rooted in mathematics and economics. The study of risk is the study of chance and choice. Probability theory is the formalism to study chance and decision theory is the formalism to study choice. Their union provides the formalism to study risk.

The importance of combining the study of chance and the study of choice was recognized by Swiss mathematician Daniel Bernoulli (1700–1782) in his 1738 essay "Exposition of a New Theory on the Measurement of Risk." In that essay, Bernoulli recognized that taking a risk is a choice to gamble on an event whose outcome is uncertain. However, a favorable or unfavorable outcome is a personal determination—one influenced by a person's view of value or worth. Bernoulli reasoned a person's wealth position influences their choice to engage in a risky prospect. His integration of chance, choice, and wealth into a mathematical theory of risk ultimately became the foundations of economic science. Bernoulli's motivation for this theory was his solution to a famous problem that became known as the St. Petersburg paradox.

A.1.1 The St. Petersburg Paradox*

The St. Petersburg paradox began as the last of five problems posed by Daniel Bernoulli's cousin Nicolas Bernoulli (1687–1759) to Pierre Raymond de Montmort (1678–1719). Montmort was a French mathematician famous for his book *Essay d'analyse sur les jeux de hazard* (1708). This work was a treatise on probability and games of chance. A second edition was published in 1713. This edition included discussions on all five problems; the last problem become known as the St. Petersburg paradox.

* The St. Petersburg paradox got its name from Daniel Bernoulli's publication of his essay in *Papers of the Imperial Academy of Sciences in Petersburg*, Vol. 5, 175–192 (1738), an academy where he held an appointment from 1725–1733. Daniel Bernoulli returned to his academic roots at the University of Basel in 1733 but remained an honorary member of the St. Petersburg Academy of Sciences, where he continued to publish many of his scientific works.

In his 1738 essay, Daniel Bernoulli's statement of his cousin's the fifth problem was as follows:

> Peter tosses a coin and continues to do so until it should land 'heads' when it comes to the ground. He agrees to give Paul one ducat if he gets 'heads' on the very first throw, two ducats if he gets it on the second, four if on the third, eight if on the fourth, and so on, so that with each additional throw the number of ducats he must pay is doubled. Suppose we seek to determine the value of Paul's expectation.

To address this question, let L_X be a lottery with possible outcomes $\{x_1, x_2, x_3, \ldots\} = \{1, 2, 4, 8, \ldots\}$, with x_n denoting the amount of ducats Paul wins from Peter on the nth coin toss. The probability that heads first appears after n tosses is $p_n = (1/2)^n$. Figure A.1 is the probability function for L_X. From Definition 3.8, the expected value of L_X is

$$E(L_X) \equiv E(X) = \frac{1}{2}(1) + \frac{1}{4}(2) + \frac{1}{8}(4) + \frac{1}{16}(8) + \ldots = \frac{1}{2} + \frac{1}{2} + \frac{1}{2} + \frac{1}{2} + \ldots = \infty \quad \text{(A.1)}$$

From Equation A.1, Paul should expect an infinite gain from this lottery. If the expected value is used as the decision rule* to enter this game, then Paul should be willing to pay any finite entry fee to play with the expectation

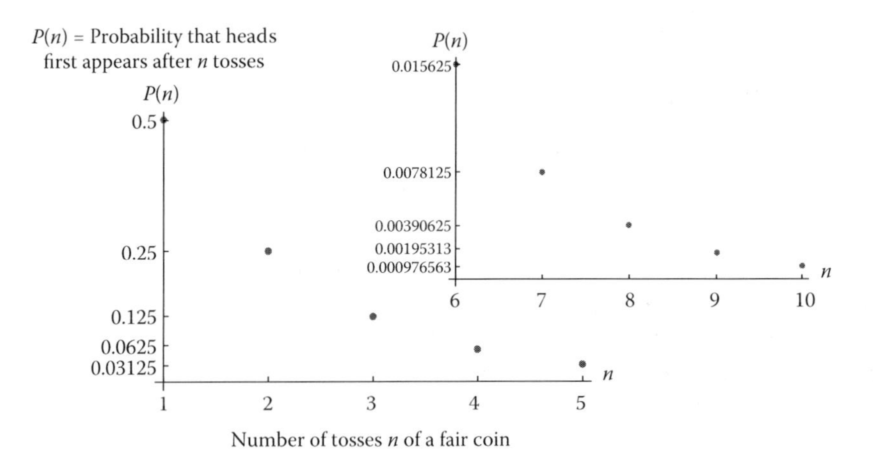

FIGURE A.1
St. Petersburg game probability function.

* Seventeenth century Dutch mathematician Christiaan Huygens (1629–1695) introduced the concept of expected value as a decision rule for whether to engage in games with risky prospects; however, in some games of chance this rule violates common sense as illustrated by the St. Petersburg paradox.

of an infinite monetary gain. However, given the probability function for this lottery (Figure A.1) the chances of realizing increasingly higher payoffs becomes rapidly improbable with each toss after the first toss of the coin. Hence, most people would be willing to pay only a small entry fee to play this game despite the expected infinite gain. This is the paradox posed in Nicolas Bernoulli's problem five.

To address this paradox, Daniel Bernoulli (1738) recognized that

> the determination of the value of an item must not be based on its price, but rather on the utility it yields. The price of the item is dependent only on the thing itself and is equal for everyone; the utility, however, is dependent on the particular circumstances of the person making the estimate. Thus there is no doubt that a gain of one thousand ducats is more significant to a pauper than to a rich man though both gain the same amount.

Bernoulli (1738) also observed that

> no valid measurement of the value of a risk can be obtained without consideration being given to its utility, that is to say, the utility of whatever gain accrues to the individual or, conversely, how much profit is required to yield a given utility. However it hardly seems plausible to make any precise generalizations since the utility of an item may change with circumstances... a poor man generally obtains more utility than does a rich man from an equal gain.

Bernoulli (1738) further recognized that

> any increase in wealth, no matter how insignificant, will always result in an increase in utility which is inversely proportionate to the quantity of goods already possessed.

With this proposition, Daniel Bernoulli identified an economic law that would become known as *diminishing marginal utility*. The stage was set for joining chance, choice, and wealth into a mathematical theory of economic decisions by individuals to engage in prospects with uncertain outcomes.

A.1.2 Use Expected Utility, Not Expected Value

With the European age of enlightment well underway, Swiss mathematician Daniel Bernoulli (1700–1782) introduced an alternative to the expected value as a way to solve the St. Petersburg paradox and other games of chance. In his 1738 essay, Daniel Bernoulli (1738) proposed using expected utility,

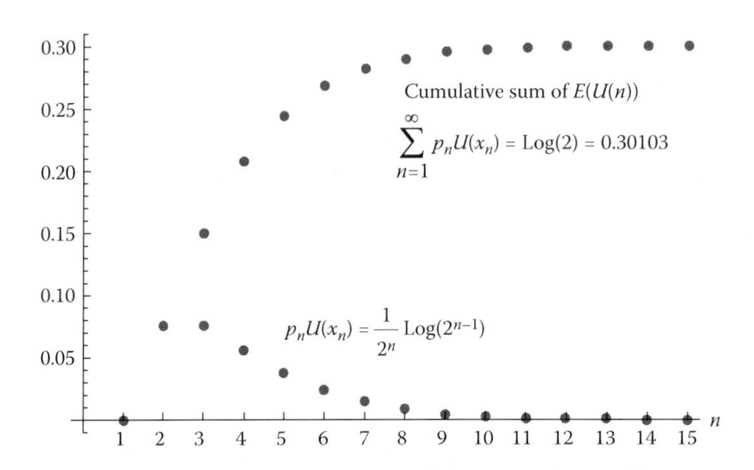

FIGURE A.2
St. Petersburg game with log utility.

not expected value, as the decision rule to engage prospects with uncertain outcomes:

> Meanwhile, let us use this as a fundamental rule: If the utility of each possible profit expectation is multiplied by the number of ways in which it can occur, and we then divide the sum of these products by the total number of possible cases, a mean utility [moral expectation] will be obtained, and the profit which corresponds to this utility will equal the value of the risk in question [Bernoulli, 1738].

Bernoulli chose a *log utility function* to demonstrate a resolution of the St. Petersburg paradox. Consider the log utility function $U(x_i) = \text{Log}(x_i)$. This function generates a cardinal utility associated to each outcome x_i from the St. Petersburg lottery L_X, where

$$L_X = (x_1, p_1; x_2, p_2; x_3, p_3; x_4, p_4; \ldots) = (1, \tfrac{1}{2}; 2, \tfrac{1}{2^2}; 4, \tfrac{1}{2^3}; 8, \tfrac{1}{2^4}; \ldots)$$

From Chapter 3, Definition 3.7, the *expected utility* of lottery L_x is

$$\begin{aligned} E(U(L_X)) &= E(U(X)) = p_1 U(x_1) + p_2 U(x_2) + p_3 U(x_3) + p_4 U(x_4) + \cdots \\ &= \tfrac{1}{2}\text{Log}(1) + \tfrac{1}{2^2}\text{Log}(2) + \tfrac{1}{2^3}\text{Log}(4) + \tfrac{1}{2^4}\text{Log}(8) + \cdots \\ &= \sum_{n=1}^{\infty} \tfrac{1}{2^n}\text{Log}(2^{n-1}) = \text{Log}(2) = 0.30103 \end{aligned}$$

Seen in Figure A.2 the above infinite series converges to Log(2). Thus, Paul obtains *a finite gain* from this lottery if the expected utility is the decision rule to enter this game instead of the expected value. With this rule, Paul should not pay an entry fee of more than $10^{\text{Log}(2)} = 2$ ducats to play *this version* of the St. Petersburg lottery.

Bernoulli's use of log utility not only solved the St. Petersburg paradox but the solution presented in his 1738 essay is considered the birth of mathematical economics and a formal theory of risk. In this work, Bernoulli introduced a mathematical theory of risk aversion and the law of diminishing marginal utility – which assumes persons have decreasing risk aversion with increasing wealth. Bernoulli chose the function $1/W$ to represent this observed human trait, where W is a person's wealth position.

Bernoulli's notion of expected utility (instead of expected value) as a rule for assessing the merits of engaging in risky prospects was innovative. It was the first time personal measures of worth were directly captured into a risk calculus. Despite Bernoulli's clever solution to this paradox, 200 years would past before mathematicians John von Neumann and Oskar Morgenstern (1944) extended these ideas into a formal theory of rational decision making under risk.

Questions and Exercises

1. In the St. Petersburg game, compute the probability of a payoff equal to $1,048,576. How many coin flips are needed to receive this payoff?
2. Solve the St. Petersburg paradox if the Bernoulli utility function is $U(W) = \sqrt{W}$ by showing the expected utility of the game converges to the finite value of $2.91.
3. Create a computer simulation of the St. Petersburg game and simulate 5000 game plays. Suppose there is an entry fee of $10 per game play. From your simulation results, answer the following:
   (A) What payoff occurred most frequently? Develop a frequency distribution of the payoffs from your simulated game and derive their occurrence probabilities.
   (B) From your simulated game, what is the occurrence probability of the maximum payoff? How many trials were needed for the maximum payoff to first appear? Given the entry fee, what was your net earnings if you kept playing the game until the maximum payoff was achieved?
   (C) Figure A.3 presents the results from a computer simulation of the St. Petersburg game played 1,000,000 times, with an entry fee of $10 per game play. The maximum payoff was $524,288. It took game play 718,550 for this payoff to occur, which meant

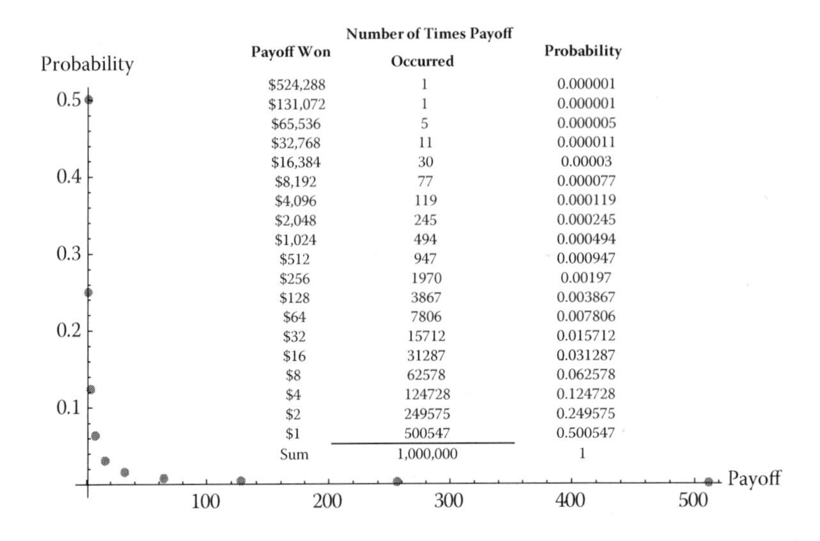

Payoff Won	Number of Times Payoff Occurred	Probability
$524,288	1	0.000001
$131,072	1	0.000001
$65,536	5	0.000005
$32,768	11	0.000011
$16,384	30	0.00003
$8,192	77	0.000077
$4,096	119	0.000119
$2,048	245	0.000245
$1,024	494	0.000494
$512	947	0.000947
$256	1970	0.00197
$128	3867	0.003867
$64	7806	0.007806
$32	15712	0.015712
$16	31287	0.031287
$8	62578	0.062578
$4	124728	0.124728
$2	249575	0.249575
$1	500547	0.500547
Sum	1,000,000	1

FIGURE A.3
St. Petersburg game simulation results: 1,000,000 game plays.

tails appeared on 19 consective coin tosses before a head first appeared on toss number 20 (with payoff = 2^{19}). If, on this game play, a head first appeared on toss number 21, then the player would have won $1,048,576. However, as shown in Figure A.3, even in a million plays of this game the million dollar payoff never occurred! Expecting such a payoff is an extremely unlikely event!

How do your simulation results in Part (B) compare to these findings?

References

Agrawal, R., Cruz, R., Okino, C., Rajan, R., 1999. "Performance Bounds for Flow Control Protocols," *IEEE/ACM Transactions on Networking*, 7(3), 310–323.

Alali, B., Pinto, C. A., 2009. "Project, Systems, and Risk Management Processes and Interaction," *Proceedings of the PICMET'09 Portland International Center for Management of Engineering and Technology*, Portland, Oregon, August 2–6.

Allen, T., Nightingale, D., Murman, E., March 2004. "Engineering Systems an Enterprise Perspective," An Engineering Systems Monograph, Engineering Systems Division, The Massachusetts Institute of Technology.

Andrews, M., 2000. "Probabilistic End-to-End Bounds for Earliest Deadline First Scheduling," *Proceedings of IEEE Infocom 2000*, Tel Aviv, 603–612.

Ang, A. H., Tang, W.H., 1984. Probability Concepts in Engineering Planning and Design: Vol. II: Decision, Risk, and Reliability, John Wiley & Sons, Inc., New York.

Arrow, K. J., 1965. "Aspects of the Theory of Risk Bearing," Yrjo Jahnsson Lectures, Helsinki, Finland: Yrjo Jahnssonin Saatio.

Ashok, E., Narayan, O., Willinger, W., 1997. "Fractal Queueing Models," *Frontiers in Queueing*, CRC Press, Inc., Boca Raton, FL, pp. 245–269.

Ayyub, B. M., 2001. *Elicitation of Expert Opinions for Uncertainty and Risks*, Chapman-Hall/CRC-Press, Boca Raton, FL.

Ayyub, B. M., McGill, W. L., Kaminsky, M., 2007. "Critical Asset and Portfolio Risk Analysis: An All-Hazards Framework," *Risk Analysis*, 27(4), 789–801.

Berger, A. W., Whitt, W., 2000. "Workload Bounds in Fluid Models with Priorities," *Performance Evaluation*, 41, 249–267.

Bernoulli, D., 1738. "Exposition of a New Theory on the Measurement of Risk," *Econometrica*, 22(1) (January 1954), 23–36 Virginia, 22060–5565, The Econometric Society, http://www.jstor.org/stable/1909829.

Bhat, U. N., Basawa, I. V. (eds.), 1992. *Queueing and Related Models*, Oxford University Press, Oxford.

Bierlant, J., Teugels, J. L., Vynckier, P., 1996. *Practical Analysis of Extreme Values*, Leuven University Press, Leuven, Belgium, 150.

Blanchard, B. S., Fabrycky W. J., 1990. *Systems Engineering and Analysis*, 2nd ed., Prentice-Hall, Englewood Cliffs, NJ.

Bolch, G., Greiner, S., de Meer, H., Trivedi, K. S., 1998. *Queueing Networks and Markov Chains Modeling and Performance Evaluation with Computer Science Applications*, John Wiley, New York.

Browning, T. R., Deyst, J. J., Eppinger, S. D., 2002. "Adding Value in Product Development by Creating Information and Reducing Risk," *IEEE Transactions on Engineering Management*, 49(4), 443–458.

Buttazzo, G. C., 1997. Hard Real-Time Computing Systems, Predictable Scheduling Algorithms and Applications, Kluwer Academic.

Castillo, E., 1988. *Extreme Value Theory in Engineering*, Academic Press, San Diego, CA.

Castillo, E., Hadi, A. S., Balakrishnan, N., Sarabia, J. M., 2005. *Extreme Value and Related Models with Applications in Engineering and Science*, Wiley-InterScience, Hoboken, NJ.

Chytka, T., Conway, B., Keating, C., Unal, R., 2004. "Development of an Expert Judgment Elicitation And Calibration Methodology for Risk Analysis in Conceptual Vehicle Design," Old Dominion University Project Number: 130012, NASA Grant NCC-1–02044, NASA Langley Research Center, Hampton, Virginia 23681.

Clemen, R. T., 1996. *Making Hard Decisions An Introduction to Decision Analysis*, 2nd ed., Brooks/Cole Publishing Company, Pacific Grove, CA.

Cox, L. A., Babayev, D., Huber, W., 2005. "Some Limitations of Qualitative Risk Rating Systems," *Risk Analysis*, 25(3), 651–662.

Cox, L. A., 2009. "Improving Risk-Based Decision Making for Terrorism Applications," *Risk Analysis*, 29(3), 336–341.

Creswell, J. W., 2003. *Research Design: Qualitative, Quantitative, and Mixed Methods Approaches*, 2nd ed., Sage University Press, Thousand Oaks, CA.

Crowther, K. G., Haimes, Y. Y., Taub, G., 2007. "Systemic Valuation of Strategic Preparedness Through Application of the Inoperability Input-Output Model with Lessons Learned from Hurricane Katrina," *Risk Analysis*, 27(5), 1345–1364.

Daigle, J. N., 1992. *Queueing Theory for Telecommunications*, Addison-Wesley, Reading, MA.

Daniels, C. B., LaMarsh, W. J., 2007. "Complexity as a Cause of Failure in Information Technology Project Management," *Proceedings of IEEE International Conference on System of Systems Engineering*, April, pp. 1–7.

de Finetti, B., 1974. *Theory of Probability*, Vol. 1, John Wiley, New York.

de Finetti, B. (author), Mura, A. (editor), 2008. *Philosophical Lectures on Probability*, Springer, Berlin.

Diakoulaki, D., Mavrotas, G., Papayannakis, L., 1995. "Determining Objective Weights in Multiple Criteria Problems: The Critic Method," *Computers and Operations Research*, 22(7), 763–770.

Dougherty, N. W., 1972. *Student, Teacher, and Engineer: Selected Speeches and Articles of Nathan W. Dougherty*, edited by W. K. Stair, University of Tennessee, Knoxville, TN.

Dyer, J. S., Sarin, R. K., 1979. "Measurable Multiattribute Value Functions," *Operations Research*, 27(4), 810–822.

Edwards, W., 1954. "The Theory of Decision Making," *Psychological Bulletin*, 41, 380–417.

Edwards, W., 1961. "Behavioral Decision Theory," *Annual Review of Psychology*, 12, 473–498.

Edwards, J. E., Scott, J. C., Nambury, R. S., 2003. *The Human Resources Program-Evaluation Handbook*, Sage University Press, Thousand Oaks, CA.

Essafi, L., Bolch, G., de Meer, H., 2001. Dynamic priority scheduling for proportional delay differentiation services. Modellierung und Bewertung von Rechen- und Kommunikations systemen, Germany.

Feller, W., 1968. *An Introduction to Probability Theory and Its Applications*, Vol. 1, 3rd ed. John Wiley, New York.

Fishburn, P. C., 1989. "Foundations of Decision Analysis: Along the Way," *Management Science*, 35(4), 387–405.

Galambos, J., 1987. *The Asymptotic Theory of Extreme Order Statistics*, Robert E. Krieger Publishing Company, Malabar, FL.

GAO: Government Accountability Office, July 2004. "Defense Acquisitions: The Global Information Grid and Challenges Facing its Implementation," GAO-04–858.

Garrett, C., Apostolakis, G., 1999. "Context in Risk Assessment of Digital Systems," *Risk Analysis*, 19(1), 23–32.

Garvey, P. R., Pinto, C. A., 2009. "An Index to Measure Risk Co-Relationships in Engineering Enterprise Systems,". *International Journal of System of Systems Engineering*, Vol. 1, Issue 3.

Garvey, P. R., 1999. "Risk Management," *Encyclopedia of Electrical and Electronics Engineering*, John Wiley, New York.

Garvey, P. R., 2000. *Probability Methods for Cost Uncertainty Analysis: A Systems Engineering Perspective*, Chapman-Hall/CRC-Press, Boca Raton, FL.

Garvey, P. R., 2001. "Implementing a Risk Management Process for a Large Scale Information System Upgrade—A Case Study," *INSIGHT*, 4(1), International Council on Systems Engineering (INCOSE).

Garvey, P. R., 2005. "System of Systems Risk Management Perspectives on Emerging Process and Practice," The MITRE Corporation, MP 04B0000054.

Garvey, P. R., 2008. *Analytical Methods for Risk Management: A Systems Engineering Perspective*, Chapman-Hall/CRC-Press, Boca Raton, FL.

Garvey, P. R., 2009. *An Analytical Framework and Model Formulation for Measuring Risk in Engineering Enterprise Systems: A Capability Portfolio Perspective*, PhD Dissertation, Old Dominion University, Virginia. August 2009, Dissertations & Theses, Old Dominion University Library, Publication No. AAT 3371504, ISBN: 9781109331325, ProQuest ID: 1863968631.

Garvey, P. R., Cho, C. C., 2003. "An Index to Measure a System's Performance Risk," *The Acquisition Review Quarterly*, 10(2).

Garvey, P. R., Cho, C. C., 2005. "An Index to Measure and Monitor a System of Systems' Performance Risk," *The Acquisition Review Journal*.

Garvey, P. R., Cho, C. C., Giallombardo, R., 1997. "RiskNav: A Decision Aid for Prioritizing, Displaying, and Tracking Program Risk," *Military Operations Research*, V3, N2.

Garvey, P. R., Pinto, C. A., 2009. "Introduction to Functional Dependency Network Analysis," *Proceedings from the Second International Symposium on Engineering Systems*, Massachusetts Institute of Technology, http://esd.mit.edu/symp09/day3.html

Gelinas, N., 2007. "Lessons of Boston's Big Dig," *City Journal*.

Gharajedaghi, J., 1999. *Systems Thinking Managing Chaos and Complexity—A Platform for Designing Business Architecture*, Butterworth-Heinemann, Woburn, MA.

Goodman, K. J., Phillips, C. V., 2005. Hill's Criteria of Causation, in *Encyclopedia of Statistics in Behavioral Science*, John Wiley, New York.

Gorney, L., 1981. *Queueing Theory: A Problem Solving Approach*, Petrocelli Books.

Greiner, S., Bolch, G., Begain, K., 1998. "A Generalized Analysis Technique for Queuing Networks with Mixed Priority Strategy and Class Switching," *Computer Communications*, 21, 819–832.

Hahn, G. J., Shapiro, S. S., 1967. *Statistical Models in Engineering*, John Wiley, New York.

Haimes, Y. Y., 1998. *Risk Modeling, Assessment, and Management*, John Wiley, New York.

Haimes, Y. Y., 2004. *Risk Modeling, Assessment, and Management*, 2nd ed., John Wiley, New York.

Haimes, Y. Y., 2009. "On the Complex Definition of Risk: A Systems-based Approach," *Risk Analysis*, 29(12), 1647–1654.

Hansson, S. O., 2008. "Risk," *The Stanford Encyclopedia of Philosophy* (Winter 2008 Edition), Edward N. Zalta (ed.), http://plato.stanford.edu/archives/win2008/entries/risk/.

Hatfield, A. J., Hipel, K. W., 2002. "Risk and Systems Theory," *Risk Analysis*, 22(6), 1043–1057.

Hildebrand, F. B., 1968. *Finite Difference Equations and Simulations*, Prentice-Hall, Englewood Cliffs, NJ.

Hillier, F. S., Lieberman, G. J., 1986. *Introduction to Operations Research*, 4th ed., Holden-Day, Inc., Oakland, CA.

Hitch, C. J., 1955. "An Appreciation for Systems Analysis," P-699, The RAND Corporation, Santa Monica, CA.

Ho, J. D., Sharma, N. K., 2000. "Multicast Performance in Shared-memory ATM Switches," *Performance Evaluation*, 41, 23–36.

Hofstetter, P., Bare, J. C., Hammitt, J. K., Murphy, P. A., Rice, G. E., 2002. "Tools for Comparative Analysis of Alternatives: Competing or Complementary Perspectives?" *Risk Analysis*, 22(5), 833–851.

Hwang, Ching-Lai, Paul, Y. K., 1995. *Multiple Attribute Decision Making: An Introduction*, Sage University Paper Series in Quantitative Applications in the Social Sciences, 07–104, Thousand Oaks, CA, Sage.

INCOSE, 2004. *Systems Engineering Handbook*, v. 2a, INCOSE-TP-2003–016–02, San Diego, CA.

INCOSE, 2010. *Systems Engineering Handbook*, v. 3.2, INCOSE-TP-2003–002–03.2, San Diego, CA.

Jackson, J. R., 1957. "Networks of Waiting Lines," *Operations Research*, 5, 518–521.

Jackson, M. C., 1991. *Systems Methodology for the Management Sciences*, Plenum, New York.

Jain, J., 1991. *The Art of Computer Systems Performance Analysis Techniques for Experimental Design, Measurement, Simulation, and Modeling*, John Wiley, New York, 10–20.

Jaiswal, N. K., 1968. *Priority Queues*, Academic Press, London.

Jaynes, E. T., 1988. "Probability Theory as Logic," Ninth Annual Workshop on Maximum Entropy and Bayesian Methods, Dartmouth College, New Hampshire, August 14, 1989. In the Proceedings Volume, *Maximum Entropy and Bayesian Methods. Edited by* Paul F. Fougere, Kluwer Academic, Dordrecht (1990).

Jiang, P., Haimes, Y. Y., 2004. "Risk Management for Leontief-Based Interdependent Systems," *Risk Analysis*, 24(5), 1215–1229.

Johnson, L. J., Kotz, S., Balakrishnan, N., 1995. *Continuous Univariate Distributions*, Vol. 2, John Wiley, New York.

Kaplan, S., 1997. "The Words of Risk Analysis," *Risk Analysis*, 4(17), 407–417.

Kaplan, S., Garrick, B., 1981. "On the Quantitative Definition of Risk," *Risk Analysis*, 1(1), 11–27.

Keating, C., Rogers, R., Unal, R., Dryer, D., Sousa-Poza, A., Safford, R., Peterson, W., Rabadi, G., 2003. "System of Systems Engineering," *Engineering Management Journal*, 15(3).

Keating, C. B., Sousa-Poza, A., Mun, Ji Hyon, 2004. "System of Systems Engineering Methodology," Department of Engineering Management and Systems Engineering, Old Dominion University, Norfolk, VA.

Keating, C., Sousa-Poza, A., Kovacic, S., 2005. "Complex System Transformation: A System of Systems Engineering (Sose) Perspective," *Proceedings of the 26th ASEM National Conference*.

Keating, C., Sousa-Poza, A., Kovacic, S., 2008. "System of Systems Engineering: An Emerging Multidiscipline," *International Journal of System of Systems Engineering*, 1(1/2), 1–17.

Keeney, R. L., Raiffa, H., 1976. *Decisions with Multiple Objectives: Preferences and Value Tradeoffs*, John Wiley, New York.
Keeney, R. L., 1992. *Value-Focused Thinking A Path to Creative Decision Making*, Harvard University Press, Cambridge, MA.
Kesidis, G., Konstantopoulos, T., 1998. "Extremal Traffic and Worst-case Performance for Queue with Shaped Arrivals," *Proceedings of Workshop on Analysis and Simulation of Communication Networks*, Toronto.
Kirkwood, C. W., 1997. *Strategic Decision Making: Multiobjective Decision Analysis With Spreadsheets*, Duxbury Press, Belmont, CA.
Kleinrock, L., 1976. *Queueing Systems: Volumes I and II Computer Applications,* John Wiley and Sons, Inc., New York.
Klungle, R., 1999. "Simulation of a Claims Call Center: A Success and a Failure," *Proceedings, Winter Simulation Conference.* Edited by P. A. Farrington, H. B. Nembhard, D. T. Sturrock, G. W. Evans, Vol 2, 1648–1653.
Kolmogorov, A. N., 1956. *Grundbegriffe der Wahrscheinlichkeitsrechnung, Ergeb. Mat. und ihrer Grenzg.*, Vol. 2, No. 3, 1933. Translated into English by N. Morrison, *Foundations of the Theory of Probability*, Chelsea, New York.
Krantz, D. H., Luce, R. D., Suppes, P., Tversky, A., 1971. *Foundations of Measurement*, Additive and Polynomial Representations, Volume 1, Academic Press, Dover Publications, New York.
Kulkarni, V. G., 1997. "Fluid Models for Single Buffer Systems," *Frontiers in Queueing: Models and Applications in Science and Engineering*, 321–338, Ed. H. Dshalalow, CRC Press Inc., Boca Raton, FL.
Leontief, W. W., 1966. *Input-Output Economics*, Oxford University Press, New York.
Lian, C., Santos, J. R., Haimes, Y. Y., 2007. "Extreme Risk Analysis of Interdependent Economic and Infrastructure Sectors," *Risk Analysis*, 27(4), 1053–1064.
Liebeherr, J., Patek, S., Burchard, A., 2001. A Calculus for End-to-End Statistical Service Guarantees, Technical Report, University of Virginia.
Malczewski, J., 1999. *GIS and Multicriteria Decision Analysis*, John Wiley, New York.
Mariampolski, H., 2001. *Qualitative Market Research: A Comprehensive Guide*, Sage University Press, Thousand Oaks, CA.
Massachusetts Turnpike Authority (MTA), *Big Dig*, retrieved from http://www.masstumpike.com /bigdig/background/facts.html
Miller, K., Bapat, V., 1999. "Case Study: Simulation of the Call Center Environment or Comparing Competing Call Routing Technologies for Business Case ROI Projection," *Proceedings, Winter Simulation Conference.* Edited by P. A. Farrington, H. B. Nembhard, D. T. Sturrock, G. W. Evans, Vol 2, 1694–1700.
MITRE, 2007. "Evolving Systems Engineering," The MITRE Corporation, Distribution Unlimited, Case Number 07–1112.
Moynihan, R. A., 2005. "Investment Analysis Using the Portfolio Analysis Machine (PALMA) Tool," The MITRE Corporation, http://www.mitre.org/work/tech_papers/tech papers_05/05_0848/ 05_0848.pdf.
Moynihan, R. A., Reining, R. C., Salamone, P. P., Schmidt, B. K., 2008. "Enterprise Scale Portfolio Analysis at the National Oceanic and Atmospheric Administration (NOAA)," *Systems Engineering*, International Council on Systems Engineering (INCOSE), September 11, 2008, Wiley Periodicals; http://www3.interscience.wiley.com/journal/121403613/references
Mucci, R., 1976. Limit theorems for extremes (Dissertation). Temple University, Philadelphia, PA.

Murphy, C., Gardoni, P., 2006. "The Role of Society in Engineering Risk Analysis: A Capabilities-Based Approach," *Risk Analysis*, 26(4), 1073–1083.

Nau, R. F., 2002. "de Finetti Was Right: Probability Does Not Exist," *Theory and Decision*, 51, 89–124.

National Transportation Safety Board, 2007. Public Meeting, July 10, 2007; "Highway Accident Report: Ceiling Collapse in the Interstate 90 Connector Tunnel," Boston, MA, NTSB/HAR-07/02.

Office of the Secretary of Defense (OSD), 2005: *Net-Centric Operational Environment Joint Integrating Concept*, Version 1.0, Joint Chiefs of Staff, October 31, 2005, Joint Staff, Washington, D.C. 20318–6000; http://www.dod.mil/cio-nii/docs/net-centric_jic.pdf.

Parsons, V. S., 2007. "Searching for Unknown Unknowns," *Engineering Management Journal*, 19(1), 43–46.

Peng, D. T., Shin, K. G., Abdelzaher, T. F., 1997. "Assignment and Scheduling of Communicating Periodic Tasks in Distributed Real-time Systems," *IEEE Transactions on Parallel and Distributed Systems*, 8(12), 745–758.

Pinto, C. A., Arora, A., Hall, D., Ramsey, D., Telang, R., 2004. "Measuring the Risk-Based Value of IT Security Solutions," *IEEE IT Professional*, 6(6), 35–42.

Pinto, C. A., Arora, A., Hall. D., Schmitz, E., 2006. "Challenges to Sustainable Risk Management: Case Example in Information Network Security," *Engineering Management Journal*, 18(1), 17–23.

Pinto, C. A., Tolk, A., Landaeta, R., 2010. "Goal Approach To Risk Scenario Identification In Systems Development," *Proceedings of the 31st National Conference of the American Society for Engineering Management*, Arkansas, October 13–16, 2010, 361–366.

Pratt, J. W., 1965. "Risk Aversion in the Small and in the Large," *Econometrica*, 32, 122–136.

Ramsey, F. P., 1931. *The Foundations of Mathematics, and other Logical Essays*. Edited by R. B. Braithwaite. Preface by G. E. Moore. International Library of Psychology, Philosophy, and Scientific Method, New York, Harcourt, Brace and Company, London, Kegan Paul, Trench, Trubner and Company, Ltd.

Rebovich, G., Jr., 2007. "Engineering the Enterprise," The MITRE Corporation; http://www.mitre.org/work/tech_papers/tech_papers_07/07_0434/07_0434.pdf.

Rebovich, G., Jr., 2005. "Enterprise Systems Engineering Theory and Practice, Volume 2, Systems Thinking for the Enterprise New and Emerging Perspectives," The MITRE Corporation; http://www.mitre.org/work/tech_papers/tech_papers_06/05_1483/05_1483.pdf.

Reilly, J., Brown, J., 2004. "Management and Control of Cost and Risk for Tunneling and Infrastructure Projects," *Proceedings of the International Tunneling Conference*, Singapore.

Rescher, N., 2006. *Philosophical Dialectics: An Essay on Metaphilosophy*, SUNY Press, Albany, New York.

Rittel, H., 1972. "On the Planning Crisis: Systems Analysis of the First and Second Generations," The Institute of Urban and Regional Development, Reprint No. 107, University of California, Berkeley, CA.

Santos, J. R., Haimes, Y. Y., 2004. "Modeling the Demand Reduction Input-Output (I-O) Inoperability Due to Terrorism of Interconnected Infrastructures," *Risk Analysis*, 24(6), 1437–1451.

Santos, J. R., Haimes, Y. Y., Lian, C., 2007. "A Framework for Linking Cybersecurity Metrics to the Modeling of Macroeconomic Interdependencies," *Risk Analysis*, 27(5), 1283–1297.

Savage, L. J., 1954. *The Foundations of Statistics*, John Wiley, New York.

Shanteau, J., Weiss, D. J., Thomas, R., Pounds, J., 2001. "Performance-based Assessment of Expertise: How to Decide if Someone is an Expert or Not," *European Journal of Operations Research*, 136, 253–263.

Stern, 2003. "*The Christian Science Monitor*," http://www.csmonitor.com

Stevens, S. S., 1946. "On the Theory of Scales of Measurement," *Science*, 103, 677–680.

Tanir, O., Booth, R. J., 1999. "Call Center Simulation in Bell Canada." *Proceedings, Winter Simulation Conference.* Edited by P. A. Farrington, H. B. Nembhard, D. T. Sturrock, G.W. Evans, Vol 2, 1640–1647.

Vasquez-Abad, F. J., Jacobson, S. H., 1994. "Application of RPA and the Harmonic Gradient Estimators to a Priority Queueing System," *Proceedings of the 1994 Winter Simulation Conference*, 369–376.

von Bertalanffy, L., 1968. *General Systems Theory, Foundations, Development, Applications*, University of Alberta, Edmonton, Canada: Published by George Braziller, New York.

von Neumann J., Morgenstern O., 1944. *Theory of Games and Economic Behavior*, Princeton University Press, Princeton, NJ.

von Winterfeldt D., Edwards, W., 1986. *Decision Analysis and Behavioral Research*, Cambridge University Press, Cambridge, New York.

Weisstein, E. W. "Graph." From MathWorld: A Wolfram Web Resource. mathworld. wolfram. com/Graph.html.

White, B. E., 2006. "Fostering Intra-Organizational Communication of Enterprise Systems Engineering Practices," The MITRE Corporation, National Defense Industrial Association (NDIA), 9th Annual Systems Engineering Conference, October 23–26, 2006, Hyatt Regency Islandia, San Diego, CA.

Index

A

a posteriori probability, 57, 59, *see also* Probability
a priori probability, 57, *see also* Probability
Additive value function (model), 74–79, 97–101, 109, 111–112, 116, 137, 160, 184, 240–241, 258
Alternatives, analysis of, 6, 33–34, 63, 69–70, 74, 76, 78, 80–81, 97–100, 102, 114–119, 257–264, 316, 424
Arrival rate, 301, 307, 310, 321, 358, 364, 376, 380–381, 387, 410
Arrow–Pratt measure of risk aversion, 30
Asymptotic
- behavior, 330
- distribution, 341–346, 353, 355, 422

Average, 78, 137–138, 140–141, 147–148, 152, 160–162, 165–166, 187, 191, 196, 198–199, 203, 210, 249, 258, 308–309, 311–313, 319–321, 324–328, 364, 373, 380, *see also* critical average; limited average; max average; weighted average
Axioms of
- expected utility, 11, 27–29, 31–32, 39, 81, 83–84
- probability, 8–9, 32–33, 44–45, 425

B

Baseline operability level (BOL), *see* FDNA
Basis of assessment (BOA), 110, 113
Bayes' rule and Bayesian approaches, 32–33, 46, 55–59, 61, 122, 295, 349, 424, *see also* Probability
Bernoulli, Daniel, 25–28, 30–31, 39–40, 179, 415–420
Best practices, 6, 36, 61–62
Birth and death process, 300, 317, 361
Boston's Central Artery/Tunnel Project, Big Dig, 2–6, 14, 19, 22, 423, 426
Bounded latency, 389–392, 401, 410, 412

C

Capability, 2, 10, 12, 23
- definition of, 23, 130
- delivery of, 12, 129
- dependencies, 13, 129, 133, 142, 144, 148, 151, 154
- portfolio risk management, 130–133, 149
  - algebra, for risk analysis, 133–148
- representation of, supplier–provider concept, 131–133
- risk management, information needs, 149–150
- risk, 133–148
  - ripple (collateral) effects of, 6, 13, 110, 134, 136, 139, 151, 154, 174, 177, 215, 222, 247, 274–275, 293, 348
- supplier risks, 133, 136, 139, 273

Capturing dependencies, 154, 157, *see also* Dependencies
Cardinal interval scale, 30, 34, 64–71, 84, 86, 112, 123, 137, 139, 160, 213, 258, 266
Cardinal risk ranking, 112–114, 152
Cardinal value function, 34, 64
Case discussions, 76, 102, 114
- case 3.1, value function theory, choosing a car, 76–81
- case 3.2, value function theory, measuring risk, 102–114
- case 3.3, utility theory, to compare designs, 114–119

Causality, *see also* Causation
Causation, *see also* Causality
- and correlation, and complexity, 349–351

Causation, *see also* Causality (*Cont.*)
physical, 37
probabilistic, 293–295, 349, 356, 423
Causes
and complexity, 349
necessary, 290, 294
necessary and sufficient. 296–298, 317–319
sufficient, 291, 294
Certain (sure) event, 9, 32, 44
Certainty equivalent, 85–89, 92–96, 123
Chain rule, multivariate, 223, 226, 229
Chance and choice, study of, 24–25, 31, 36, 81, 415, *see also* Probability
Child node, 131–132, 134, 140–141, 179, *see also* Node
Compartmentation, 246
Compensatory model, 184, 258–259
Competing alternatives, 74, 81, 102, 257–260, 263
Complement of an event, 42
Complexity
complex systems, systems of systems, enterprise systems, 20, 125–129, 279, 324
complexity and extreme events, 347–349
and correlation, 349
in queuing, 358
Compound event, 42–43
Compromise model, 259, *see also* TOPSIS
Conditional probability, 46–47, 49–50, 55–57, 61, 294–295, 349
Bayes' rule and Bayesian approaches, 32–33, 46, 55–59, 61, 122, 295, 349, 424, *see also* Probability
independent events, relationship to, 50–51
Condition–if–then construct, for risk statement, 61–62, 292
Conditioning event, for risk statement, 61–62, 102–103
Consequence assessment (analysis), 1, 6–10, 13–14, 33, 36–38, 41, 49, 61–62, 103–104, 110–112, 129–130, 133, 135–136, 149–150, 158–159, 163, 272, 274, 277–278, 285, 292, 294, 324–325, 347–348
Constituent node, 239
Constructed scale, 66–67
examples of, 104–105, 107–108, 137, 139, 212–213, *see also* Measurement scales
Cost function (queue) minimization, 382, 395, 401, 403–404, 411–412
Criteria, decision, 10, 33, 63, 70, 73–78, 81, 97, 102, 109, 114–115, 257–262, 264–265, 269–270, 351, 356, 422–423, 425
Criteria weights, 262
Critical average, 140–141, 144, 147, 152
Criticality of dependency (COD), 189, 192–193, 214–215
Cutting edge, of engineering risk management, 12, 20–21, 150–151

D

Data analysis
basic statistics, 327
of extreme and rare events, 329–330, 348, 351
eyeballing, 325
Decision (performance) matrix, 77–78, 260–265
Decision theory, foundations of, 15–16, 24–26, 28, 32, 34–36, 38, 41, 63–123, 240, 259, 262–263, 415, 422, 425, 427
analysis, scientific method, spirit of, 11
applications to engineering risk management, 101–119
Bernoulli, Daniel, log utility, 30
cardinal interval scale, 30, 34, 64–71, 84, 86, 112, 123, 137, 139, 160, 213, 258, 266
cardinal value function, 34, 64
certainty equivalent, 85–89, 92–96, 123
chance and choice, study of, 24–25, 31, 36, 81, 415, *see also* Probability
criteria, 10, 33, 63, 70, 73–78, 81, 97, 102, 109, 114–115, 257–262, 264–265, 269–270, 351, 356, 422–423, 425
expected utility, 84

expected value, 84
exponential utility function, 34, 91–94, 97, 116, 123
exponential value function, 70–75, 77, 91, 95, 97, 106, 115, 123, 182
lottery (gamble), 1, 25, 27–29, 31, 39, 82–89, 92, 95–96, 99, 123, 318, 416–419
measurable value function, 33–34, 64, 66
monotonically decreasing utilities, 71–72, 77, 89, 91, 94, 98–99, 101, 123, 181, 259
monotonically increasing utilities, 70–71, 77, 86, 88–92, 94–95, 97–100, 106, 115–116, 123, 181, 259
multiattribute risk tolerance, 34–35, 98, 115–116
multiattribute utility, 33, 97
mutually preferentially independent, 75–76, 258
piecewise linear single dimensional value function, 67, 69, 77, 104
preference differences, 30, 34, 65–66
preferences, as primitive concept, 28, 32, 34, 66
ratio method, for sensitivity analysis, 79
risk and utility functions, 81–119
 certainty equivalent, 85–89, 92–96, 123
 expected utility 11, 27–29, 31–32, 34–35, 39, 81, 84, 86–87, 89, 94–95, 116, 119, 123, 258–259, 417–419
 exponential utility function, 34, 91–94, 97, 116, 123
 lotteries and risk attitudes, 1, 25, 27–29, 31, 34, 39, 81–89, 91–93, 95–99, 123, 318, 416–419
 multiattribute utility, 33, 97
 power–additive utility function, 34–35, 97–100, 123
risk attitude, 29, 34, 81–82, 85–86, 91–93, 97–98, 123
 risk–averse, 1, 27, 29–31, 34, 39, 85–93, 98, 115–116, 419, 426
 risk–neutral, 29–30, 34, 85–86, 89–92, 98–99
 risk–seeking, 29, 85–86, 89–92, 98
risk, definition of, 1, 8
risk impact dimensions, 102, *see also* Consequence assessment
risk tolerance, 34–35, 91, 98, 115–116, 161, 400
St. Petersburg paradox, 25, 31, 415–420
value function, theory of, 63–81
 additive value function, 74–79, 97–101, 109, 111–112, 116, 137, 160, 184, 240–241, 258
 cardinal (interval) scale, 30, 34, 64–71, 84, 86, 112, 123, 137, 139, 160, 213, 258, 266
 constructed scale, 66–67, examples of, 104–105, 107–108, 137, 139, 212–213, *see also* Measurement scales
 developing, value increment approach, 64, 67–70, 105, 109, 122–123
 exponential value function, 70–75, 77, 91, 95, 97, 106, 115, 123, 182
 measurable value function, 33–34, 64, 66
 measurement scales, 65–67
  constructed scale, 66–67
  interval scale (cardinal), 30, 34, 64–71, 84, 86, 112, 123, 137, 139, 160, 213, 258, 266
  nominal scale, 65
  ordinal scale, 65
  ratio scale, 66
Deficit, degree of (in queues), 397
Definitions
 additive value function, 75
 axioms, of expected utility theory, 27–29
 axioms, of probability, 8–9
 baseline operability level (BOL), 192
 birth and death process, 300
 capability, 130
 capability portfolio, 130
 capability risk, 133
 causality, necessary and sufficient causes, 289–290
 causation, probabilistic, 295
 cause, necessary and sufficient, 296

Definitions (*Cont.*)
- certain (sure) event, 9
- certainty equivalent, 85
- class, priority level of, 359
  - relative risk between classes, 398
  - risk of, 383
- classes, number of, 359
- complement, of an event, 42
- complex system, 20–24
- compound event, 42
- conditional probability, 46–47
- constituent node, and operability level, 239, 241
- constructed scale, 66
- critical average, 140
- criticality of dependency (COD), 192–193, 214
- degradation index, and computing, 223–224
- degradation tolerance level, leaf node, 228
- dependence, probabilistic, 50–51
- deterministic process, 283
- domain of attraction, 335
- dynamic priority system, 365
  - state–dependent, 366
  - time–dependent, 371
- elementary events, 42, *see also* Probability
- enterprise, system, 20–24
- equally likely interpretation, of probability, 43
- event, probability, 42
- exceedance probability, 329
  - nonexceedance, 381
- expected utility, 84
- expected value, 84
- exponential distribution, 304
- exponential utility function, 91
- exponential value function, 70–72
- extreme and rare events, 323
- FDNA (α,β) weakest link rule, 191
- FDNA general weakest link rule (GWLR), 186
- FDNA limited average weakest link rule (LAWLR), 187
- frequency interpretation, of probability, 43
- gamble, lottery, 82
- increment function, 360
- intersection, of an event, 42
- latency threshold, 382
  - cost due to late entities, 402
  - cost due to refused entities, 402
- leaf node degradation tolerance level (LNDTL), 228
- limited average, 187
- lottery, gamble, 82
- marginal probability, 46
- max average, 137
- maximum criticality of dependency, 189
- maximum strength of dependency, 196
- measurable value function, 64
- measure of belief interpretation, of probability, 9, 288–289
- measurement scales, 65
- midvalue, 71
- minimum effective operability level (MEOL), 216
- multiattribute risk tolerance, 34, 98
- multiplication rule, 49
- mutually exclusive (disjoint), 51
- null event, 42
- objective probability, 43
- order statistics, 330
- poisson distribution, 304
- portfolio (capability), 130
- power–additive utility function, 97–98
- preferential independence, mutual, 74–75
- priority levels, set of, 359
- probability, 9, 44, 288
- probability of exceedance, 329
- probability of nonexceedance, 381
- random event, 285
- random process, 285
- random variable, 51
- random variable, continuous, 54
- random variable, discrete, 52
- risk, 8
- risk inheritance, risk corelationship (RCR) index, 158
  - capability node risk corelationship (RCR) index, 162
  - capability node risk score, 161

node risk score (inheritance), 161
node risk score (noninheritance), 160
program node risk corelationship (RCR) index, 161
program node risk score (mixed case), 161
risk score, 112, 137, 160–162
sample space, 42
saturated priority system (PS), supersaturated, 363
simple (elementary) event, 42
static priority system, 364
strength of dependency (SOD), 192
subjective probability, 45
subset, of an event, 42
sure (certain) event, 9
system–of–systems, 20
tolerance level, for risk, in priority system, 396
degree of deficit, 397, 400
differentiation, 400
tolerance, network, 228
total probability law, 55
uncertainty, 8
aleatory, 8, 286
epistemic, 8, 286
union, of an event, 42
utility, axioms of expected, 11, 27–29, 31–32, 39, 81, 83–84
expected utility, 11, 27–29, 31–32, 34–35, 39, 81, 84, 86–87, 89, 94–95, 116, 119, 123, 258–259, 417–419
exponential utility function, 91
utility function, 81
value function, 63
additive, 75
cardinal, 64
direct rating, 69
exponential constant, 71
exponential value function, 70–71
measurable value function, 64
single dimensional, 63
value increment approach, 67
waiting entities, in priority system, 361
waiting lines, 361
finite, 361
infinite, 361
length of, 361
Departure rate, 301, 307, 358
of class n (queuing), 386
Dependencies, consideration of, 13, 22, 129, 133, 142, 144, 148, 151, 154, 157, 162, 174, 177–256, 272–275, 277, 280, 423, 427, *see also* Ripple effects
Determinism, 282, 316, 318
mathematical, 283
philosophical, 284
Deterministic process, 282–284
Differentiation (queue), 367, 373, 375, 379, 395
relative risks, 398–399
tolerance level, 400
Disjoint (mutually exclusive) events, 42, *see also* Probability
Domain of attraction, 335–336, 340, 345–346, 386–387, 389, 394, 405–411
graphical method, 341–342
inverse function, 336
plotting position, 341
Dynamic priority systems (DPS), 365, 379
definition of, 365
state—dependent DPS, 365, *see also* SDPS
time—dependent DPS, 371, *see also* TDPS

E

Elementary events, 42, *see also* Probability
Emerging behavior, 19– 20, 22, 347– 348, 351, 356
Emerging discipline, 12, 23, 125–126, 271, 423– 424, 426
Engineer, ideal, 38
Engineering management, 1, 6, 12, 24, 26, 107, 125, 135, 150, 155, 165, 247, 272, 277
Engineering risk management, general considerations, 1–12, *see also* Engineering risk management, enterprise systems; Engineering risk management, special analytical topics

Engineering risk management, general considerations (*Cont.*)
intellectual groundwork, modern, 36–38
new challenges, 12–13
objectives, goals of, 6–7
process and practice, general, 10–12
resource allocation decisions, for, 6, 11, 13, 36, 102, 113, 129–130, 133, 150, 171–173, 247–248, 257–258, 272, 275–276, 279, 300
risk score, 112, 137, 160–162
risk statement, 60–61, 292
steps, risk management process, 10
Engineering risk management, enterprise systems, 125–155, *see also* Engineering risk management, general considerations; Engineering risk management, special analytical topics
capabilities-based approach, 129–155
algebra, for risk analysis, capability portfolio, 133
capability portfolio view, 132
capability risks, defined, 133
critical average, 140–141, 144, 147, 152
max average, 137–138, 140, 147–148, 152, 160–162, 165–166
ripple-in-the-pond effect, 134, 136, 139, 151, 154, 174, 177, 215, 222, 247, 274–275, 293, 348
risk corelationship index (RCR), 157–170, 174–175
risk score, 112, 137, 160–162
supplier-provider concept, 132–134, 150, 154, 157–158, 177, 248, 259, 265, 274–275, 280
supplier risks, 133, 136, 139, 273
consequence (risk impact) measurement, 133–150, *see also* Consequence assessment
cutting edge, 12, 20–21, 150–151
definition of enterprise, 20–24
dependencies, consideration of, *see* Dependencies
enterprise problem space, 126–129, 274
enterprises, nested nature of, 126
holistic view, 2, 126, 129, 272
risk score, 112, 137, 160–162
Engineering risk management, special analytical topics
enterprise risk analysis algebra, 133
enterprise risk model, 271
extreme event theory, 323, 379
functional dependency network analysis (FDNA), 177
measuring risk corelationships, 157
prioritization systems, 357
ranking risk criticality, 112–114, 152, 257
risk score, 112, 137, 160–162
supplier-provider concept, 132–134, 150, 154, 157–158, 177, 248, 259, 265, 274–275, 280
Engineering systems, 1–14, 15–24, 36, 125, 177, 271–280, 324, 357, 421, 423, 425–427, *see also* Engineering risk management, enterprise systems
Enterprise systems, xv–xvii, 2–3, 6, 12–13, 17–18, 20–24, 36, 38–39, 125–155, 274, *see also* Complex system
Entropy method, objective weights, 262–264, 269–270, 422, 424
Expected utility, 11, 27–29, 31–32, 34–35, 39, 81, 84, 86–87, 89, 94–95, 116, 119, 123, 258–259, 417–419
Expected value, 84
Exponential
distribution, 304
domain of attraction, 335
Kendall notation, 310
utility function, 91
value function, 70–72
Extreme and rare event, 323
complex system, 347
in engineering system, 324
order statistics, 330
Extreme event, *see also* Rare event
order statistics, 330
probability distributions, 345
theory, 323–351
Extreme value analysis, 327
Extreme value theory (EVT), 330, 340

F

Feeder node, 185, *see also* Node
Frechet
  distribution, 336–338
  plot, 339
Functional dependency network analysis (FDNA), 177–256, *see also* Dependencies
  baseline operability level (BOL), 192
  chain rule, multivariate, 223, 226, 229
  compartmentation, FDNA node, 246
  constituent nodes, 239
  critical node analysis and degradation index, 222
  criticality constraint, 188, 192–193
  criticality of dependency (COD), 189, 192–193, 214–215
  criticality of dependency, maximum, 189, 215
  cycle dependencies, 245
  degradation index, 223
  expressing dependence, operability, 179
  FDNA dependency function, forming the, 204
  FDNA fundamentals, 177–186
  FDNA research areas, 248–249
  leaf node degradation tolerance level (LNDTL), 228
  Leontief, Wassily, input–output model, 249
  maximum criticality of dependency, 189, 215
  maximum strength of dependency, 196
  minimum effective operability level (MEOL), 216
  network operability, tolerance analysis, 215
  operability function composition, 183–184
  operability function regulation, 237
  operability levels, as value functions, 181–182
  postulates, of FDNA, 185–186
  properties, of FDNA, 188, 197, 202–203, 210
  receiver node with multiple feeder nodes, 197
  reference point operability level (RPOL), 193
  single component node, 239
  single feeder–receiver node pair, 195
  special topics, 237
  strength of dependency, 192
  strength of dependency, maximum, 196
  strength of dependency protocol, 210–211
  strength of dependency (SOD) constructed scale, 213
  tolerance analysis, network operability, 215–237
    FDNA (α,β) weakest link rule, 191
    FDNA general weakest link rule (GWLR), 186
    FDNA limited average weakest link rule (LAWLR), 187
    weakest link rules, 186–187, 191

G

Gamble, *see* Lottery
General systems theory, 15
GEVD (generalized extreme value distributions), 335, 345, *see also* Asymptotic distributions
Gumbel types distribution, 315, 386–387, *see also* Asymptotic distributions
  distribution, 336 –338
  plot, 346

H

Human judgment, 1, 35, 38–39, 41, 46, 173
Hypothesis analysis, Bayesian inference, 57–60

I

Ideal engineer, 38
Impact assessment, risk, *see* Consequence assessment
Interpretations, axioms, of probability, 8–9, 32–33, 43–45, *see also* Probability

Intersection of events, 42, *see also* Probability
Interval scale, 30, 34, 64–71, 84, 86, 112, 123, 137, 139, 160, 213, 258, 266
  value differences, meaning of, 33–34, 64, 66, 70
Intuition, unaided, 46

J

James river, 325
Joint/Coalition C2 Enterprise, 128

K

Kendall notation, 303, 309, 314, *see also* Queuing systems
Knapsack model, 171–173
Kolmogorov's axioms, *see* Axioms of probability

L

Latency, 308–313
  bounded, 389–394
  of a class, 367, 372–374
  extreme, 379
  threshold, 382
  unbounded, 386–388
Leaf node, 133–134, 185, *see also* Node
Lessons learned, Big Dig, 2–6
Limit distributions, 334, *see also* Asymptotic distributions
Limited average, 187
Linear combination, *see* Additive value function
Lottery, 1, 25, 27–29, 31, 39, 82–89, 92, 95–96, 99, 123, 318, 416–419

M

Marginal probability, 46
Marginal utility, law of diminishing, 27, 417, 419
Markov, *see also* Birth and death process
  characteristic of queuing systems, 303–304
  process, 298–299
Max average, 137–138, 140, 147–148, 152, 160–162, 165–166
Maxima (in extreme value analysis), 327, 330, 336–337, 340–346, 353, 355
Maximum criticality of dependency, 189, *see also* FDNA
Maximum strength of dependency, 196, *see also* FDNA
Measurable value function, 33–34, 64, 66
Measure of belief, 9, 45, 288–289, *see also* Probability
Measure of effectiveness, 186, 204
Measure of performance, 185, 204
Measurement scales, 65–67
  constructed scale, 66–67
  interval scale, 30, 34, 64–71, 84, 86, 112, 123, 137, 139, 160, 213, 258, 266
  nominal scale, 65
  ordinal scale, 65
  ratio scale, 66
Measuring risk corelationships, 157–174
Minima (in extreme value analysis), 330, 336, 338, 340–342, 344, 346, 353, 355
MITRE Corporation, 22, 126, 187, 423, 425–427
Model formulation, for measuring enterprise risk, 271–279
Multiattribute risk tolerance, 34–35, 98, 115–116
Multiattribute utility, 33, 97
Multiplication rule, 49
Multivariate chain rule, 223, 226, 229
Mutual independence, 50–51, *see also* Probability
Mutual preferential independence, 75–76, 258

N

Necessary causes, 290–295, *see also* Sufficient
Network analysis, 215–237, 357–377, 421, 423–426
  degradation tolerance analysis, 227–237
  functional dependency network analysis (FDNA), 177–256
  network operability analysis, 215–227

Network Operations Capability Portfolio, 130–132, 134, 146–147, 152–153, 274, 277
Networked environments, 19–22, 38, 102, 126, 357–377
Networks, 2, 12, 19–22, 38, 67, 102, 125–126, 130–132, 134, 152–154, 178
Node, types of, 133–134, 185–186
Nominal scale, 65–66, 123
Nonexceedance Probability
  curve, 383
  definition, 381
  example, 406–410
Null event, *see also* Probability
  description, 42
  probability of, 44

O

Objective probabilities, 42–43, *see also* Probability
Objective weighting, entropy, 262–264, 269–270, 422, 424
Old Dominion University, 422–424
Operability, 185–186
Operations research, knapsack model and, 171
Optimistic decision (on queues), 406
Order statistics, *see also* Extreme event
  common domain of attraction, 340
  definition, 330
  example of, 332
  limit distributions of, 334
Ordinal scale, 65
Outcome and uncertainty, *see also* Causes, necessary and sufficient

P

Parent node, *see* Node
Performance
  (decision) matrix, 77–78, 260–265
  measure of, 185, 204
  (queue) derived, 365, 373, 395
Perspectives on theories of systems and risk, 15–36
Pessimistic decision (on extremes), 404–409
Philosophical foundations, probability, risk and decision theory, 31–35
Piecewise linear single dimensional value function, 67
  along a constructed scale, example of, 104–105
  graph, example of, 77
Poisson
  distribution, 303–304
  process, 303
  property, 309, 364
  variance–to–mean ratio (VMR) of, 264, 269
Portfolio risk management, 134, 129–150, *see* Capability
Postulates, properties, and propositions
  postulates, for
    functional dependency network analysis (FDNA), 185–186
    risk corelationship (RCR) index, for, 158–159
  properties, for
    functional dependency network analysis (FDNA), 188, 197, 202–203, 210
    risk corelationship index (RCR), 164, 174
  propositions, for
    bounded latency times, 390
    bounded latency times is a saturated SDPS, 392
    risk of a priority system, 384
    state–dependent dynamic priority system (SDPS), 367
    unbounded latency times, 388
Power–additive utility function, 97–101
Preference difference, primitive concept, 28, 32, 34, 66
Preferences, strength of, 28, 32–34, 64, 66, 69, 84, *see also* Measurable value function
Preferential independence, mutual, 75–76, 258
Primitive, value measure as, 28
Prioritization systems, highly networked environments, 357–375
Prioritizing risks, *see* Risk prioritization

Priority systems (queue)
class of entities, 357
initial priority levels, 358
length of waiting line, 360–361
notation, 358
number of classes, 358
priority increment function, 358–360
saturated, 363
super-saturated, 363
types
dynamic (DPS), 365–371, *see also* DPS
state-dependent (SDPS), 365, *see also* SDPS
static (SPS), 363–364, *see also* SPS
time-dependent (TDPS), 371, *see also* TDPS
Probabilistic
causality, 284, 289, *see also* Probabilistic causation
causation, 291–296, 349–350
condition-if-then, protocol, 61–62, 292
event and risk, 7
FDNA, 248
independence, 50–51
Probability scale, ordinal, example of, 104
Probability theory, elements of, 41–62
a posteriori probability, 57, 59
a priori probability, 57
axioms, of, 8–9, 32–33, 44–45, 425
Bayes' rule and Bayesian approaches, 32–33, 46, 55–59, 61, 122, 295, 349, 424
certain (sure) event, 9, 32, 44
complement, 42
compound event, 42–43
conditional probability, 46–47, 49–50, 55–56, 294–295, 349
condition-if-then construct, for risk statement, 61–62, 292
independent events, relationship to, 50–51
conditioning event, for risk statement, 61–62, 102–103
density function, uniform, 96, 116–117, 123, 340, 342
elementary events, 42
equally likely, interpretation, of probability, 43
event, definition of, 42
expected value, 84
frequency interpretation, of probability, 43–44, 288
independent events, 50–51
interpretations, axioms, of probability, 8–9, 32–33, 43–45, *see also* Probability
intersection, 42
lottery (gamble), *see* Lottery
marginal (or unconditional) probability, 46
multiplication rule, 49
mutually exclusive (disjoint), 42
null event, 42
objective probabilities, 42–43
personal (subjective, measure of belief) probabilities, 9, 45, 57, 288–289
philosophical foundations, probability, risk and decision theory, 31–35
risk versus uncertainty, 8, 45–46
sample space, 9, 42–44, 47, 51–52, 55, 60, 120, 287, 289
simple (elementary) event, 42
subjective probabilities, 9, 45, 57, 288–289
subset, 42
sure (certain) event, 9
theorems, on, 44–45
total probability law, 55
union, of an event, 42
venn diagram, 48
Probability, value function for, 74, 104
Program management, 6, 13
Programmatic impacts, from risks, 9, 103, 107–110, 135, 152

Q

Queue, 301, *see also* Priority system
bounded, 363, 389–401
finite, 310–312
infinite, 310

Kendall variable, 319
latency times, 308
parameters, 302–315
performance measures, 310–313
priority, *see* Priority system
saturated, 363, 371
service rate, 307–310, 363–364, 381
supersaturated, 363
unbounded, 386–389, 395–401
waiting entities, 308
Queuing systems, complex, risk of extreme events, 379–410
Queuing theory, 281–315

R

Random
event, 9, 82, 285–289
processes and queuing theory, 281–321
property, 285, 289
Ranking risks, 112–114, 152, 257–265, *see also* Risk prioritization
Rare event, 323, *see also* Extreme event
complex systems, and, 347
in engineering systems, 324
Ratio method, sensitivity analysis, 79
Ratio scale, 66
Receiver node, 185, *see also*, Node
Ripple effects, *see also* Dependencies, 6, 13, 110, 134, 136, 139, 151, 154, 174, 177, 215, 222, 247, 274–275, 293, 348
Risk analysis, *see* Engineering risk management
Risk analysis framework
analytical and model formulation, 277
for engineering enterprise systems, 125–155
Risk and decision theory, 24
philosophical foundations, 31–35
Risk attitude, families of, 29
risk–averse, 1, 27, 29–30, 31, 34, 39, 85–93, 98, 115–116, 419, 426
risk–neutral, 29–30, 34, 85–86, 89–92, 98–99
risk–seeking, 29, 85–86, 89–92, 98
Risk corelationship index, 157–170, 174–175
properties of, 164, 174
Risk dependencies, *see* Dependencies
Risk, extreme events in complex queuing systems, 379–410
Risk impact, *see* Consequence assessment
Risk management, *see* Engineering risk management, enterprise systems; Engineering risk management, general considerations; Engineering risk management, special analytical topics
Risk prioritization, 112–114, 152, 257–265, *see also* Engineering risk management
Risk score, 112, 137, 160–162
Risk statement, writing of, 61–62, 292
Risk tolerance, 34–35, 91, 98, 115–116, 161, 400
Risk, utility functions and, 81–119
attitude towards, *see* Risk attitude
certainty equivalent, 85–89, 92–96, 123
definition of, 1, 8
expected utility, 11, 27–29, 32, 34–35, 39, 81, 84, 86–87, 89, 94–95, 116, 119, 123, 258–259, 417–419
exponential utility function, 34, 91–94, 97, 116, 123
lotteries, 1, 25, 27–29, 31, 39, 82–89, 92, 95–96, 99, 123, 318, 416–419
multiattribute utility, 33, 97
power–additive utility function, 34–35, 97–101, 123
risk tolerance, 34–35, 91, 98, 115–116, 161, 400
utility and utility functions, 81–119
utility independence, 33, 98, 116
Root node, 133

S

Sample space, 9, 42–44, 47, 51–52, 55, 60, 120, 287, 289
Saturated (queue), 363, *see also* Queue
Scales, *see* Measurement scales
Scientific analysis, spirit of, 11, 289

SDPS (State-dependent Dynamic Priority Systems)
  bounded latency, 392
  class latency, 367
  definition of, 366
  derived performance measures, 395–403
  difference with TDPS, 371
  increment function, 366
Server (in queue), 300
  multiple, 304, 310–313
  single, 301, 304, 308–312
Service rate, 307–310, 363–364, 381
Simple event, 42, *see also* Probability
  Markov property, and, 299
Single component node, 186, 239
Single dimensional value function, 63
Sociopolitical considerations, in engineering systems, 6, 17, 19, 27, 126, 129
SPS (Static Priority Systems)
  definition of, 364
  FCFS model, 369
  nonpreemptive, 369
St. Petersburg paradox, 25, 31, 415–420
Strength of dependency, (SOD) 192, 196, 210–211, 213
Strength of preference, 28, 32–34, 64, 66, 69, 84, *see also* Measurable value function
Subjective probabilities, 9, 45, 57, 288–289, *see also* Probability
Subjective weighting, 262–263, *see also* Objective weighting, entropy; Weighting
Subset, 42, *see also* Probability
Sufficient cause, *see* Necessary cause
Supersaturated queue, 363
Supplier–provider concept, 132–134, 150, 154, 157–158, 177, 248, 259, 265, 274–275, 280
Supplier risks, 133, 136, 139, 273
Sure (certain) event, 9, 32, 44, *see also* Probability
System–of–systems, 20–24, *see also* Engineering systems
System safety, experiences, Big Dig, 3
Systems engineer, ideal, 38
Systems engineering, new challenges, 12–13, *see also* Engineering systems

T

TDPS (Time-dependent Dynamic Priority Systems)
  class differentiation, 373
  class latency, 372
  compared to FCFS, 372
  definition of, 371
  difference with SDPS, 371
Theorems, properties, and propositions, on
  bounded latency times, 390
  bounded latency times in saturated SDPS, 392
  functional dependency network analysis (FDNA), 188, 197, 202–203, 210
  probability theory, 9, 44–45, 330, 385, 389, 393–394, 425
  risk corelationship index (RCR), 164, 174
  risk of a priority system, 384
  state–dependent dynamic priority system (SDPS), 367
  unbounded latency times, 388
  utility theory, 84, 94, 99
  value function theory, 63–81
Theories of systems and risk, 15
Threshold, latency, 382
Tolerance level, for risk, in priority system, 396
  degree of deficit, 397, 400
  differentiation, 400
  differentiation level (priority system), 400
  for risk (priority system), 396
Tolerance, network, 228
TOPSIS, 259
Total probability law, 55–57, 385, *see also* Probability
Traditional systems engineering, 12
True zero, 66, 264

U

Unbounded latency, 386, *see also* Queue
Uncertainty, 4, 8, 20, 27, 29, 33–35, 41, 45–46, 81, 84, 248, 262, 264, 281–282, 284, 286–289, 294, 298, 315–318, 347–348, 421, 423

aleatory, 8, 286–287
causality, 284, 289–290
epistemic, 8, 286–287
versus risk, 8, 45–46
Unifying risk–analytic framework, 271
model formulation, enterprise risk analysis, 272
Union of events, 42
Utility theory, 81–119
axioms, of expected utility, 11, 27–29, 31–32, 39, 81, 83–84
Bernoulli utility, 25, 27, 30, 415–419
certainty equivalent, 85–89, 92–96, 123
expected utility, 11, 27–29, 32, 34–35, 39, 81, 84, 86–87, 89, 94–95, 116, 119, 123, 258–259, 417–419
exponential utility function, 34, 91–94, 97, 116, 123
lotteries and risk attitudes, 82, 25, 27–29, 31, 39, 82–89, 92, 95–96, 99, 123, 318, 416–417
marginal utility, law of diminishing, 27, 417, 419
multiattribute utility and power–additive utility function, 33, 34–35, 97–101, 123
power–additive utility function, 97–101
risk tolerance, 34–35, 91, 98, 115–116, 161, 400
St. Petersburg paradox, 25, 31, 415–420
utility functions, 81–119
utility independence, 33, 98, 116

V

Value function theory, 11, 63–81
additive value function (model), 74–79, 97–101, 109, 111–112, 116, 137, 160, 184, 240–241, 258
cardinal interval scale, 30, 34, 64, 66–71, 112, 123, 137, 139, 160, 258, 266
cardinal risk ranking, 112–114, 152, 257–265
cardinal value function, 34, 64
constructed scale, 66–67, *see also* Measurement scales
examples of, 104–105, 107–108, 137, 139, 212–213
developing, value increment approach, 67–70
direct rating, 69
exponential constant, 71
exponential value function, 70–75, 77, 91, 95, 97, 106, 115, 123, 182
measurable value function, 33–34, 64, 66
measurement scales, 65–67
constructed scale, 66–67
interval scale, 30, 34, 64–71, 84, 86, 112, 123, 137, 139, 160, 213, 266
nominal scale, 65
ordinal scale, 65
ratio scale, 66
piecewise linear single dimensional value function, 67, 77, 104
preference difference, primitive concept, 28, 32, 34, 66
preference differences, 30, 34, 65–66
preferences, as primitive concept, 28, 32, 34, 66
preferences, strength of, 28, 32–34, 64, 66, 69, 84, *see also* Measurable value function
primitive, value measure as, 28
probability, value function for, 74, 104
ratio method, for sensitivity analysis, 79
value differences, meaning of, 33–34, 64, 66, 70
Value models, scientific, objective, 11
Variance–to–mean ratio (VMR), 264, 269
Venn diagram, 48
Von Neumann–Morgenstern utility theory, 81–85
Arrow–Pratt measure of risk aversion, 30
axioms of expected utility theory, *see* Axioms

W

Weakest link rules, 186–187, 191, *see also* FDNA
FDNA (α,β) weakest link rule, 191

Weakest link rules (*Cont.*)
FDNA general weakest link rule (GWLR), 186
FDNA limited average weakest link rule (LAWLR), 187
Weibull distribution, 336–338
plot, 346
Weight function, example of, 140
Weighted average, 78, *see also* Additive value function
Weighted average, compensatory model, 258
Weighted average, constrained, 187
Weighting, *see* Objective weighting, entropy; Subjective weighting
Weights, for criteria, 262
Writing a risk, *see* Risk statement

Z

Zero, true, 66, 264

Printed in the United States
by Baker & Taylor Publisher Services